ISOZYMES IN PLANT BIOLOGY

ISOZYMES IN PLANT BIOLOGY

edited by
Douglas E. Soltis
and
Pamela S. Soltis

Department of Botany, Washington State University
Pullman, Washington

Introduction by G. L. Stebbins

Advances in Plant Sciences Series
VOLUME 4
Theodore R. Dudley, Ph.D., General Editor

DIOSCORIDES PRESS
Portland, Oregon

ELMHURST COLLEGE LIBRARY

© 1989 by Dioscorides Press (an imprint of Timber Press, Inc.)

All rights reserved
ISBN 0-931146-13-5
Printed in Hong Kong

DIOSCORIDES PRESS
9999 SW Wilshire
Portland, Oregon 97225

Library of Congress Cataloging-in-Publication Data

Isozymes in plant biology / edited by Douglas E. Soltis and Pamela S.
 Soltis ; introduction by G.L. Stebbins.
        p.   cm. -- (Advances in plant sciences series ; v. 4)
   Includes bibliographies and index.
   ISBN 0-931146-13-5
   1. Plant isozymes.   I. Soltis, Douglas E.   II. Soltis, Pamela S.
 III. Series.
 QK898.I8I86   1989
 581.19'25--dc20   89-11739
                                                                  CIP

Contents

Introduction
G. L. Stebbins 1

CHAPTER 1
Visualization and Interpretation of Plant Isozymes
Jonathan F. Wendel and Norman F. Weeden 5

CHAPTER 2
Genetics of Plant Isozymes
Norman F. Weeden and Jonathan F. Wendel 46

CHAPTER 3
Isozyme Analysis of Plant Mating Systems
A. H. D. Brown, J. J. Burdon and A. M. Jarosz 73

CHAPTER 4
**Isozymes and the Analysis of Genetic Structure
in Plant Populations** J. L. Hamrick 87

CHAPTER 5
Isozyme Variation in Colonizing Plants
Spencer C. H. Barrett and Joel S. Shore 106

CHAPTER 6
**Physiological and Demographic Variation Associated
With Allozyme Variation** Jeffry B. Mitton 127

CHAPTER 7
Enzyme Electrophoresis and Plant Systematics
Daniel J. Crawford 146

CHAPTER 8
Isozymic Evidence and the Evolution of Crop Plants
John Doebley 165

CHAPTER 9
Isozyme Analysis of Tree Fruits
Andrew M. Torres 192

CHAPTER 10
**Isozymes as Markers for Studying and Manipulating
Quantitative Traits** Charles W. Stuber 206

CHAPTER 11
Bryophyte Isozymes: Systematic and Evolutionary Implications
Robert Wyatt, Ann Stoneburner and Ireneusz J. Odrzykoski 221

CHAPTER 12
**Polyploidy, Breeding Systems, and Genetic
Differentiation in Homosporous Pteridophytes**
Douglas E. Soltis and Pamela S. Soltis 241

Index 259

Introduction

G. L. Stebbins

Since their discovery by Hunter and Markert in 1957, isozymes have played a key role in many branches of biology. To date, they have become the most widely recognized links between the organismal and molecular approach to our science. Isozymes were originally defined by Markert and Moller (1959) as different variants on the same enzymes, having identical or similar functions, and present in the same individual. As such, their importance for understanding gene action in development and differentiation was exploited during the 1960s in both animals and plants. A review of this early work was made by Scandalios (1969) on six different kinds of enzymes. For some of them he noted differences in different parts of the same plant: both presence vs. absence, and quantitative differences in concentration.

Nevertheless, isozymes played a minor role in research on plant biochemistry until after the dramatic discovery in 1966, independently by H. Harris in humans (Harris, 1966) and by Lewontin and Hubby (1966) in *Drosophila*, of genetic polymorphism for isozymes within the same population. At last, population geneticists could make precise quantitative estimates of genetic variability based upon one parameter of the molecular structure of the primary products of the genes themselves. For many workers in the field, optimism evoked the feeling: "The sky's the limit".

Plant population geneticists were not long in following their zoological colleagues. Polymorphism in both cultivars of barley and natural populations of *Avena* were reported by R. W. Allard and his associates (Kahler and Allard, 1970, Marshall and Allard, 1970a, b). A plant geneticist (Gottlieb, 1971) reviewing both plant and animal populations pointed out the inestimable value of isozymes for gaining knowledge about processes of evolution. From then on, investigations of both animals and plants increased explosively. One after another, new ways of looking at population genetics, evolutionary processes, and the phylogenetic relationships of populations opened up, expanding vistas.

The technique of extracting and analyzing the electric mobility of soluble proteins, particularly enzymes, has many advantages, but at the same time, some disadvantages. Consider first examples similar to the original one of lactate dehydrogenase, in which the different isozymes are tetrameric molecules containing two different polypeptide chains in different proportions, all of them in the same individual (Hunter and Markert, 1957). This situation raises the question of differentiation: Why should an organism have different isozymes of the same enzyme in different tissues, such as flight muscles and heart muscles? Is each isozyme structured so that it functions most efficiently in the cellular environment in which it exists? In plants, we can ask the same question about examples reviewed by Scandalios (1969), with respect to tissue specificity of leucine aminopeptidase (LAP) in maize. Such questions, though interesting, are not reviewed in the present volume since, particularly on the part of population geneticists, evolutionists, systematists and plant breeders, they have been overshadowed by the greater opportunities afforded by research on populations.

For population studies, isozymes (or allozymes, as they are sometimes called) make possible comparisons between individuals and populations on the basis of several gene loci, rather than just one or two. Moreover, if the analysis is accompanied by investigation of progenies of the organisms analyzed, Mendelian segregation ratios can be obtained without the trouble of isolating parents and making crosses. Two parameters have been extensively used: the proportion of enzyme loci for which the population is polymorphic (P), and the mean number of loci for which individuals are heterozygous (H). A review by Selander (1976) comparing these parameters in various populations has been followed by many other studies. In the present volume, J. B. Mitton has used H to evaluate the importance of heterozygosity in natural populations. The degree of polymorphism expressed by P, has been used in several contributions to approach various problems of population genetics, particularly breeding structure and mating systems by Hamrick, Barrett and Shore, Brown, Burdon and Jarosz, as well as Soltis and Soltis, and Wyatt, Stoneburner, and Odrzykoski. New knowledge derived from these investigations has strengthened a point of view already stressed by Darwin: evolution takes place in a complex environment, that can be constantly changing over long periods of time, or can alternate between long periods of relative stability and cycles of rapid change. The most successful plant species become adjusted to these vagaries in several ways, including shifts in heterozygosity, polymorphism and mating systems. The strength of isozyme analysis for testing hypotheses is well illustrated by the contribution of the Soltises, who have shown clearly that a previously held hypothesis, predicting self fertilization fortified by polyploid genetic segregations in ferns, must be rejected.

The contribution of Wyatt, Stoneburner and Odrzykoski adds a dimension, since it deals for the first time with predominantly haploid multicellular organisms, and with two subphyla, liverworts and mosses, that are the most ancient of widespread modern land plants. The presence of polymorphism in populations of haploids is of general interest to population geneticists, whereas the difference in genetic diversity between two subphyla that are almost equally ancient is highly relevant to hypotheses of adaptation, neutrality and the molecular clock.

In their contribution, Barrett and Shore present valuable data relevant to the genetics of weeds and other colonizing species, especially self-fertilizing allopolyploids.

The most important theme of this volume as a whole is that, regardless of what isozymes can tell us about Darwinian evolution, or the lack of it as some claim to be so, isozymes as marker genes still continue to grow in value as aids to solving a variety of problems in systematics, evolution, and plant breeding. Among the systematic problems, a critical one is whether two sets of populations that are sympatric and closely similar in morphological characteristics do or do not share the same gene pool, i.e., do or do not belong to the same species. In his contribution, Crawford reports examples in *Salicornia* and *Chenopodium* for which electrophoresis and isozyme comparisons have provided positive answers. A second question: is the ranking of entities as different subspecies or distinct species, if based entirely upon morphology, a reflection of actual degree of genetic relationship? In many instances it is answered in the positive by isozyme comparisons. The probable answer to a third question: how much genetic differentiation is necessary for species to become distinct genetic systems?, has raised a new, previously unexpected series of situations. Although the mean genetic identity between related plant species (0.67) (Gottlieb, 1977a) is much less than between populations of the same species (0.90 or higher), nevertheless the values for different species pairs that are reproductively isolated but regarded by taxonomists as nearest neighbors vary greatly.

For well defined species of perennial herbs in the genus *Heuchera* (Saxifragaceae) identity values are 0.83 to 0.85, whereas among herbaceous ferns living in similar habitats, identity values are as low as 0.272 to 0.514 (Soltis and Soltis, this volume). A possible explanation, cited by Crawford, is provided by Witter's analysis of the silversword alliance in Hawaii. Identity values of species pairs are correlated with the geologically determined recency of the island on which they occur. This result suggests the possibility that, in terms of the geological time scale, speciation is always a rapid process, but that after the species have become distinct, they may continue to diverge more with respect to isozymes than in morphology. Several comparisons of morphological and isozyme patterns of variability in both plants and animals have shown that they are often poorly correlated with each other.

Isozymes also provide valuable information with respect to hybridization and gene duplication, including polyploidy. With respect to hybrid origin of diploid species, one isozyme comparison (in *Stephanomeria diegensis*) confirmed a hybrid origin hypothesis based upon morphology, whereas another (in *Lasthenia burkei*), rejected it.

Isozyme studies have shown in one instance, *Clarkia*, that duplication of gene loci may exist in species that on the basis of all criteria are diploid (Gottlieb, 1977b), and in another extensive study (Soltis and Soltis, this volume), that genera previously believed to be of polyploid origin at least do not have the genetic attributes of polyploids. Crawford cites other examples of the value of isozymes, always in conjunction with other techniques, for unravelling plant relationships above the species level.

The use of isozymes as markers has been a new approach of considerable value to those plant breeders who have employed it. In this volume, Doebley has shown that it can aid greatly our understanding of crop plant evolution. The careful research of Stuber has used the method for attacking one of the most important and difficult problems in plant breeding: the location, number and nature of the genes that contribute to patterns of quantitative inheritance. These results should improve greatly efforts to increase grain yield. Dealing with an entirely different type of economic plant, fruit trees, in which the problems of improvement by breeding are intensified by the great length of their generations, Torres has shown that using marker isozymes provides valuable short cuts.

In spite of this brilliant succession of valuable biological discoveries, the difficulties and limitation of isozymes must be recognized. One of them is the ongoing problem of whether or not differences in isozymes are neutral or adaptive with respect to evolution. This problem has no clear solution. One can hardly doubt that adaptive differences exist with respect to isozymes that are cell- or tissue-specific within a single organism, such as those of lactate dehydrogenase in birds and some enzymes in germinating seeds. Moreover, several comparisons of animal isozymes, particularly in marine forms, have compared climatic distribution with intensive biochemical studies of extracted and purified enzyme preparations, and in this way have provided strong evidence for adaptive differences. On the other hand, investigations of some different allelic pairs have failed to provide such evidence. At present, the best answer to this question is that both adaptiveness and neutrality exist with respect to different isozymes and different pairs of isozymic alleles; and the degree of adaptiveness or neutrality, in many instances, varies with the environment.

Two other drawbacks to the technique are first, that the enzymes extracted and subjected to electrophoresis are a tiny and probably non-representative sample of the total array of proteins present in an organism, and second, that electrophoretic differences are only one kind of difference, and on the whole, a minor one, of the various differences that exist even between genetically related proteins. Although not discussed in the present volume, these difficulties should be recognized, and should warn theorists to be conservative in extrapolating from isozyme evidence to generalizations about major evolutionary and genetic problems. This field of investigation will continue to flourish in the future, but will do so only to the extent that it is coordinated with other techniques.

The authors of this volume, and particularly the editors, Doug and Pam Soltis, are to be heartily congratulated on their achievement in putting together a seminal volume that constitutes a further step ahead in the development of contemporary plant science.

LITERATURE CITED

Gottlieb, L. D. 1971. Gel electrophoresis: new approach to the study of evolution. *BioScience* 21: 939–944.

―――. 1977a. Electrophoretic evidence and plant speciation. *Ann. Missouri Bot. Gard.* 64: 161–180.

―――. 1977b. Evidence for duplication and divergence of the structural gene for phosphoglucose isomerase in diploid species of *Clarkia*. *Genetics* 86: 289–307.

Harris, H. 1966. Enzyme polymorphism in man. *Proc. Roy. Soc. Ser. B,* 164: 298–310.

Hunter, R. L., and C. L. Markert. 1957. Histochemical demonstration of enzymes separated by zone electrophoresis in starch gels. *Science* 125: 1294–1295.

Kahler, A. L., and R. W. Allard. 1970. Genetics of enzyme variants in barley. I. Esterases. *Crop Sci.* 10: 444–448.

Lewontin, R. C., and J. L. Hubby. 1966. A molecular approach to the study of genetic heterozygosity in natural populations. II. Amount of variation and degree of heterozygosity in natural populations of *Drosophila pseudoobscura*. *Genetics* 54: 595–609.

Markert, C. L., and F. Moller. 1959. Multiple forms of enzymes: tissue, ontogenetic and species specific patterns. *Proc. Natl. Acad. Sci. USA* 45: 753–763.

Marshall, D. R., and R. W. Allard. 1970a. Isozyme polymorphism in natural populations of *Avena fatua* and *A. barbata*. *Heredity* 29: 373–382.

———, and ———. 1970b. Maintenance of isozyme polymorphisms in natural populations of *Avena barbata*. *Genetics* 66: 393–399.

Scandalios, J. 1969. Genetic control of multiple molecular forms of enzymes in plants: a review. *Biochem. Genet.* 3: 37–79.

Selander, R. K. 1976. Genic variation in natural populations. In F. J. Ayala [ed.], *Molecular evolution*, pp. 21–45. Sinauer Associates, Sunderland, MA.

CHAPTER 1

Visualization and Interpretation of Plant Isozymes

Jonathan F. Wendel

Department of Botany
Bessey Hall
Iowa State University
Ames, Iowa 50011

Norman F. Weeden

Department of Horticultural Sciences
New York State Agricultural Experiment Station
Cornell University
Geneva, New York 14456

Gel electrophoresis of proteins has become a standard and powerful research tool for application in a multitude of biological disciplines. One form of protein electrophoresis, isozyme analysis, has become particularly prominent in systematic and evolutionary biology as well as agronomy (Tanksley and Orton, 1983). Isozymes, or multiple molecular forms of enzymes, are enzymes that share a common substrate but differ in electrophoretic mobility (Markert and Moller, 1959). They are revealed when tissue extracts are subjected to electrophoresis in various types of gels and subsequently submersed in solutions containing enzyme-specific stains. Genetic analysis may indicate that some of the variant electromorphs are encoded by alternate alleles at a single locus, in which case the allelic products are termed allozymes (Prakash et al., 1969). Data retrieved from electrophoretic gels consist of the number and relative mobilities of various enzyme products, which with appropriate genetic analyses become transformed into single or multilocus genotypes for each individual analyzed. Reasons are many for the popularity of electrophoretic data (Avise, 1975; Gottlieb, 1977; Crawford, 1983), but foremost among these is that isozymes provide a series of readily scored, single-gene markers.

This chapter focuses on two interdependent considerations that promote the efficient development of isozyme markers: selection of the electrophoretic methodology and genetic interpretation. Emphasis is placed on starch gel electrophoresis of enzymes from simply prepared tissue homogenates, because these techniques are the most practical and frequently employed for applications requiring the scoring of several loci in large numbers of individuals. Many of the technical aspects of starch gel electrophoresis have been addressed elsewhere, including preparation of extracts, gels, electrophoretic buffer systems, and staining assays (Brewer and Sing, 1970; Shaw and Prasad, 1970; Gabriel, 1971; Harris and Hopkinson, 1976; Siciliano and Shaw, 1976; O'Malley et al., 1980; Conkle et al., 1982; Cardy et al., 1983; Soltis et al., 1983; Vallejos, 1983; Heeb and Gabriel, 1984; Werth, 1985; Micales et al., 1986; Morden et al., 1987). Although these are valuable sources of information, most are either directed toward isozyme analysis of specific plant groups, or do not provide adequate coverage of the staining assays and electrophoretic buffer systems that have proven useful in plants. In addition, many minor matters of technique have either remained widely acknowledged but unpublished, or have been recently developed. Consequently, much of this chapter is devoted to procedural details and methodological recommendations, with the aim of providing a reference manual for starch gel electrophoresis of isozymes for use by plant scientists in a variety of fields. Although the equation between phenotype and genotype is often simple for isozyme characters, several phenomena commonly complicate genetic interpretation of electrophoretic banding patterns. These problems are briefly discussed to facilitate correct genotypic assignments.

ISOZYME ELECTROPHORESIS

Electrophoretic separation of complex mixtures of proteins can be accomplished in several types of support media, including starch, polyacrylamide, and agarose gels, and cellulose acetate membranes. The latter two typically lack sufficient resolving power, and for this and other reasons they are generally not employed for the study of enzyme polymorphisms (see, however, Womack and Moll, 1986). Biochemists and others requiring maximum resolving power often prefer polyacrylamide gel electrophoresis (PAGE) (see review by Blackshear, 1984). A valuable property of PAGE is that the stringency of molecular sieving can be varied by altering the acrylamide concentration, thus increasing flexibility in the range of protein molecular weights or dimensions that are separable (Leaback, 1976). Other advantages of polyacrylamide gels include their uniformity and transparency, facilitating densitometric quantification of product, inertness of components allowing broad assay compatibility, usually rapid run times, and the ability to concentrate proteins into very thin starting zones in stacking gels prior to their entrance into the resolving gels (Chrambach and Rodbard, 1971; Chrambach, 1980; Hames and Rickwood, 1981). Some enzymes, such as ribonuclease (Isola and Franzoni, 1981), pyruvate decarboxylase (Zehender et al., 1983), and amylase (Kiang, 1981; Zimniak-Przybylska et al., 1985), are analyzed with PAGE because the assays involve production of white precipitates that are difficult to observe on starch gels.

Starch gel electrophoresis (Smithies, 1955), however, rather than PAGE, continues to be preferred for most studies involving the analysis of large numbers of individuals for several to many different enzymes, despite the greater resolving power and flexibility of PAGE. Reasons for this are several: simplicity of starch gel preparation, the non-toxic nature of the material used (the acrylamide monomer used for PAGE is a neurotoxin), relative costs of equipment, and the ease of loading samples onto the gels. Samples for starch gels are usually crude uncentrifuged homogenates, but PAGE requires clarified samples. Probably the most compelling reason for the continued popularity of starch gels is the difference in amount of data generated per gel. Starch gels can be sliced horizontally into replicate slices, allowing multiple (up to six) enzyme systems to be assayed per gel, whereas polyacrylamide gels are usually stained for a single enzyme. In studies where 20 or more enzymes and large numbers of plants are assayed, the efficiency and cost-effectiveness of starch more than compensate for its lower resolution relative to acrylamide.

There is considerable evidence that electrophoresis is an imperfect detector of allozyme polymorphisms (Shaw, 1970; Marshall and Brown, 1975), with many potential variants remaining hidden under any single set of electrophoretic conditions (reviewed by Coyne, 1981). An important question for some users of starch gels concerns the nature and degree of cryptic variability that might be revealed using additional electrophoretic techniques. This question has been addressed in a number of empirical studies, mostly with *Drosophila*, but also with plant enzymes (Gottlieb, 1979, 1984; Shumaker et al., 1982; Lowrey and Crawford, 1985). Although there is great variability from locus to locus, most available evidence indicates that polymorphic loci become more highly polymorphic under additional electrophoretic conditions, whereas monomorphic loci remain invariant or nearly so. Hidden heterogeneity is unlikely to be a problem for the majority of studies, however, as its magnitude seems to have little impact on systematic or phylogenetic inferences (Crawford, 1983) and its presence may be irrelevant in other applications (e.g., when selected allozyme polymorphisms are used as marker loci for measuring mating systems or tracking linked factors in breeding programs). For studies employing exclusively starch gel electrophoresis, probably the most important lesson to be learned regarding hidden heterogeneity is that willingness to modify experimental procedures is likely to increase levels of polymorphism or alter relative migration rates of allozymes.

SAMPLE PREPARATION

Choice of Tissue

Electrophoresis can be performed successfully on extracts from a wide variety of plant tissues, the proper choice of which depends on several variables that must be considered and optimized in each study. Availability of material is often a foremost consideration, and for this reason extracts are usually obtained from vegetative tissues, such as the lamina or petioles of leaves, succulent stems, terminal portions of roots, or where seed supply is not limiting, portions of or entire germinating seedlings. Several parts of seeds have been used, including entire embryos, endosperm (e.g., Nickrent, 1986), imbibed cotyledons, hypocotyls, radicles, or in grasses, coleoptiles (e.g., Cardy et al., 1983). Imbibed cotyledons, germinating seedlings, or young metabolically active leaves typically contain the highest enzyme activities and often are the tissue of choice when readily available. Pollen can also be a rich source of enzymes (Brewbaker, 1971; Loukas et al., 1983; Sari-Gorla et al., 1986) and has the additional advantages of being easily extracted (Weeden and Gottlieb, 1980) and assisting genetic interpretation through comparisons with diploid somatic tissues (Weeden and Gottlieb, 1979). In exceptional cases, fruit parts (Torres et al., 1978), single pollen grains (Mulcahy et al., 1979), or, remarkably, even herbarium specimens (Ranker and Werth, 1986) may be used. Most analyses in conifers are performed on extracts derived from the enzyme-rich haploid megagametophyte or its embedded embryo, although appropriately extracted vegetative tissues may also yield usable homogenates (Cheliak and Pitel, 1984; Neale et al., 1984; Pitel and Cheliak, 1984). In ferns, sporophyte tissue is ordinarily extracted, but gametophytes (Gastony and Gottlieb, 1982, 1985; Haufler and Soltis, 1984; Soltis, 1986; Wolf et al., 1987) or megaspores of heterosporous pteridophytes (Soltis and Soltis, 1986) may also serve as the source of enzymes.

Whichever tissue is chosen, care should be taken to assure that different samples are in the same physiological and ontogenetic condition. Although it is widely acknowledged that the spectrum of isozymes expressed may vary in different plant tissues, it is important to recognize that physiological state and environment may also influence electrophoretic phenotypes of the same tissue. Manifestations range from changes in intensity or relative intensity of bands to appearance or disappearance of bands, and may be correlated with such factors as leaf position or developmental stage (Endo, 1981; Tyson et al., 1985), season (Kelley and Adams, 1977b; Oncelay et al., 1979), or processes occurring during seed storage and germination (Tanksley, 1980).

Enzyme Extraction

For most purposes, tissues to be analyzed are prepared as crude buffered extracts, with no elaborate protein purification or concentration steps required. Each taxon and tissue poses its own set of problems regarding the difficulty of cell breakage and associated problems caused by endogenous tannins, phenols, phenoloxidases, and other (mostly unidentified) cellular constituents. Soft tissues, such as imbibed seeds or succulent vegetative parts, are often readily ground with an appropriate buffer in a chilled mortar and pestle, spotglass, weighing boat, or other mechanical contraptions (e.g., Cardy et al., 1983; Fig. 1.1). More forceful methods are required for some tissues (particularly coriaceous leaves or conifer needles), such as grinding in liquid nitrogen or the use of a power homogenizer (e.g., Soltis et al., 1980; Wendel and Parks, 1982). The primary objective, whatever the grinding procedure, is to effect adequate cell breakage in a minimum amount of time with no warming of the extract.

An additional consideration for obtaining extracts with high enzyme activity is the optimal tissue to volume ratio. Although it is generally true that samples should be ground in the minimum volume of extraction buffer necessary to provide sufficient extract for electrophoresis, the proper ratio should be empirically determined. With some tissues, samples homogenized in minimal volumes are inferior to those ground in larger volumes, perhaps due to insufficient buffering or incomplete complexing of phenolics.

Following homogenization, extracts may be briefly centrifuged to remove cellular debris (typically in microfuge tubes in a refrigerated microfuge). This necessitates an additional step in sample preparation, but often results in zymograms with sharper resolution and less streaking. Tissue

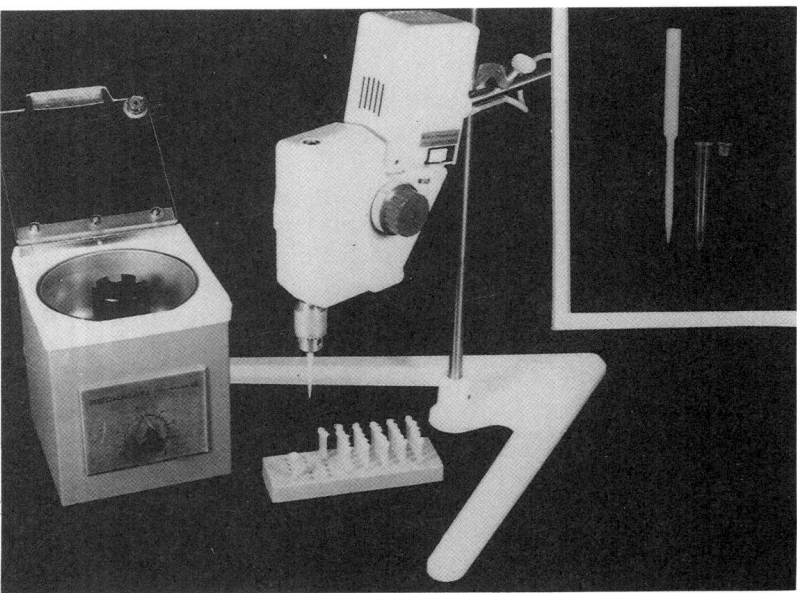

Fig. 1.1. Sample preparation equipment. Various types of equipment may be used for tissue homogenization. Shown is a power driven acetal pestle that is designed to fit microfuge tubes (Cardy et al., 1983). Samples are centrifuged for 1 minute in a microfuge subsequent to homogenization.

availability and laboratory workloads often make it desirable to prepare homogenates and freeze them days or even months in advance of electrophoresis. This is conveniently accomplished with microfuge tubes. It must be determined in each case, however, whether extracts can be frozen without unacceptable loss of enzyme activity. In general, superior results are obtained from freezing extracts rather than tissue samples, and from freezing in an ultracold (−70° C) rather than conventional (−20° C) freezer. Extracts from some tissues retain high activity levels even after a few years of storage in an ultracold freezer, as long as they are not subjected to repeated freeze-thaw cycles. Some loss of enzyme activity is inevitable with freezing, but it is often apparent only for a few enzymes. In maize, for example, freezing causes little deterioration of extract quality other than for the products of one of the two hexokinase loci (Wendel et al., 1986) and for all of the aconitate hydratase isozymes (Wendel et al., 1988).

Of paramount importance is the type of grinding buffer to be used in preparation of the extract. If the tissue to be homogenized is relatively free from phenolics and other interfering substances, best results are usually obtained with comparatively simple extraction buffers. Many plant tissues, however, contain high levels of phenolics that complex with enzymes following cellular disruption. Probable consequences include loss of enzyme activity and/or generation of artifacts. Mechanisms by which phenolics interfere with protein extraction and methods that overcome or retard this interference have been discussed by Loomis and Battaile (1966), Pierpoint (1966), Anderson and Rowan (1967), Anderson (1968), Loomis (1969, 1974), King (1971), and Kelley and Adams (1977a). An assortment of buffer additives is available that may improve extract quality, including phenol-complexing agents [borate, germanate, polyvinylpyrrolidone (PVP) or polyvinylpolypyrrolidone (PVpP), bovine serum albumin (BSA)], phenoloxidase inhibitors [PVP, PVpP, diethyldithiocarbamate (DIECA)], and various antioxidants and reducing agents (mercaptoethanol, dithiothreitol, ascorbate, bisulfite). Other common buffer additives include enzyme-stabilizing osmotica such as sucrose or glycerol and detergents such as Triton X-100 or Tween-80. No single extraction buffer will be optimally effective in protecting all enzymes from any given tissue or any given enzyme from all taxa, as demonstrated in several experiments designed to test the effects of different extraction buffers on isozyme expression

(Kelley and Adams, 1977a; Wilson and Hancock, 1978; Soltis et al., 1980; Pitel and Cheliak, 1984). In addition, genotype-specific responses to different extraction buffers may occur (Watson and Cook, 1982), and some buffer additives may inhibit or reduce quantitative recovery of certain enzymes [e.g., PVP is reported to inhibit glutamine synthetase (Sadler and Shaw, 1978); DIECA inhibits the Cu-Zn isozymes of superoxide dismutase (Jaaska, 1982)]. Few investigators have the resources to extract each sample repeatedly in several different grinding buffers; therefore, a compromise extraction buffer that yields the sharpest bands for the broadest spectrum of isozymes is typically chosen. Empirical testing of extraction buffers may be time-consuming and laborious, but inevitably becomes effort well spent.

The trial and error approach has led to the development of a large number of highly useful extraction buffers, many of which differ only in minor detail. Those given below represent modal types derived and modified from numerous sources, and should be considered illustrative. There is a great deal of variation in constituent molarities reported for each of the following buffers, which represent intermediate stages in a continuum of possibilities. A guiding principle in the selection and development of extraction buffers should be to use only those additives that are necessary, and to use no greater buffering capacity than is needed.

1. *Extraction buffer for tissues low in phenolics or other interfering substances*

50 mM	Phosphate or Tris-HCl buffer, pH 7.5
5 %	Sucrose (w/v)
14 mM	Mercaptoethanol (0.1% v/v)

   Modifications: Glycerol may substitute for sucrose. Small quantities of detergent may be added (0.1% Tween-80 or Triton X-100). Dithiothreitol (DTT) or dithioerythritol (DTE) (1–3 mM) may substitute for mercaptoethanol.

2. *Extraction buffer for tissues with moderate levels of interfering substances*

75 mM	Phosphate or Tris-HCl buffer, pH 7.5
5 %	Sucrose (w/v)
5 %	PVP-40 (w/v)
14 mM	Mercaptoethanol (0.1% v/v)
50 mM	Ascorbic acid, Na-salt
10 mM	Diethyldithiocarbamate
0.1 %	Bovine serum albumin (w/v)

   Modifications: Small quantities of detergent (0.1 %) or sodium metabisulfite (10 mM) may be helpful. Glycerol may substitute for sucrose. Millimolar quantities (e.g., 10 mM) of DTT or DTE may be included. 10 mM KCl or $MgCl_2$ are often added.

3. *Extraction buffer for tissues with high levels of interfering substances*

100 mM	Tris-HCl or Tris-maleate buffer, pH 7.5
7 %	Sucrose (w/v)
10 %	PVP-40 (w/v)
14 mM	Mercaptoethanol (0.1% v/v)
250 mM	Ascorbic acid, Na-salt
20 mM	Diethyldithiocarbamate
1.0 %	Bovine serum albumin (w/v)
20 mM	Sodium metabisulfite
200 mM	Sodium tetraborate

   Modifications: Glycerol may substitute for sucrose. Detergents such as Tween-80, Triton X-100 or PEG (up to 2%) are sometimes included. Millimolar quantities (e.g., 10 mM) of DTT or DTE and 10 mM KCl or $MgCl_2$ are often added.

Subcellular Localization of Isozymes

Many enzymes routinely examined in electrophoretic studies have complex phenotypes resulting from the expression of more than a single structural gene, that is, more than one isozymic form may be present. These isozymes may or may not reside in the same subcellular compartment (see Gottlieb, 1981). Genetic analyses and organelle extractions conducted in many species indicate that alternate isozymes capable of catalyzing similar reactions often occur in more than one compartment (e.g., cytosol, chloroplast, mitochondrion, glyoxysome, and peroxisome). For some applications it is desirable or even essential to identify the portion of the phenotype contributed by each subcellular isozyme or isozymes. Interpretation of complex phenotypes involves a series of experimental steps, including formal genetic analyses and determination of subcellular compartmentalization of alternate isozymes. Most organellar isozymes are located in the chloroplasts and mitochondria, reviews of which have been recently published by Weeden (1983) and Newton (1983). This section is concerned with methodology involved in determining compartmentalization of plant isozymes.

One of two techniques is generally employed to demonstrate the presence of a particular isozyme in organelles: analysis of pollen extracts, and organelle isolation following subcellular fractionation. Weeden and Gottlieb (1980) demonstrated that a variety of buffers was effective in eluting cytoplasmic isozymes from small quantities (e.g., 5 mg) of pollen from several species, but that plastid, mitochondrial, and microbody isozymes were not eluted as rapidly. Organellar isozymes are readily extracted if the pollen is crushed, however, facilitating interpretation of zymograms through comparison. This procedure alone does not allow assignment of particular isozymes to specific organelles, but has found wide acceptance as a first step in the interpretation of complex zymograms (e.g., Warwick and Gottlieb, 1985; Bayer and Crawford, 1986). Some precautions should be observed in the interpretation of isozyme patterns derived from pollen. Absence of an isozyme in soaked-pollen extracts does not necessarily indicate that the isozyme is organellar, because it may simply not be expressed or may not be leachable. Moreover, it should be noted that the suite of isozymes expressed in pollen grains may not be identical to that expressed in somatic tissues (Tanksley et al., 1981; Sari-Gorla et al., 1986). Some isozyme loci may have expressions restricted to either gametophytic or sporophytic tissues.

A more direct approach to the determination of subcellular localization of isozymes is organelle isolation and electrophoretic analysis of purified organellar extracts. In most cases, organelles are isolated by cell fractionation and various combinations of differential and gradient centrifugation (techniques reviewed by Quail, 1979). A large number of protocols exist, and species vary widely in the degree of difficulty of organelle isolation. The problems posed are several, and include selection of the appropriate osmoticum and buffer additives (again, phenolics are often a problem), homogenization techniques, and the centrifugation medium and procedures that will allow intact organelles to be isolated. Detailed protocols are found in many of the references cited by Quail (1979), Newton (1983), and Weeden (1983). Examples of organelle purification for the purpose of clarifying subcellular localization of plant isozymes are provided in Baum and Scandalios (1979), Gastony and Darrow, (1983), Tanner et al. (1983), Odrzykoski and Gottlieb, (1984); Pichersky et al. (1984), Tanksley and Kuehn (1985), Arulsekar et al. (1986), Doebley et al. (1986), Weeden and Robinson (1986), Wendel et al. (1986), and Morden et al. (1987). Once organelles have been isolated, their purity may be checked by light microscopy and by assaying for the activity of organelle-specific marker isozymes. Among the more easily visualized markers are: plastid-specific forms of glucosephosphate isomerase, aspartate aminotransferase, triosephosphate isomerase, phosphoglucomutase, 6-phosphogluconate dehydrogenase, and superoxide dismutase; mitochondrial-specific isozymes of malate dehydrogenase, superoxide dismutase and aspartate aminotransferase; and for microbodies, catalase and peroxidase. Many of these same enzymes have counterparts which may serve as appropriate cytosolic markers.

Despite repeated efforts, the tissues of some plant species are particularly intractable with respect to isolation of intact organelles. In some species, chloroplast envelopes are very sensitive to cellular disruption, and even extensive periods of destarching (in the dark) and gentle homogenization fail to yield intact chloroplasts. High endogenous levels of phenolic compounds have long been recognized as a problem in organelle isolation (Anderson, 1968). However, through the inclusion of some of the grinding buffer additives that improve extract quality (see above), it is often possible to generate organelle-enriched subcellular pellets. While these are unlikely to yield definitive evidence regarding intracellular localizations of isozymes, they often provide clues in the form of altered band intensities (e.g., in comparisons of crude whole-cell extracts with organellar pellets). These clues may be of particular value in generating hypotheses that can be tested through the design of appropriate crossing experiments.

GEL PREPARATION

Proper preparation of starch gels is necessary to obtain desirable gel handling properties and adequate isozyme separation and clarity. Important factors include the buffers used in the gel and the electrode reservoirs, the electrophoretic apparatus, and the cooking method employed in gel preparation. While all of these are consequential, the most critical influence on the quality of results is often the electrophoretic buffers chosen.

Gel and Electrode Buffers

Most buffer systems for starch gel electrophoresis were developed in the 1960s for use with animal tissues and have been summarized in the widely cited and influential compilations by Brewer and Sing (1970) and Shaw and Prasad (1970). Many of the formulations presented in those publications have been adapted and modified by numerous investigators, both in the plant and animal sciences, so that the literature currently contains an almost endless variety of recipes. Unfortunately, publication of a novel buffer system did not necessarily make it useful, particularly for plant scientists. The novice is thus faced with an overwhelming array of options, with little available guidance beyond that which could be gleaned from miscellaneous technical bulletins and advice from the more experienced.

This past decade has witnessed an explosion in the number of electrophoretic investigations of plant isozymes, with a concomitant increase in our understanding of which electrophoretic buffer systems are likely to be useful. Many methods manuals have been published in recent years for specific plants or plant groups, including maize (Cardy et al., 1983), conifers (O'Malley et al., 1980; Conkle et al., 1982), ferns (Soltis et al., 1983), and sorghum (Morden et al., 1987). Through the accumulated experience of numerous investigators, it has become clear that only a few gel and electrode buffer types, each with several common modifications, are being used fruitfully for the resolution of a broad spectrum of enzymes. These basic types are presented below, with the hope of expediting progress for those initiating isozyme studies. The buffer systems listed are intended to serve as starting points. None of these systems is likely to be ideal for all taxa or for every enzyme that can be electrophoretically resolved, nor even for every variant allele at a single locus (e.g., Stuber and Goodman, 1983). For this reason, many common modifications are also presented. As with extraction buffers, optimization of electrophoretic buffers requires an empirical approach, and laboratory and workload constraints are likely to force compromise selections.

1. *Histidine-citrate, pH 5.7 (Stuber et al., 1977)*

    Electrode Buffer: 0.065 M L-Histidine, free base
                              0.019 M Citric acid, monohydrate
                              Dissolve L-Histidine (10.09g/l) in H_2O. Lower pH to 5.7 with citric acid
                              (approx. 3.0 g/l).
    Gel Buffer: 0.009 M L-Histidine, free base
                              0.006 M Citric acid, monohydrate
                              pH 5.7
                              Prepared by dilution of 1 part electrode buffer with 6 parts H_2O.

Modifications: Frequently reported modifications include varying the pH (range = 5.0 to 6.5) by adjusting the citric acid molarity, and altering the ionic strength of the gel buffer (dilutions ranging from 1:3 to 1:12).

2. *Morpholine-citrate, pH 6.1 (Clayton and Tretiak, 1972)*

    Electrode Buffer: 0.040 M Citric acid, monohydrate
                                  Approx. 0.068 M N-(3-Aminopropyl)-morpholine
                                  Dissolve citric acid (8.41 g/l) in H_2O. Raise pH to 6.1 with N-(3-Aminopropyl)-morpholine (approx. 10 ml/l).
    Gel Buffer: Prepared by dilution of 1 part electrode buffer with 19 parts H_2O.

Modifications: Electrode and gel buffer pH are often changed (to between 5.5 and 6.5) by adjusting the amount of titrant. Higher ionic strength gel buffers are sometimes used (smaller dilution of electrode buffer, e.g., 1:9).

3. *Tris-citrate, pH 7.0 (Meizel and Markert, 1967)*

    Electrode Buffer: 0.135 M Tris (16.35 g/l)
                                    0.043 M Citric acid, monohydrate (9.04 g/l)
                                  pH 7.0
    Gel Buffer: 0.009 M Tris
                                  0.003 M Citric acid, monohydrate
                                  pH 7.0
                                  Prepared by dilution of 1 part electrode buffer with 14 parts H_2O.

Modifications: Frequently reported modifications include pH adjustment (range = 7.0 to 8.5) either by increasing the concentration of tris, decreasing the citric acid molarity, or a combination of these two methods. Gel buffer dilutions are occasionally increased up to 1:27.

4. *Citrate/histidine, pH 7.0 (Fildes and Harris, 1966)*

    Electrode Buffer: 0.410 M Citric acid, Na_3 salt (120.59 g/l), titrated to pH 7.0 with free citric acid.
    Gel Buffer: 0.005 M L-Histidine, free base (0.78 g/l), adjusted to pH 7.0 with NaOH.

Modifications: Electrode buffer titrant is sometimes HCl, or NaOH if free citrate is used instead of its sodium salt. Either histidine or histidine-hydrochloride may be included in the gel buffer, with molarities ranging from 0.005 to 0.020.

5. *Sodium-borate, pH 8.0/Tris-citrate, pH 8.6 (Poulik, 1957)*

    Electrode Buffer: 0.300 M Boric acid (18.55 g/l), adjusted to pH 8.0 with NaOH.
    Gel Buffer: 0.076 M Tris (9.20 g/l), titrated to pH 8.6 with citric acid (final citrate molarity = approximately 0.005).

Modifications: Most modifications of the electrode buffer involve changing the pH (range = 7.7–8.6) by adjusting the amount of NaOH. Gel buffer recipes vary considerably, both with respect to constituent molarities (molarity of tris is typically in the range of 0.015 to 0.100) and pH (through titration with citric acid).

6. *Lithium-borate, pH 8.3/Tris-citrate, pH 8.3 (Ashton and Braden, 1961)*

    Electrode Buffer: 0.192 M Boric acid (11.87 g/l), titrated to pH 8.3 with Lithium hydroxide (approx. 1.6 g/l; final molarity approximately 0.038).
    Gel Buffer: 0.019 M Boric acid
                                  0.004 M LiOH
                                  0.047 M Tris
                                  0.007 M Citric acid

This buffer is prepared by adding 1 volume of electrode buffer to 9 volumes of Tris-citrate buffer, pH 8.3 (0.052 M Tris-0.008 M citric acid).

Modifications: Most modifications are minor and entail adjustments of gel and/or electrode buffer pH (range = 7.6–8.5) through alterations in constituent molarities. An often used and more substantial modification is that of Ridgeway et al. (1970), who introduced a buffer system that bears similarities to both systems 5 and 6 (electrode buffer = 0.300 M boric acid, titrated to pH 8.1 with LiOH; gel buffer = 1 part electrode buffer to 99 parts 0.030 M tris-0.005 M citric acid, pH 8.5).

7. Tris-Borate-EDTA, pH 8.6 (Markert and Faulhaber, 1965)

Electrode Buffer: 0.180 M Tris (21.80 g/l)
0.100 M Boric acid (6.18 g/l)
0.004 M EDTA, Na_2 salt (1.34 g/l)
pH 8.6
Gel Buffer: 0.045 M Tris
0.025 M Boric acid
0.001 M EDTA, Na_2 salt
pH 8.6
Prepared by dilution of 1 part electrode buffer with 3 parts H_2O.

Modifications: The majority of TBE systems are similar to the one given, with alterations usually minor and consisting of small changes in buffer pH and/or constituent molarities. More substantial modifications are occasionally reported (e.g., Siciliano and Shaw, 1976).

Preparing Starch Gels

Commercially available starch is generally obtained from one of three sources (Connaught Laboratories, Willowdale, Ontario; Electrostarch Co., Madison, Wisconsin; Sigma Chemical Co., St. Louis, Mo.) or their distributors. It is a biological product that is subject to an alarming amount of variation from lot to lot, both in cooking and electrophoretic qualities. For this reason it is generally advisable to acquire sufficient quantities of a single lot to last an entire study. Gels are generally prepared with approximately 10–12% starch (w/v), but usable gels may be formed containing as little as 9 or as much as 14% starch. Sucrose (3.5%, w/v) is occasionally added. Sucrose adds resistance, and thus increases voltage (and hence migration rate) for any given wattage relative to gels that lack sucrose. Sucrose may also be a necessary gel constituent for optimal resolution of some isozymes [e.g., *Gossypium* MDH isozymes are significantly sharper with the inclusion of sucrose (J. F. Wendel, unpublished data)].

Starch gels are generally prepared by one of two methods. The first, using a microwave oven, is preferred because it gives the greatest reproducibility of results. Approximately three-fourths of the buffer required for a single gel (varies according to the specific gel mold used) is heated in an Erlenmeyer flask in a microwave oven until boiling, while the remaining volume of buffer is mixed with the starch (and sucrose, if used) in another flask at room temperature and swirled to form an even suspension. The boiling buffer is poured into the starch suspension, vigorously mixed for a few seconds, and returned to the microwave oven, where it is heated until it begins to boil. The solution is then degassed with an aspirator and poured into an acrylic gel mold (which can be treated with 0.5% Photoflo to prevent sticking). The gel is allowed to cool until it reaches room temperature (approximately 30 min.), covered with polyethylene film or Saran Wrap to prevent dehydration, and left (often overnight) until the samples are loaded for electrophoresis. Several factors influence the required cooking times, including buffer volumes and the brand and wattage of the microwave oven used. Once the appropriate times have been determined, they should be adhered to for maximum reproducibility of results. If a microwave oven is not available, the solution of starch and buffer (using all of the buffer) may be heated over a flame (with continuous and vigorous swirling) or on a hot plate with magnetic stirrer until a clear boiling solution is obtained. This solution is then degassed and treated as above.

Gel molds are commercially available in a variety of sizes from several sources, but are often locally manufactured. They generally range in volume from 300 to 600 ml. A particularly useful type of gel mold (Johnson and Schaffer, 1974; see Fig. 1.2) provides direct contact between the gel and electrode buffer (Fig. 1.3) thus obviating the need for the conventional sponge electrode wicks. Additional diagrams or photographs of apparatus and gel cooking methods used in starch gel electrophoresis are presented in Smith (1976), Conkle et al. (1982), Cardy et al. (1983), Werth (1985), and Micales et al. (1986).

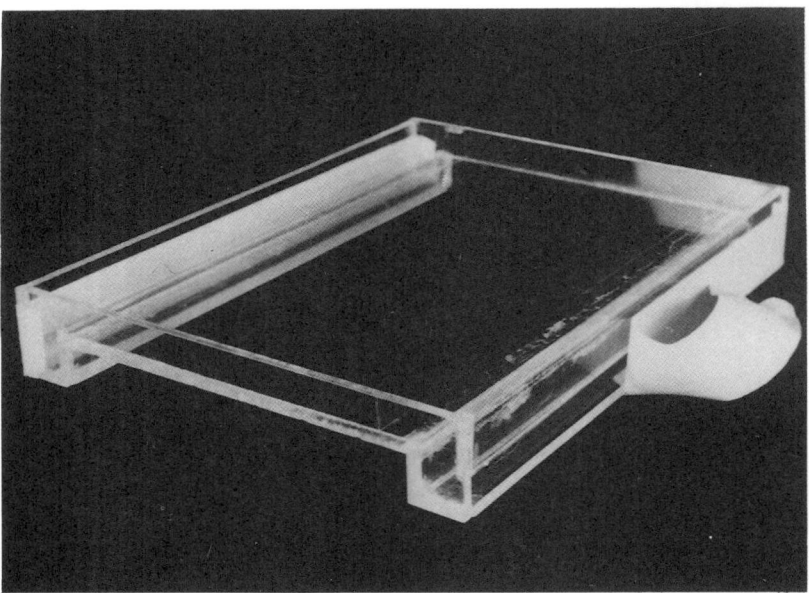

Fig. 1.2. Gel mold design. The acrylic gel tray shown is designed to provide direct contact between the gel and the electrode buffers, and hence avoid the use of conventional sponge electrode wicks (Johnson and Schaffer, 1974). During gel preparation the cathodal and anodal ends of the mold are sealed by masking tape, which is removed after sample loading.

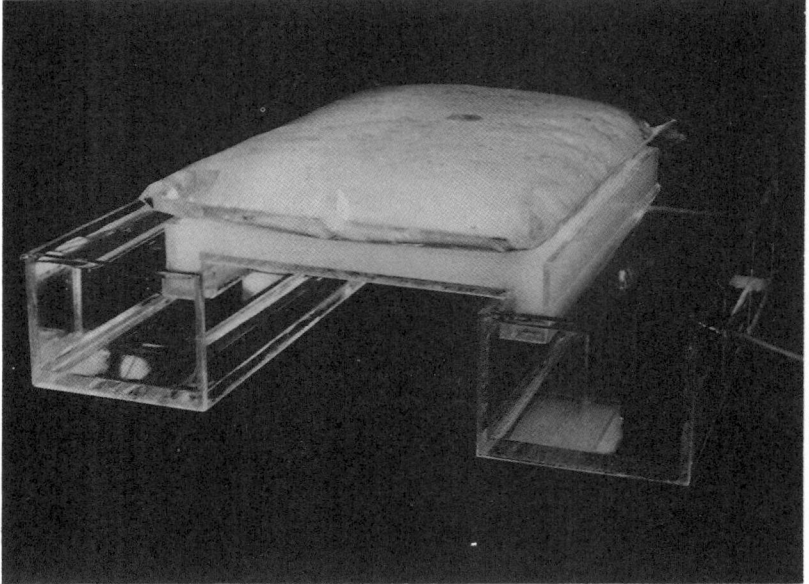

Fig. 1.3. Electrophoretic apparatus. The gel is placed on the electrode buffer tanks and covered with a chilled water pack. Electrodes are 24-gauge platinum wire.

Fig. 1.4. Gel slicing apparatus. Gel slices are cut by pulling a tightened wire horizontally through the gel slab on an acrylic slicing bed. The slicing wire is elevated by strips of formica (1.25 mm thick) placed on both sides of the gel. Additional strips, secured by corner posts, are added for each subsequent slice.

ELECTROPHORESIS

Prior to application of the samples, the gels are cooled to approximately 4° C in a refrigerator or on trays with ice. This procedure generally takes about 20 minutes, during which time final preparations for gel loading are made and electrode buffer trays are filled with the appropriate buffers. Once the gel is cool, the polyethylene covering or Saran Wrap is trimmed with scissors or a razor blade so that it covers only the surface of the gel. The wrap is pulled back from the cathodal edge of the gel and a slit is cut perpendicular to the gel surface approximately 3 cm from the edge of the gel to serve as the origin. The cathodal strip of gel is gently pulled back to allow ease of sample insertion.

Samples are applied to the gel as extracts absorbed onto wicks that are typically cut or punched from filter paper (often Whatman 3MM, although other grades are sometimes used) but may be composed of cellulose acetate membranes, which are reported to have a number of desirable properties (Werman, 1986). Wick size varies among laboratories, but most fall within the range of 2–4 mm (width) × 8–15 mm (length) depending on the dimensions of the gel mold and the desired sample number per gel. The gel mold illustrated in Figs.1.2–1.4 allows 30 wicks, 2 mm × 11 mm, to be inserted approximately every 3 mm. Excess sample homogenate is blotted from each wick to prevent contamination of neighboring lanes. The gel slit is gently opened and the sample wicks inserted. Control samples representing reproducible genotypes run on all gels throughout a study, and wicks containing bromophenol blue or food coloring as tracker dyes are often interspersed with sample wicks. Air bubbles introduced between the underside of the gel and the gel mold are squeezed out, and the polyethylene or Saran Wrap is replaced on the surface of the gel. Wicks are often removed from gels after approximately 15 minutes of electrophoresis, but resolution is generally of equivalent quality if the cathodal and anodal sides of the origin are pressed together by insertion of a 3 mm thick beveled section of acrylic between the cathodal end of the mold and the cathodal end of the gel.

Contact between the gel and electrode buffer may either be direct (Fig. 1.3) or be established by cloth or sponge wicks that serve as bridges between the electrode reservoirs and the ends of the gel. It is important that gels be kept cool during electrophoresis, and for this reason chilled water bags (Fig. 1.3) or equivalent (Blue Ice packs) may be placed on top of the gels for the duration of the run. Bands are usually sharpest if electrophoresis is performed at 4° C. Any type of refrigerated chamber may house the gels during electrophoresis; the only essential criterion is effective and stable cooling. An electrical potential is applied across the gel by a power supply for a period of time and wattage that must be empirically determined. Because of resistance changes that often occur during an electrophoretic run, constant wattage is preferred over constant voltage or amperage (Schaffer and Johnson, 1973). However, for some enzymes and some electrophoretic buffer systems, these more expensive power supplies do not appear to be essential.

After electrophoresis, the gel is removed from the refrigeration chamber and a rectangular slab is cut by trimming away excess starch from the sides of the gel. The cathodal strip should also be sliced and stained unless previous results have demonstrated that cathodal activity is lacking. A predetermined corner of the gel is cut or notched in some way so that proper orientation can be determined during later manipulations. Gel slices are cut by pulling a tightened wire or nylon thread horizontally through the gel slab on an acrylic slicing bed. Numerous apparati for gel slicing have been devised. In the method illustrated in Fig. 1.4, a strip of formica or acrylic is placed on each side of an acrylic bed to elevate the slicing wire. The thickness of these strips varies among laboratories, but is usually in the range of 1–2 mm. Additional strips, either secured by corner posts or stacked on top of each other, are sequentially added until the desired number of slices have been cut. Many investigators use a simpler system consisting of an acrylic sheet with a single strip securely fastened or glued to each side. After each slice is cut, the gel is turned over, the slice is removed, and the gel is returned to the sheet for additional slices. Whichever method is used for gel slicing, the gel slices are placed in staining trays or boxes for staining.

Several problems may be encountered that hamper effective use of starch gels, including some that arise from improper preparation or electrophoresis of gels. A useful diagnostic "troubleshooting" guide is presented in Micales et al. (1986).

ENZYME VISUALIZATION

Gel Staining

Following electrophoresis, isozymes are detected *in situ* through the use of specific activity stains. General principles of enzyme activity staining and the chemistry of enzyme localization in gels have been reviewed by Gabriel (1971), Ostrowski (1983), and Vallejos (1983). Following immersion of the gel slice into the staining solution, the substrates and other required reagents diffuse into the gel where they are acted upon by the enzyme under study. Detection is primarily based upon precipitation of soluble indicator dyes, which become insoluble and colored in zones of enzyme activity. Dyes may be various, but most stains employ either soluble tetrazolium salts (e.g., MTT, NBT), which become insoluble blue formazans upon reduction by enzymatically generated reduced pyridine cofactors (NAD or NADP), diazonium salts (e.g., Fast Blue BB, Fast Black K, Fast Garnet GBC), which precipitate as colored azo dyes upon chemical coupling of the salt with an enzymatically cleaved substrate, or various redox dyes (e.g., dichlorophenol indophenol, 3-amino-9-ethylcarbazole) that undergo changes in color and/or solubility upon oxidation. Natural or artificial fluorochromes and ultraviolet light are used for the detection of some enzymes; these methods rely on either generation of fluorescence upon enzymatic cleavage of a non-fluorescent substrate (4-methylumbelliferyl derivatives) or loss of fluorescence upon enzymatic oxidation (e.g., oxidation of NADH).

Several publications contain valuable protocols and lists of reagents for enzyme staining (Brewer and Sing, 1970; Shaw and Prasad, 1970; Gabriel, 1971; Harris and Hopkinson, 1976; Siciliano and Shaw, 1976; Conkle et al., 1982; Cardy et al., 1983; Soltis et al., 1983; Vallejos, 1983; Heeb and Gabriel, 1984). Some of these publications focus on isozyme detection in specific groups of plants and are consequently limited in their coverage. Others are too inclusive, in that they contain staining protocols for enzymes that have either not been detected in plants or that are sub-optimal for the detection of plant isozymes. Whereas staining protocols exist for over 200 different enzymes, only about 40 of these have proven useful in plants. To provide a single source for visualization of the majority of useful plant isozymes, annotated recipes for their detection are presented below. These recipes are derived from numerous sources, including those listed above. Most, however, have been substantially modified from their original form by numerous investigators. Several versions of many enzyme stains have proven effective in plants, and for this reason modifications for optimization are suggested below each recipe. In addition, appropriate electrophoretic buffer systems are recommended for many enzymes, particularly for those frequently studied. Suggested electrophoresis buffers are listed below each stain. These recommendations should be interpreted as starting points, however; optimal resolution in each taxon may require additional experimentation.

For most of the following recipes, quantities of reagents have been given that are appropriate for 50 ml of staining solution, which is a sufficient volume for the majority of commonly used stain boxes and gel sizes. These volumes and reagent quantities may be adjusted as necessary as long as constituent molarities remain similar. In some cases, intermediate products of the histochemical reaction are highly diffusable and must be localized to maintain maximum band sharpness. This is usually accomplished by applying the stain as an agar or agarose overlay, in which case volumes are less than 50 ml. A popular alternative to the use of agar overlays is to prepare stains in minimal volumes, which are applied dropwise and allowed to soak into the surface of the gel slices. Both of these alternative methods also reduce cost.

Procedures for preparing the stains vary among laboratories, but are often simplified by dispensing frequently used reagents as aqueous solutions in a convenient volume. Nitro blue tetrazolium (NBT), tetrazolium thiazolyl blue (MTT), and cofactors such as NAD and NADP are often refrigerated as 10 mg/ml solutions (5 mg/ml for NADP), and used in 1 ml volumes. Similar solutions (often in other concentrations, however) may be prepared for other indicator dyes and many salts and substrates. Buffers are usually stored as stock solutions of 10–20 times the concentrations used in the stains. Some coupling enzymes, such as glucose-6-phosphate dehydrogenase, are purchased as lyophilized powders, which may be diluted and frozen in vials or tubes in the appropriate quantities for the assays. For each of the following stains, reagents that are often used as aqueous solutions are indicated by listing an appropriate volume along with the recommended amount. Shelf life of reagents stored in this manner varies widely, and it is generally advisable to prepare only enough reagent solution for a couple of weeks of lab work. Many stains are sensitive to light and/or begin to deteriorate shortly after they are made. For this reason stains are generally made immediately prior to their addition to the gel. All incubations listed are for 37° C in the dark unless otherwise indicated.

Although phenazine methosulfate (PMS) is listed as the intermediary catalyst for most assays involving tetrazolium stains, an alternative compound, meldolablue (MLB), may substitute for PMS in many assays (Turner and Hopkinson, 1979). Stains using MLB in some cases may have slower reaction times than stains using PMS, but result in gels with less background staining.

Some staining procedures require the addition of glucose-6-phosphate dehydrogenase (G6PDH) as a coupling or "helper" enzyme in assays where endogenous glucose-6-phosphate can be produced (e.g., GPI, PGM). G6PDH dehydrogenates glucose-6-phosphate and in the process reduces a pyridine nucleotide, which, coupled to the phenazine-tetrazolium system, results in the precipitation of blue formazans. Most of the original staining schedules called for a nicotinamide dinucleotide phosphate (NADP)-dependent form of G6PDH and as a consequence required NADP as the pyridine nucleotide cofactor. Modified enzyme stains that substitute nicotinamide dinucleotide (NAD) and NAD-dependent G6PDH for NADP and NADP-dependent G6PDH, with no resultant loss of staining intensity, have been presented by Buth and Murphy (1980). Because NADP is approximately 10 times as expensive as NAD, use of the NAD and NAD-dependent G6PDH alternatives will result in considerable monetary savings. Some commercial preparations of G6PDH are active with either cofactor (NAD/NADP-dependent G6PDH), and use of this type of G6PDH with NAD can also be recommended. To indicate this preference, enzyme stains that require G6PDH carry the notation "(NAD)" following the G6PDH.

Nomenclature of the enzymes listed follows the guidelines and recommendations of the nomenclature committee of the International Union of Biochemists (I.U.B., 1984). For each enzyme, the recommended colloquial name, systematic name (I.U.B.:), and enzyme commission (E.C.) number are given. Commonly used synonyms and acronyms are parenthetically listed after the recommended name.

N-Acetyl-β-glucosaminidase (NAG). I.U.B.: N-Acetyl-β-D-glucosaminide N-acetylglucosaminohydrolase. E.C. 3.2.1.30

Solution A:	
100 mM Na-citrate buffer, pH 4.5	15 ml
Agar or agarose	200 mg
Solution B:	
100 mM Na-citrate buffer, pH 4.5	15 ml
4-Methylumbelliferyl-N-acetyl-β-D-glucosaminide	10 mg

Procedure: Prepare solutions A and B. Bring solution A to a boil. Cool to 60° C. Gently mix in solution B and pour on gel. Allow agar to set and view under long wave UV (370 nm). 4-Methylumbelliferone diffuses rapidly, so gel should be photographed or scored immediately upon band development.

Comments: Alternatively, use only solution B and apply dropwise to the gel with a pasteur pipette. NAG is rarely studied, and reports of variation are few (Weeden and Marx, 1987).

Buffer systems: Too few species have been examined to allow specific recommendations. Weeden and Marx (1987) used a modification of buffer system 6.

Acid phosphatase (ACP). I.U.B.: Orthophosphoric-monoester phosphohydrolase. E.C. 3.1.3.2

50mM Na-acetate buffer, pH 5.0	50 ml
Na-α-napthyl acid phosphate	50 mg
$MgCl_2$	50 mg (1 ml)
Fast Garnet GBG salt	50 mg (1 ml)

Procedure: Dissolve reagents in buffer and pour on gel. Incubate in the dark until desired staining intensity has occurred. Rinse, and store in water or fix.

Comments: Plants contain a variety of enzymes with phosphatase activity, some of which may have broad and overlapping substrate specificities in vitro. For any species and tissue it is good practice to experiment with stain modifications to visualize the products of other or additional loci. Common modifications include altering the pH (usually between 4.5 and 6.0), using a different color reagent (e.g., o-dianisidine, Fast Black K salt) or employing other artificial substrates [e.g., Na-β-napthyl acid phosphate, napthol-AS-BI phosphate with hexazotized pararosaniline (Barka, 1960)]. Caution should be exercised in interpreting "alkaline phosphatase" (E.C. 3.1.3.1) or "fructose bisphosphatase" (E.C. 3.1.1.11) banding patterns as distinct isozymes without checking for comigration of products with "acid phosphatase" as stained above. In maize coleoptiles, for example, all three stains reveal the products of the same locus (J. F. Wendel, unpublished data), despite the fact that the pH of the former two stains is typically 8.0 (e.g., Shaw and Prasad, 1970).

Buffer systems: Because of the heterogeneity of isozymes capable of cleaving the substrate, it is difficult to state with confidence which buffer systems will yield optimal resolution. In many species, however, the lower pH buffer systems seem to give the clearest patterns.

Aconitate hydratase (ACO) [= Aconitase]. I.U.B.: Citrate (isocitrate) hydro-lyase. E.C. 4.2.1.3

Solution A:	
50 mM Tris-HCl, pH 8.0	15 ml
Agar or agarose	200 mg
Solution B:	
50 mM Tris-HCl, pH 8.0	15 ml
$MgCl_2$	50 mg (1 ml)
NADP	5 mg (1 ml)

Neutralized (pH 7.0) cis-aconitic acid	75 mg (1 ml)
Isocitrate dehydrogenase	60 units
MTT	10 mg (1 ml)
PMS	2 mg (0.4 ml)

Procedure: Prepare solutions A and B. Bring solution A to a boil. Cool to 60° C. Gently mix in solution B and pour on gel. Once agar has solidified, incubate until blue bands appear. Stained gel may be refrigerated unfixed for a few days or fixed.

Comments: cis-Aconitic acid is usually prepared as a neutralized solution (25mg/ml) and kept refrigerated for up to three weeks. In taxa with very active aconitase it may be possible to lower expenses by using reduced quantities of isocitrate dehydrogenase.

Buffer systems: Buffer system 3.

Adenylate kinase (ADK). I.U.B.: ATP:AMP phosphotransferase. E.C. 2.7.4.3

Solution A:
50mM Tris-HCl, pH 8.0	15 ml
Agar or agarose	200 mg

Solution B:
50mM Tris-HCl, pH 8.0	15 ml
$MgCl_2$	50 mg (1 ml)
NAD	10 mg (1 ml)
Adenosine diphosphate (ADP)	20 mg (1 ml)
Glucose	90 mg
Glucose-6-phosphate dehydrogenase (NAD)	40 units (1 ml)
Hexokinase (HEX)	170 units (1 ml)
MTT	10 mg (1 ml)
PMS	2 mg (0.4 ml)

Procedure: G6PDH and HEX, if purchased as lyophilized powders, are dissolved in H_2O and stored as frozen aqueous solutions in 1 ml vials. Prepare solutions A and B. Bring solution A to a boil. Cool to 60° C. Gently mix in solution B and pour on gel. Once agar has solidified, incubate until blue bands appear. Stained gel may be refrigerated unfixed for a few days or fixed.

Comments: Band intensity may be significantly affected by the concentration of stain constituents, which accordingly should be experimentally optimized.

Buffer systems: ADK has been studied in relatively few species (Wendel et al., 1988). Buffer systems 1, 2, and 3 have proven most useful.

Alanine aminotransferase (ALT) [= Glutamate-pyruvate transaminase (GPT)]. I.U.B.: L-Alanine:2-oxoglutarate aminotransferase. E.C. 2.6.1.2

200 mM Tris-HCl, pH 8.0	5 ml
NADH	10 mg
L-Alanine	20 mg
α-Ketoglutarate	10 mg
Lactate dehydrogenase	100 units

Procedure: Combine stain components in a small flask. Apply to gel dropwise with a pasteur pipette and allow to soak into gel surface. View under longwave UV (370 nm). Zones of activity are revealed as bands of non-fluorescence in an evenly fluorescent background. Development times may vary widely among tissues and taxa, but maximum clarity and visibility are usually obtained within 30 minutes. Gel cannot be preserved and must be photographed immediately through a yellow filter.

Comments: Care should be taken to minimize the volume of stain; excess solution will mask the zones of defluorescence. If higher molarities of substrates are used, as indicated in some published recipes, the stain pH should be checked and adjusted if necessary.

Buffer systems: Although few species have been studied (e.g., Wheeler and Guries, 1982; Wheeler et al., 1983; Strauss and Conkle, 1986) best resolution has been obtained with buffer systems 6 or 7.

Alcohol dehydrogenase (ADH). I.U.B.: Alcohol; NAD+ oxidoreductase. E.C. 1.1.1.1

50 mM Tris-HCl, pH 8.0	50 ml
NAD	10 mg (1 ml)
Ethanol	0.2 ml
NBT or MTT	10 mg (1 ml)
PMS	2 mg (0.4 ml)

Procedure: Combine ingredients and pour over gel. Incubate until bands are optimally developed. Rinse and store in water (if NBT is used) or fixative.

Comments: Many published staining methods for ADH use higher concentrations of substrate than suggested here. This may be unnecessary, and in some cases may even be inhibitory. Plant ADHs may have broad substrate specificities and/or be inadvertently revealed through contaminating levels of alcohols in other enzyme stains. Accordingly, other dehydrogenase zymograms should be checked for comigration with ADH as stained above.

Buffer systems: ADH is often easily resolved, and many different buffer systems provide adequate resolution. Bands are often sharpest with systems 5 and 6, but variants are often maximally separated with lower pH buffer systems.

Aldolase (ALD). See Fructose-bisphosphate aldolase.

Aminopeptidase (AMP), including leucine aminopeptidase (LAP) I.U.B.: α-Aminoacyl-peptide hydrolase. E.C. 3.4.11.1

200mM Tris-200mM maleate, pH 3.7	20 ml
0.2 N NaOH	15 ml
H_2O	15 ml
$MgCl_2$	50mg (1 ml)
L-leucyl-β-napthylamide·HCl	20 mg
N,N-dimethylformamide	2 ml
Fast Black K salt	20 mg

Procedure: Combine first three ingredients (should result in an assay pH of 6.0). Dissolve substrate in the dimethylformamide and add, along with remaining ingredients, to the stain buffer. Pour over gel and incubate until bands are optimally developed. Rinse and either fix or store refrigerated in water.

Comments: Numerous stain recipes have been published for plant aminopeptidase isozymes. In our experience, inclusion of $MgCl_2$ and use of a tris-maleate stain buffer maximize activity. Some authors obtain better results by omitting N,N-dimethylformamide. Many tissues and taxa contain a number of different aminopeptidases of varying substrate specificities, and activity level and isozyme number can vary dramatically according to which aminoacyl-napthylamide is included in the assay (Vodkin and Scandalios, 1981; Murray and Waters, 1985; Ott and Scandalios, 1985; Harry, 1986). Two others that are often useful are L-alanyl-β-napthylamide·HCl and L-arginyl-β-napthylamide·HCl.

Buffer systems: One of the lower pH buffer systems (1, 2, 3 or 4) usually gives the best results.

Amylase (AMY). I.U.B.: 1,4-α-D-Glucan glucanohydrolase (α-amylase); 1,4-α-D-Glucan maltohydrolase (β-amylase). E.C. 3.2.1.1 and 3.2.1.2

Procedure: Revealed as translucent areas or bands in the opaque starch matrix on gels stained for other enzymes, particularly those involving tetrazolium salts or Fast Black K.

Comments: Gels may be left overnight at room temperature to enhance band development. Regions of amylase activity are translucent or "clear", and should not be confused with achromatic regions resulting from the activity of superoxide dismutase. Amylases are usually studied on polyacrylamide gels (Kiang, 1981; Zimniak-Przybylska et al., 1985) using an iodine-potassium iodide assay (Vallejos, 1983). As discussed by Vallejos (1983), α- and β-amylases may be distinguished by differential sensitivity to heat, pH, and heavy metal ions.
Buffer systems: Best results are usually obtained with buffer system 6.

Aspartate aminotransferase (AAT) [= Glutamate oxaloacetate transaminase (GOT)]. I.U.B.: L-Aspartate:2-oxoglutarate aminotransferase. E.C. 2.6.1.1

AAT substrate solution*	50 ml
Fast Blue BB salt	50 mg (1 ml)

AAT Substrate Solution (pH 7.4)

H_2O	800 ml
α-Ketoglutaric acid	292 mg
L-Aspartic acid	1.07 g
PVP-40 (polyvinylpyrrolidone)	4.00 g
EDTA, Na_2 salt	400 mg
Sodium phosphate, dibasic	11.36 g

Procedure: Add Fast Blue BB to substrate solution and incubate at room temperature in the dark, until blue bands appear. Rinse and fix.
Comments: Substrate solution can be refrigerated for up to three weeks without noticeable loss of activity. Specific AAT isozymes can have relatively narrow pH optima, and adjusting the pH of the assay solution may change their relative staining intensities. Some published stain recipes use Fast Violet B instead of Fast Blue BB and/or include pyridoxal-5'-phosphate in the substrate solution. Pyridoxal-5'-phosphate is apparently a tightly bound prosthetic group, however, and we have found that its inclusion results in no increase in activity.
Buffer systems: Buffer systems 5 and 6.

Catalase (CAT). I.U.B.: Hydrogen peroxide:hydrogen peroxide oxidoreductase. E.C. 1.11.1.6

Hydrogen peroxide, 0.01%	50 ml
H_2O	50 ml
Ferric chloride	500 mg
Potassium ferricyanide	500 mg

Procedure: Pour H_2O_2 on gel slice and leave for 5 minutes (time may vary according to activity level). Meanwhile, mix remaining ingredients. Pour off peroxide and add stain solution. Swirl gently until bands are developed. Catalase activity is revealed as achromatic zones on a green background. Rinse three or four times and store in water.
Comments: Gel may be kept in water for a few days, but it is best to photograph it or score it shortly after staining. Results with this stain are usually superior to those obtained with the more common thiosulfate-potassium iodide method (Thorup et al., 1961).
Buffer systems: Systems 3, 5, and 6.

Diaphorase (DIA). See NAD(P)H Dehydrogenase.

Endopeptidase (ENP). E.C. 3.4.-.-

200 mM Tris-200 mM Maleate pH 3.7	25 ml
0.2 N NaOH	10 ml
H_2O	15 ml
$MgCl_2$	50 mg (1 ml)

α-N-Benzoyl-DL-arginine-β-Napthylamide·HCl	25 mg
N,N-Dimethylformamide	2 ml
Fast Black K salt	20 mg

Procedure: Mix together first three ingredients (pH should be 5.55). Dissolve substrate in solvent and add to buffer. Stir in Fast Black K salt and pour over gel. Incubate until dark brown bands appear. Rinse and fix.

Comments: It is not clear which protease(s) is revealed by this stain, and consequently no systematic name can be given. If activity levels are low, gel may be left in stain overnight at room temperature in the dark.

Buffer systems: Buffer systems 5 and 6.

Esterase (EST). E.C. 3.1.1.-

Method 1 (Colorimetric):

100 mM Na-phosphate, pH 6.0	50 ml
α-Napthyl acetate	25 mg
β-Napthyl acetate	25 mg
Fast Garnet GBC or Fast Blue RR salt	50 mg (1 ml)

Procedure: Substrates are usually added as solutions in acetone (e.g., 25mg/ml). Combine all ingredients and pour on gel. Incubate until red or brown bands appear. Rinse and fix.

Method 2 (Fluorescent):
Solution A:

| 50 mM Na-acetate buffer, pH 5.0 | 15 ml |
| Agar or agarose | 200 mg |

Solution B:

| 50 mM Na-acetate buffer, pH 5.0 | 15 ml |
| 4-Methylumbelliferyl acetate dissolved in 3 ml acetone | 10 mg |

Procedure: Prepare solutions A and B. Bring solution A to a boil. Cool to 60° C. Gently mix in solution B and pour on gel. Allow agar to set and view under long wave UV (370 nm). 4-Methylumbelliferone diffuses rapidly, so gel should be photographed or scored immediately upon band development.

Comments: It is not surprising that a large number of variations on the above staining methods exist. Plants contain numerous enzymes capable of hydrolyzing esters, some of which may have overlapping substrate specificities and different pH optima (see Chapter by Weeden and Wendel in this volume). Both of the above methods are non-specific and may result in the products of numerous loci being visualized. Experimentation with the stl, rather than the more commonly used filter paper overlay, considerably improves the fluorescent stain. Some of the common modifications of the colorimetric assay involve the use of other substituted α- and β-napthols (e.g., propionate, valate, butyrate) and changes in buffer composition (e.g., inclusion of 2 ml N-propanol), molarity, and pH (almost always in the range of 5.0–6.5, however).

Buffer systems: Due to the heterogeneity of the products visualized, it is difficult to generalize about appropriate electrophoretic buffers. Buffer systems 5 and 6 often provide the best resolution in many taxa, but all buffer systems employed should be tested.

Formate dehydrogenase (FDH). I.U.B.: Formate; NAD^+ oxidoreductase. E.C. 1.2.1.2

50 mM Tris-HCl, pH 8.0	50 ml
Formic acid, sodium salt	100 mg (1 ml)
NAD	10 mg (1 ml)
NBT or MTT	10 mg (1 ml)
PMS	2 mg (0.4 ml)

Procedure: Mix together all ingredients and pour over gel. Incubate until blue bands appear. Rinse and store in water (if NBT is used) or fix.

Comments: This rarely studied plant isozyme can be visualized in many species using the above assay (Wendel and Parks, 1982; Farinelli et al., 1983; Ellstrand and Marshall, 1985).

Buffer system: Buffer system 7.

Fructose-bisphosphate aldolase (FBA) [= Aldolase (ALD)]. I.U.B.: D-Fructose-1,6-bisphosphate D-glyceraldehyde-3-phosphate lyase. E.C. 4.1.2.13

50 mM Tris-HCl, pH 8.0	50 ml
NAD	10 mg (1 ml)
Arsenic acid, sodium salt	75 mg (1 ml)
Fructose-1,6-diphosphate, sodium salt	200 mg
Glyceraldehyde-3-phosphate dehydrogenase	100 units
NBT or MTT	10 mg (1 ml)
PMS	2 mg (0.4 ml)

Procedure: Combine ingredients and pour over gel. Incubate until bands are optimally developed. Rinse and store in water (if NBT is used) or fixative.

Comments: Caution should be used in interpreting all bands as isozymes of FBA. Enzymatic cleavage of fructose-1,6-diphosphate creates dihydroxyacetone phosphate and glyceraldehyde-3-phosphate, substrates for triosephosphate isomerase, which is usually abundant and highly active in plant extracts.

Buffer systems: Resolution is often poor; Weeden and Marx (1984) report relatively good bands using a modification of system 1.

Fructose-bisphosphatase (FBP) [= Fructose-1,6-diphosphatase (FDP)]. I.U.B.: D-Fructose-1,6-bisphosphate 1-phosphohydrolase. E.C. 3.1.3.11

50 mM Tris-HCl, pH 8.0	50 ml
MgCl$_2$	50 mg (1 ml)
NAD	10 mg (1 ml)
Fructose-1,6-diphosphate, sodium salt	120 mg
Glucosephosphate isomerase	40 units
Glucose-6-phosphate dehydrogenase (NAD)	40 units (1 ml)
MTT	10 mg (1 ml)
PMS	2 mg (0.4 ml)

Procedure: Mix together all ingredients and pour on gel. Zones of activity are revealed as blue bands. Rinse and fix when desired intensity of staining has been reached.

Comments: Costs may be reduced by preparing stain as an agar overlay in a smaller volume. An alternative staining method, based on precipitation of phosphate with calcium, has been published by Nimmo and Nimmo (1982). Reports of FBP isozymes in plants are few (Wheeler et al., 1983; Millar, 1985; Strauss and Conkle, 1986).y be non-specific and reveal the products of enzymes other than FBP.

Buffer systems: The few existing reports do not allow specific recommendations to be made.

Fumarate hydratase (FUM) [= Fumarase]. I.U.B.: (S)-Malate hydro-lyase E.C. 4.2.1.2

50 mM Tris-HCl, pH 8.0	50 ml
NAD	10 mg (1 ml)
Fumaric acid, sodium salt	200 mg
Malate dehydrogenase	200 units
MTT	10 mg (1 ml)
PMS	2 mg (0.4 ml)

Procedure: Mix together all ingredients and pour on gel. Zones of activity are revealed as blue bands. Rinse and fix when desired intensity of staining has been reached.

Comments: FUM isozymes are infrequently reported from plants (Guries and Ledig, 1982; Wheeler and Guries, 1982; Wheeler et al., 1983; Strauss and Conkle, 1986). The stain given may not be optimal, and modifications may prove worthwhile.

Buffer systems: No buffer system can be considered superior, due to the paucity of studies.

β-Galactosidase (GAL). I.U.B.: β-D-Galactoside galactohydrolase. E.C. 3.2.1.23

Method 1 (Colorimetric):

100 mM Na-phosphate buffer, pH 6.0	50 ml
β-Napthyl-β-D-galactopyranoside	25 mg
Fast garnet GBC or Fast blue RR salt	50 mg (1 ml)

Procedure: The substrate is usually added as a solution in acetone (e.g., 25mg/ml). Combine all ingredients and pour on gel. Incubate until red or brown bands appear. Rinse and fix.

Comments: A number of staining alternatives exist, with most alterations consisting of varying the buffer composition and pH (e.g., Wallner and Walker, 1975).

Method 2 (Fluorescent):
Solution A:

100 mM Na-phosphate buffer, pH 6.0	15 ml
Agar or agarose	200 mg

Solution B:

100 mM Na-phosphate buffer, pH 6.0	15 ml
4-Methylumbelliferyl-β-D-galactoside	10 mg

Procedure: Prepare solutions A and B (dissolve substrate in acetone). Bring solution A to a boil. Cool to 60° C. Gently mix in solution B and pour on gel. Allow agar to set and view under long wave UV (370 nm). 4-Methylumbelliferone diffuses rapidly, so gel should be photographed or scored immediately upon band development.

Comments: Stain buffer composition and pH vary widely in different reports [100 mM phosphate, pH 7.0 (Hughes, 1981); 50 mM citrate-phosphate, pH 4.5 (Yazdani and Rudin, 1982; Weeden and Robinson, 1986); 100mM Tris-HCl, pH 8.0 (Rieseberg and Soltis, 1987)]. It is probable that in many taxa there exist isozymes with low pH optima and others with high pH optima (Weeden, 1985). Gels should also be stained for glucosidase activity (Stuber et al., 1977), because some plant glucosidases have substrate specificities broad enough to cleave the above galactoside (J. F. Wendel, unpublished data).

Buffer systems: All buffer systems employed should be tested.

Glucose-6-phosphate dehydrogenase (G6PDH). I.U.B.: D-Glucose-6-phosphate:NADP$^+$1-oxidoreductase. E.C. 1.1.1.49

50 mM Tris-HCl, pH 8.0	50 ml
NADP	5 mg (1 ml)
MgCl$_2$	50 mg (1 ml)
Glucose-6-phosphate, Na$_2$-salt	50 mg
NBT or MTT	10 mg (1 ml)
PMS	2 mg (0.4 ml)

Procedure: Mix ingredients and pour over gel. Incubate until blue bands have appeared. Rinse and store in water (if NBT is used) or fixative.

Comments: There are occasional reports of G6PDH activity or band clarity being enhanced by inclusion of millimolar levels of NADP in the extraction and electrophoretic buffers (Brewer and Sing, 1970).

Buffer systems: Buffer systems 3, 5, 6, or 7.

Glucose-6-phosphate isomerase (GPI) [= Phosphoglucoisomerase (PGI)]. I.U.B.: D-Glucose-6-phosphate ketol-isomerase. E.C. 5.3.1.9

50 mM Tris-HCl, pH 8.0	50 ml
NAD	10 mg (1 ml)
Fructose-6-phosphate, Na$_2$-salt	20 mg
Glucose-6-phosphate dehydrogenase (NAD)	20 units
MTT	10 mg (1 ml)
PMS	2 mg (0.4 ml)

Procedure: Combine ingredients and pour over gel. Incubate until blue bands are optimally developed. Rinse with water and fix.

Comments: Most staining methods call for the inclusion of magnesium, which is not required for either the GPI or the coupling reaction.

Buffer systems: GPI activity is often easily detected with many electrophoretic buffers. Sharp resolution of the plastid isozymes, however, often requires a buffer system similar to number 5.

β-Glucosidase (GLU). I.U.B.: β-D-Glucoside glucohydrolase. E.C. 3.2.1.21

Method 1 (Colorimetric):

50 mM Na-phosphate buffer, pH 6.5	50 ml
6-Bromo-2-napthyl-β-D-glucopyranoside	50 mg
Fast Blue BB salt	50 mg (1 ml)
PVP-40 (polyvinylpyrrolidone)	1.0 g

Procedure: Add together all ingredients and pour on gel. Incubate until red bands appear. Rinse and fix.

Comments: This method is modified from Stuber et al. (1977). Band development may be slow, but gel slice can be left in the stain overnight if necessary.

Buffer systems: Systems 1, 2, 3, and 4.

Method 2 (Fluorescent):
Solution A:

50 mM Na-phosphate buffer, pH 6.5	15 ml
Agar or agarose	200 mg

Solution B:

50 mM Na-phosphate buffer, pH 6.5	15 ml
4-Methylumbelliferyl-β-D-glucopyranoside	10 mg

Procedure: Prepare solutions A and B. Bring solution A to a boil. Cool to 60° C. Gently mix in solution B and pour on gel. Allow agar to set and view under long wave UV (370 nm). 4-Methylumbelliferone diffuses rapidly, so gel should be photographed or scored immediately upon band development.

Comments: There are still too few reports using this method to specify the pH range that will allow activity to be visualized. Yeh and Layton (1979) report activity with a pH 4.0 citrate-phosphate buffer. In maize, isozymes are apparent within 5 minutes using the listed fluorescent method, whereas a few hours are required using the colorimetric method.

Buffer systems: Systems 1, 2, 3, and 4.

Glutamate-ammonia ligase (GS) [= Glutamine synthetase]. I.U.B.: L-Glutamate:ammonia ligase (ADP forming). E.C. 6.3.1.2

Reaction solution:

100 mM Tris-acetate, pH 6.5	100 ml
EDTA	25 mg
Arsenic acid, sodium salt	600 mg
Adenosine diphosphate	25 mg (1 ml)

L-Glutamine	780 mg
Hydroxylamine·HCl	180 mg
MnCl$_2$	200 mg (1 ml)
Staining solution:	
Trichloroacetic acid	5.0 g
Ferric chloride	10.0 g
Hydrochloric acid, 2 N	100 ml

Procedure: Combine all ingredients of reaction solution. Adjust pH to 6.5 with NaOH. Pour over gel and incubate 1–3 hours. Pour off the reaction mixture and rinse once with water. Add staining solution dropwise with a pasteur pipette and allow to soak into gel. Zones of GS activity will appear within a few minutes as brown bands on a yellow background. Photograph or score immediately.

Comments: Glutamine synthetase can be detected by either a glutamyl transfer reaction (above) or a biosynthetic reaction, as described by Barratt (1980). The reaction mixture listed above is modified from the transfer reaction mixture presented by Barratt (1980) and gives superior results in all taxa studied (Rayapati, unpublished data). The TCA-ferric chloride staining solution is stable and may be kept indefinitely at room temperature. Polyvinylpyrrolidone (PVP) and polyvinylpolypyrrolidone (PVpP), common components of enzyme extraction buffers, reportedly inhibit GS in some taxa (Sadler and Shaw, 1978), although inclusion of PVP has no detectable effect on GS activity in *Gossypium* (J. F. Wendel, unpublished data).

Buffer systems: In all taxa studied, best results have been obtained with buffer systems 5 and 7.

Glutamate dehydrogenase (GDH). I.U.B.: L-Glutamate:NAD+ oxidoreductase (deaminating). E.C. 1.4.1.2

50 mM Tris-HCl, pH 8.0	50 ml
NAD	10 mg (1 ml)
CaCl$_2$	50 mg (1 ml)
L-Glutamate, sodium salt	200 mg
NBT or MTT	10 mg (1 ml)
PMS	2 mg (0.4 ml)

Procedure: Mix together ingredients and pour over gel. Incubate until blue bands appear. Rinse and store in water (if NBT is used) or fixative.

Comments: Band development may be slow. If necessary, the gel slice may be left in the stain overnight at room temperature in the dark. Calcium is not included in the majority of GDH recipes, but is reported to be an activator (Scheid et al., 1980).

Glutamate oxaloacetate transaminase (GOT). See Aspartate aminotransferase.

Glutamine synthetase (GS). See Glutamate-ammonia ligase.

Glyceraldehyde-3-phosphate dehydrogenase (G3PDH) [= Triosephosphate dehydrogenase]. I.U.B.: D-Glyceraldehyde-3-phosphate:NAD(P)+ oxidoreductase (phosphorylating). E.C. 1.2.1.12 (NAD-dependent form); E.C. 1.2.1.9 (NADP-dependent form).

50 mM Tris-HCl, pH 8.0	50 ml
Arsenic acid, sodium salt	75 mg (1 ml)
NAD(P)	10(5) mg (1 ml)
Fructose-1,6-diphosphate	100 mg
Aldolase	50 units
NBT or MTT	10 mg (1 ml)
PMS	2 mg (0.4 ml)

Procedure: Combine ingredients and pour over gel. Incubate until blue bands appear. Rinse and store in water (if NBT is used) or fixative.
Comments: Plant tissues contain different NAD-dependent and NADP-dependent G3PDHs (Cerff, 1982). Separate staining solutions should be prepared for both forms. Costs may be reduced by preparing in a minimal volume or as an agar overlay.
Buffer systems: Systems 1, 2, 3, or 4.

Hexokinase (HEX). I.U.B.: ATP: D-Hexose-6-phosphotransferase. E.C. 2.7.1.1

Solution A:

50mM Tris-HCl, pH 8.0	15 ml
Agar or agarose	200 mg

Solution B:

50mM Tris-HCl, pH 8.0	15 ml
$MgCl_2$	50 mg (1 ml)
NAD	10 mg (1 ml)
Glucose	200 mg
Adenosine triphosphate	125 mg
Glucose-6-phosphate dehydrogenase (NAD)	40 units
MTT	10 mg (1 ml)
PMS	2 mg (0.4 ml)

Procedure: G6PDH is added as a 1-ml, frozen, aqueous solution. Prepare solutions A and B. Bring solution A to a boil. Cool to 60° C. Gently mix in solution B and pour on gel. Allow agar to set, and incubate until blue bands appear. Stained gel may be refrigerated unfixed for a few days or fixed.
Comments: Other hexose-phosphorylating enzymes exist in plants, and staining methods have been developed for some. For fructokinase (E.C. 2.7.1.4) use the same recipe but substitute fructose for glucose and include 40 units of glucosephosphate isomerase. In maize, both stains are utilized to facilitate locus assignment of variants, because fructose is a more specific phosphate acceptor than is glucose (Wendel et al., 1986).
Buffer systems: Buffer systems 1 or 3.

Isocitrate dehydrogenase (IDH). I.U.B.: Isocitrate:NAD(P)$^+$ oxidoreductase (decarboxylating). E.C. 1.1.1.42 (NADP form); E.C. 1.1.1.41 (NAD form)

Solution A: 50mM Tris-HCl, pH 8.0	15 ml
Agar or agarose	200 mg
Solution B:	
50mM Tris-HCl, pH 8.0	15 ml
$MgCl_2$ or $MnCl_2$	50 mg (1 ml)
NAD(P)	10(5) mg (1 ml)
Isocitric acid, Na_3-salt	100 mg
NBT or MTT	10 mg (1 ml)
PMS	2 mg (0.4 ml)

Procedure: Prepare solutions A and B. Bring solution A to a boil. Cool to 60° C. Gently mix in solution B and pour on gel. Once agar has solidified, incubate until blue bands appear. Stained gel may be refrigerated unfixed for a few days or the agar overlay may be gently removed and the gel fixed.
Comments: Both NAD and NADP specific IDHs occur in plants, and separate stains should be made for each. The use of an agar overlay, although not essential, considerably improves the clarity of bands.
Buffer systems: Best results are likely to be obtained with buffer systems similar to numbers 1, 2, 3, or 4.

Lactate dehydrogenase (LDH). I.U.B.: (S)-Lactate:NAD$^+$ oxidoreductase. E.C. 1.1.1.27

50 mM Tris-HCl, pH 8.0	50 ml
NAD	10 mg (1 ml)
Lactic acid, lithium salt	100 mg
NBT or MTT	10 mg (1 ml)
PMS	2 mg (0.4 ml)

Procedure: Combine all ingredients and pour over gel. Incubate until blue bands appear. Rinse and store in water (if NBT is used) or fixative.

Comments: Most published stain assays for LDH use neutralized 85% DL-lactate as the substrate (e.g., Siciliano and Shaw, 1976; Vallejos, 1983). McLeod et al. (1981), however, report no LDH activity in *Capsicum* with this substrate and recommend lithium lactate as an alternative.

Buffer systems: There are too few reports of plant LDH isozymes to allow specific recommendations to be made.

Malate dehydrogenase (MDH). I.U.B.: (S)-Malate:NAD$^+$ oxidoreductase. E.C. 1.1.1.37

50 mM Tris-HCl, pH 8.5	50 ml
NAD	10 mg (1 ml)
Malic acid	150 mg (1 ml)
NBT or MTT	10 mg (1 ml)
PMS	2 mg (0.4 ml)

Procedure: Malic acid (L- or DL-, which is less expensive) is usually added as a neutralized (with NaOH) aqueous solution. Ingredients are combined and poured over gel. Incubate until blue bands appear. Rinse and store in water (if NBT is used) or fixative.

Comments: MDH isozymes are often sharper if the starch gel is prepared with the inclusion of 4% sucrose. In some taxa, homogenization of tissue with ascorbic acid selectively eliminates the cytosolic forms of MDH, facilitating identification of the mitochondrial forms (Goodman et al., 1980).

Buffer systems: Although MDH is one of the more abundant and active plant isozymes and can be visualized on practically any gel, best resolution is nearly always obtained with buffer system 1 or a modification thereof.

Malate dehydrogenase (oxaloacetate-decarboxylating) (NADP$^+$) [= Malic enzyme (ME)]. I.U.B.: (S)-Malate:NADP$^+$ oxidoreductase (oxaloacetate-decarboxylating). E.C. 1.1.1.40

50 mM Tris-HCl, pH 8.0	50 ml
MgCl$_2$	50 mg (1 ml)
NADP	5 mg (1 ml)
Malic acid	150 mg (1 ml)
NBT or MTT	10 mg (1 ml)
PMS	2 mg (0.4 ml)

Procedure: Malic acid (L- or DL-, which is less expensive) is usually added as a neutralized (with NaOH) aqueous solution. Ingredients are combined and poured over gel. Incubate until blue bands appear. Rinse and store in water (if NBT is used) or fixative.

Comments: A parallel slice should be stained for MDH, as some MDH isozymes may be weakly active with NADP as a cofactor and hence be visualized with this stain.

Buffer systems: ME activity has been reported from all seven buffer systems, but results do not allow predictions for optimal electrophoretic resolution.

Malic enzyme (ME). See Malate dehydrogenase (oxaloacetate-decarboxylating) (NADP$^+$).

Mannose phosphate isomerase (MPI). I.U.B.: D-Mannose-6-phosphate ketol-isomerase. E.C. 5.3.1.8

Solution A:
50mM Tris-HCl, pH 8.0	15 ml
Agar or agarose	200 mg

Solution B:
50mM Tris-HCl, pH 8.0	15 ml
$MgCl_2$	50 mg (1 ml)
NAD	10 mg (1 ml)
Mannose-6-phosphate	30 mg
Glucosephosphate isomerase	50 units
Glucose-6-phosphate dehydrogenase (NAD)	40 units
NBT or MTT	10 mg (1 ml)
PMS	2 mg (0.4 ml)

Procedure: Prepare solutions A and B. Bring solution A to a boil. Cool to 60° C. Gently mix in solution B and pour on gel. Once agar has solidified, incubate until blue bands appear. Stained gel may be refrigerated unfixed for a few days or the agar overlay may be gently removed and the gel stored in water (if NBT is used) or fixed.

Comments: This is an infrequently studied plant enzyme (Wheeler and Guries, 1982; Wheeler et al., 1983; Gottlieb, 1984; Strauss and Conkle, 1986). Costs of this expensive stain may be reduced by using minimal volumes or preparing as an agar overlay.

Buffer systems: A modification of buffer system 7 typically has been used.

Menadione reductase (MNR). See NAD(P)H dehydrogenase.

NAD(P)H dehydrogenase (NAD(P)DH), including diaphorase (DIA), menadione reductase (MNR), and others. I.U.B.: NAD(P)H: (acceptor) oxidoreductase. E.C. 1.6.99.- and possibly others

Method 1:
Solution A:
50 mM Tris-HCl, pH 8.0	15 ml
Agar or agarose	200 mg

Solution B:
50 mM Tris-HCl, pH 8.0	15 ml
NADH	10 mg
MTT	10 mg (1 ml)
2,6-Dichlorophenol-indophenol	2 mg

Procedure: Bring solution A to a boil to dissolve agar. Cool to 60° C. Add solution B and pour on gel slice. Allow agar to set, and incubate until dark blue bands form on a light blue background. Store refrigerated.

Method 2:
50 mM Tris-HCl, pH 7.0	50 ml
NADH	10 mg
Menadione	25 mg
NBT	10 mg (1 ml)

Procedure: Mix ingredients and pour solution on gel. Incubate until pink to purple bands appear on a white background. Rinse and store in water or fix.

Comments: A diverse group of plant flavoproteins is capable of using reduced pyridine cofactors to reduce acceptor substrates, and these stains may have little specificity. In many plants, both methods reveal the same isozymes, whereas in others there may be overlap but not identity. Our present understanding of the homologies and specificities of the proteins revealed by these stains is limited.

Buffer systems: Because of the heterogeneity of enzymes that may be visualized with this stain, no specific recommendations can be offered.

Peptidase (PEP). E.C. 3.4.-.-

50 mM Tris-HCl, pH 7.0	50 ml
L-Amino acid oxidase	20 units
Peroxidase	40 units
Di- or Tri-peptides	20 mg
3-Amino-9-ethylcarbazole	25 mg
N,N-Dimethylformamide	3 ml

Procedure: Dissolve 3-amino-9-ethylcarbazole in the N,N-dimethylformamide. Add along with remaining ingredients to buffer and pour over gel. Incubate until red bands appear.

Comments: No systematic name can be given because it is not known which plant peptidases are visualized with this staining solution. Activity may be enhanced by preparation in a minimal volume of buffer or as an agar overlay. Many variations of this staining method have been published. The most common alterations involve choice of buffer composition and pH (e.g., 50 mM Na-phosphate, pH 6.5; 100 mM Tris-HCl, pH 8.0; 50 mM Na-borate, pH 7.0) and the specific dipeptides and/or tripeptides included in the incubation mixture. Commonly used substrates include glycyl-L-leucine, L-leucyl-L-tyrosine, L-leucyl-L-alanine, L-leucyl-L-leucine, L-leucyl-L-glycyl-L-glycine, and L-leucyl-L-leucyl-L-leucine. Different isozymes may have different substrate specificities, and experimentation is required to optimize the stain recipe for each tissue and species examined. Gels should also be stained for aminopeptidase (described above) to determine if AMP and PEP represent identical or non-identical products.

Buffer systems: Because homologies across taxa are uncertain, it is not possible to recommend a specific buffer.

Peroxidase (PRX). I.U.B.: Donor:hydrogen peroxide oxidoreductase. E.C. 1.11.1.7

50 mM Na-acetate buffer, pH 5.0	50 ml
$CaCl_2$	50 mg (1 ml)
Hydrogen peroxide, 3%	0.25 ml
3-Amino-9-ethylcarbazole	25 mg
N,N-Dimethylformamide	2 ml

Procedure: Dissolve 3-amino-9-ethylcarbazole in the N,N-dimethylformamide. Add along with remaining ingredients to buffer and pour over gel. Incubate at room temperature until red bands appear.

Comments: Peroxidases are sensitive to peroxide concentration, and failure to detect PRX often results from the inclusion of too high a concentration of H_2O_2. Some peroxidases are also inactivated by too high a concentration of reducing agents in extraction buffers.

Buffer systems: Peroxidase isozymes have been resolved with a wide variety of electrophoresis buffers. Systems 1, 2, 3, and 4 are the most likely to yield clear banding. In many plants, however, some of the peroxidases migrate cathodally at these pHs.

Phosphoglucoisomerase (PGI). See Glucose-6-phosphate isomerase.

Phosphoglucomutase (PGM). I.U.B.: α-D-Glucose 1,6-phosphomutase. E.C. 5.4.2.2 (formerly 2.7.5.1)

50 mM Tris-HCl, pH 8.5	50 ml
$MgCl_2$	50 mg (1 ml)
Glucose-1-phosphate, Na_2-salt	150 mg

NAD	10 mg (1 ml)
Glucose-6-phosphate dehydrogenase (NAD)	20 units
MTT	10 mg (1 ml)
PMS	2 mg (0.4 ml)

Procedure: Combine ingredients and pour over gel. Incubate until blue bands are optimally developed. Rinse well and fix.

Comments: PGM activity is dependent upon the presence of millimolar concentrations of α-D-glucose-1,6-diphosphate, which is present in appropriate quantities in many commercial preparations of glucose-1-phosphate.

Buffer systems: PGM is activated by histidine (N. F. Weeden, unpublished data); best resolution is typically achieved using a buffer system similar to numbers 1 or 4.

Phosphogluconate dehydrogenase (PGD). I.U.B.: 6-Phospho-D-gluconate: NADP$^+$ 2-oxidoreductase (decarboxylating). E.C. 1.1.1.44

50 mM Tris-HCl, pH 8.0	50 ml
$MgCl_2$	50 mg (1 ml)
6-Phosphogluconic acid, (Na or Ba salt)	20 mg (1 ml)
NADP	5 mg (1 ml)
NBT or MTT	10 mg (1 ml)
PMS	2 mg (0.4 ml)

Procedure: Combine ingredients and pour over gel. Incubate until blue bands appear. Rinse and store in water (if NBT is used) or fixative.

Comments: Isozymes of PGD may be particularly sensitive to variation in starch lots and electrophoretic running conditions (e.g., Stuber and Goodman, 1984).

Buffer systems: Best resolution is usually achieved with buffer system 2 or a modification thereof. Alternative systems include numbers 1, 3, and 4.

Ribulose-bisphosphate carboxylase (RBC) = [rubisco]. I.U B.: 3-phospho-D-glycerate carboxy-lyase (dimerizing). E.C. 4.1.1.39

Amido Black	25 mg
Fixative (5 parts methanol, 5 parts water, 1 part acetic acid)	50 ml

Procedure: Dissolve the amido black in the fixative. Pour over gel and stain for 15–30 minutes. Pour off stain and rinse three times with fixative, for 15 minutes each, to remove background staining.

Comments: This is a non-specific protein stain, and will reveal the most abundant proteins in the sample extracts. The major band present from leaf tissue extracts is the rubisco holoenzyme (Weeden, 1984). Band number and intensity may be altered by varying the staining time or by using other protein staining reagents, such as coomassie blue or ponceau S.

Buffer systems: Systems 5, 6, and 7.

Shikimate dehydrogenase (SKD). I.U.B.: Shikimate: NADP$^+$ 3-oxidoreductase. E.C. 1.1.1.25

50 mM Tris-HCl, pH 8.5	50 ml
NADP	5 mg (1 ml)
Shikimic acid	50 mg
NBT or MTT	10 mg (1 ml)
PMS	2 mg (0.4 ml)

Procedure: Combine ingredients and pour over gel. Incubate until blue bands appear. Rinse and store in water (if NBT is used) or fixative.

Comments: SKD has a pH optimum of 9.1 (Lourenco and Neves, 1984), but background staining is greater at pH 9.1 than at pH 8.5. Shikimate is a moderately strong acid, so if substrate concentration is increased buffer pH should be adjusted as necessary.
Buffer systems: Best resolution is typically observed with buffer system 2. Other useful systems are likely to be numbers 1, 3, and 4.

Superoxide dismutase (SOD). I.U.B.: Superoxide: superoxide oxidoreductase. E.C. 1.15.1.1

50 mM Tris-HCl, pH 8.0	50 ml
Riboflavin	2 mg
EDTA	1 mg
NBT	10 mg (1 ml)

Procedure: Combine ingredients and pour over gel. Incubate for 30 minutes (times may vary depending on activity). Remove from incubator and illuminate on a light box. Zones of SOD activity are revealed as achromatic regions on a dark blue to black background. Rinse with water and fix.
Comments: The stain recipe given is modified from Beauchamp and Fridovich (1971) and Baum and Scandalios (1979), which should be consulted for alternatives. In many plants it is not necessary to stain specifically for SOD, as activity may be manifested as white bands against the blue background of other stains containing tetrazolium salts ("tetrazolium oxidase"). It should be noted that diethyldithiocarbamate (DIECA), a common component of enzyme extraction buffers, has been reported to inhibit the Cu-Zn containing isozymes (Jaaska, 1982).
Buffer systems: Optimal buffer systems vary among taxa.

Triose-phosphate isomerase (TPI). I.U.B.: D-Glyceraldehyde-3-phosphate ketol-isomerase. E.C. 5.3.1.1

50 mM Tris-HCl, pH 8.0	50 ml
Arsenic acid, sodium salt	75 mg (1 ml)
Dihydroxyacetone phosphate	10 mg
NAD	10 mg (1 ml)
Glyceraldehyde-3-phosphate dehydrogenase	300 units
NBT or MTT	10 mg (1 ml)
PMS	2 mg (0.4 ml)

Procedure: Combine ingredients and pour over gel. Incubate until blue bands appear (may be left in the stain overnight at room temperature in the dark if necessary). Rinse and store in water (if NBT is used) or fixative.
Comments: Dihydroxyacetone phosphate (DHAP) is commercially available as a lithium salt or as a cyclohexyl ammonium salt. Use of the latter is less expensive, but requires hydrolysis (according to manufacturers' instructions) prior to stain preparation. Agar overlays in minimal volumes of buffer can be effective in reducing costs and sharpening bands in some cases. A more complicated but frequently used assay involves enzymatic production of DHAP: Add 300 mg DL-α-glycerophosphate, 150 mg pyruvic acid (Na salt), 10 mg NAD, 50 units α-glycerophosphate dehydrogenase, and 70 units lactate dehydrogenase to 25 ml Tris-HCl, pH 8.0 (50 mM). Incubate for 2 hours at 37° C. Lower pH to 2.0 with concentrated HCL. Raise to pH 7.0 with Tris crystals (or a saturated solution of Tris). Add 20 ml of 50 mM Tris-HCl (pH 8.0), 75 mg arsenic acid, 10 mg NAD, 300 units glyceraldehyde-3-phosphate dehydrogenase, 10 mg NBT (or MTT), and 2 mg PMS.
Buffer systems: Best results are usually obtained with buffer system 6, with systems 5 and 7 being acceptable alternatives.

Additional Enzyme Assays

Several additional plant enzymes have been electrophoretically studied in one or more taxa.

Illustrative recipes for their detection in starch gels are referenced below. For the most part these enzymes are relatively poorly characterized, both genetically and with respect to stain assay parameters. In our experience they are less likely to be useful than those listed above.

Catechol oxidase (polyphenol oxidase, phenolase, laccase, tyrosinase). E.C. 1.10.3.-. Shaw and Prasad (1970); Schwennesen et al. (1982); Vallejos (1983).
Glycerate dehydrogenase. E.C. 1.1.1.29. Siciliano and Shaw (1976).
α-Glycerophosphate dehydrogenase. E.C. 1.1.1.8. Shaw and Prasad (1970); Siciliano and Shaw (1976).
Nitrate reductase. E.C. 1.6.6.2. Vallejos (1983).
Nitrite reductase. E.C. 1.6.6.4. Vallejos (1983).
Phosphoenolpyruvate carboxykinase (PEP carboxylase). E.C. 4.1.1.38. Brown et al. (1978); Vallejos (1983).
Phosphoglycerate kinase. E.C. 2.7.2.3. Siciliano and Shaw (1976).
Pyruvate kinase. E.C. 2.7.1.40. Brewer and Sing (1970); Harris and Hopkinson (1976); Siciliano and Shaw (1976).
Sorbitol dehydrogenase. E.C. 1.1.1.14. Shaw and Prasad (1970); Millar (1985).
Transketolase. E.C. 2.2.1.1. Tanksley (1984).
Xanthine dehydrogenase. E.C. 1.1.1.204. Brewer and Sing (1970); Shaw and Prasad (1970); Vallejos (1983).

Post-staining Treatment

When the isozyme bands are sufficiently developed the staining reaction is stopped, usually by pouring off the staining solution and fixing the gels. Common fixatives include methanol:water:acetic acid (5:5:1), ethanol:water:acetic acid:glycerol (5:4:2:1), and glycerol:water (1:1). Some stains, particularly those that use NBT, may be left in water at room temperature for a few days, or refrigerated for a week or two without fixation. Most banding patterns, however, lose some clarity with fixation or storage, and for this reason they are often scored and/or photographed a few minutes after stopping the staining reaction. Many types of photographic equipment and films are satisfactory. We routinely use a 35 mm SLR camera with ASA 100 color print film or Kodak technical pan 2415, and photograph the gels with illumination provided by a light box from underneath and floodlamps from above. Prints so obtained often are as easily and accurately scored as the original gel slices. For those without access to photographic equipment, fixed gels may be wrapped in plastic wrap and stored indefinitely in the refrigerator.

INITIATING ISOZYME STUDIES

Effective development of techniques for starch gel electrophoresis of isozymes requires that many different aspects of the procedure be simultaneously optimized, a prospect that some may consider daunting due to the number of options and amount of detail. Novices considering their first trial runs and experienced practitioners that switch to new groups of organisms both face similar technical demands, namely, selection of the tissue to use, choice of an extraction procedure and grinding buffer, and determination of the enzymes that can be electrophoretically resolved. Fortunately, it has become clear that not all methodological details affect the final results equally, so that initial efforts can be efficiently focused on the most critical elements of the procedure, namely, the extraction protocol and the electrophoretic buffer systems. This is not to deny the importance of other aspects, such as proper gel preparation and optimization of stain recipes. Rather, it merely emphasizes that no amount of "fine tuning" can overcome enzyme extracts of poor quality or selection of inappropriate electrophoretic buffers. Concentration on these two factors, consequently, often allows for the most efficient and rapid progress.

Several screening runs generally are required prior to final selection of protocols. If the objectives of the study include the development of techniques for a broad spectrum of enzymes, these initial screening runs should be designed to determine the extraction procedure that yields a maximally active homogenate. This is often accomplished by selecting several different tissues and grinding buffers (given above) and staining for isozymes that are often easily resolved (e.g., GPI, AAT, IDH, PGM, ADH, MDH) on a few of their recommended electrophoretic buffer systems (see above). Advances may also be made by grinding the tissue several different ways (e.g., with a chilled mortar and pestle, with liquid nitrogen, with and without added insoluble PVpP) and with several different tissue to volume ratios. With the caveat that no single homogenizing technique is likely to be optimal for all enzymes, it is generally true that the procedure that yields the most active extract during these screening runs will ultimately prove the most useful when additional enzymes and buffer systems are added.

Once the provisional extraction protocol has been selected, subsequent screening runs may focus on additional enzymes and on ascertaining which buffer systems to use for their resolution. An effective procedure for accomplishing this is to load replicate samples of a few genotypes several times on each of the buffer systems presented above, so that a few strips can be cut from every gel slice, each one containing an identical array of samples. This allows numerous enzymes to be assayed on each gel, facilitating both determination of the enzymes that can be resolved on each gel and comparisons of band number and clarity across gels. After the initial selection of sample preparation technique, electrophoresis buffers, and enzymes to assay, it is generally advisable to conduct additional experiments with some of the other important variables, such as starch percentage, run time and wattage, and buffer system modifications. Such experimentation is often time consuming and tedious, but is almost always rewarded with superior results.

GEL INTERPRETATION

Genetic Bases of Isozyme Phenotypes

When electrophoretic gels are submersed in enzyme-specific staining solutions one or more regions of enzyme activity may be revealed. The resulting banding pattern is the electrophoretic phenotype, which usually consists of one or more colored bands for each individual analyzed. This phenotype varies greatly in its complexity, depending on numerous factors, including the organism, tissue, and enzyme assayed. In some cases, it may be simple, and consist of a single invariant band in all samples. In contrast, some enzymes may be encoded by several genes and display complex phenotypes with 15 or more bands per individual. Although the relative migration and number of bands observed in various samples may be adequately informative for some types of studies, the majority of applications requires that electrophoretic phenotypes be translated into genotypes for the underlying genetic loci. Correct interpretation of banding patterns in genetic terms requires the proper "framework for analysis" (Gottlieb, 1977), i.e., cognizance of the pertinent factors that influence the electrophoretic phenotype, and in many cases requires some amount of formal genetic documentation.

Several factors may be considered primary determinants of the number of bands observed on a gel, including: (1) the number of coding genes; (2) their allelic states (homozygous or heterozygous); (3) quaternary structure of the protein products; and (4) their subcellular compartmentalization (see additional discussion in chapter by Weeden and Wendel in this volume). Evidence regarding these factors is often easily derived from simple crosses or progeny tests, as illustrated in Figure 1.5.

The simplest case involves a single region of staining with variant electromorphs (allozymes) observed in different individuals. Because allozymes are usually codominantly inherited, crosses between individuals bearing different electromorphs will result in F_1 progeny that display both parental bands. In addition, the F_1 may possess "hybrid" bands not observed in either parent, the presence and number of which depend on the number of polypeptide subunits contained in the active enzyme. Thus, for a dimeric enzyme, three bands are observed, the two parental homodimers and an additional product of intermediate mobility, or heterodimer, composed of one polypeptide specified by each of the two parental alleles. If, on the other hand, the enzyme is tetrameric, five bands will be observed in heterozygotes, two homotetramers (AAAA, BBBB) and three heterotetramers (AAAB, AABB, ABBB). Random association of subunits into multimeric proteins results in unequal numbers

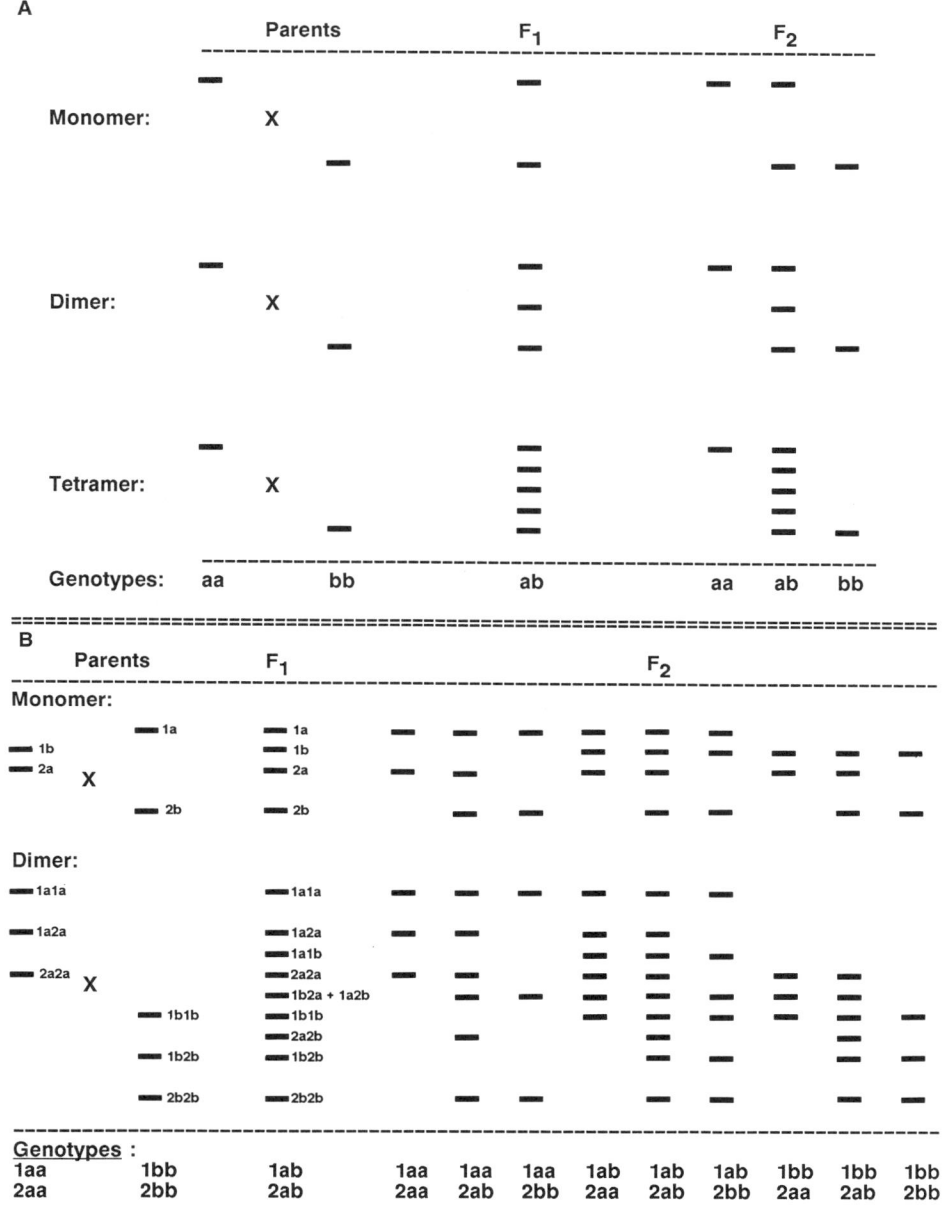

Fig. 1.5. Schematic illustration of idealized relationships between electrophoretic phenotypes, enzyme subunit number, and allelic state of the encoding gene or genes. **A.** The number of bands observed in a heterozygote formed from crossing individuals that are homozygous for alleles that specify different allozymes is dependent on enzyme quaternary structure. Novel bands not observed in either parent arise from random association of the polypeptide subunits specified by the alternate parental alleles. Evidence regarding isozyme subunit structure is often gained from segregation data, as shown in the righthand column. **B.** Electrophoretic phenotypes can become complex if there is more than a single coding gene. The crosses illustrated involve parents that are homozygous for different alleles at two loci. Enzyme subunit composition is given next to the bands in the parents and the F_1 individuals. For monomeric enzymes, codominant inheritance leads to the presence of all four parental bands in the F_1, which then segregate in the F_2 generation. If the enzyme is dimeric, F_1 plants produce ten different protein products, formed from random association of four variant polypeptides into four homodimers, two intragenic heterodimers, and four intergenic heterodimers. Fewer than ten bands are usually observed, however, due to comigration of some products. Selfing of dihybrids results in an F_2 with nine progeny classes, as illustrated (genotypes given below).

of each type of molecule [given by the coefficients of the binomial expansion $(A + B)^n$, where A and B are polypeptides specified by alleles a and b and n is the number of polypeptides in the active enzyme] and hence unequal staining intensity of bands, typically 1:2:1 and 1:4:6:4:1 for dimeric and tetrameric enzymes, respectively. Irrespective of subunit composition, only three classes of progeny are expected (in a 1:2:1 ratio) upon selfing the F_1 or intercrossing individuals heterozygous for the same two alleles (Fig. 1.5).

Many enzymes are encoded by more than a single gene, i.e., more than one isozyme is present. If the enzymes are located in the same subcellular compartment and if polypeptides specified by all encoding alleles randomly associate into multimeric products, intergenic heteromultimers will be observed as well as homomultimers and intragenic heteromultimers. Predicted band number (n) for multigenic systems depends on the number of different types of subunits (S) and the number of subunits in the active enzyme (Q), as modified from Shaw (in Shannon, 1968) as:

$$n = \frac{(S + Q - 1)!}{(Q!)(S - 1)!}$$

The complexity of the phenotype rises dramatically with increasing gene number. In a double heterozygote, for example, random association of four different polypeptide products into dimers will result in 10 bands, four homodimers, two intragenic heterodimers, and four intergenic heterodimers (Fig. 1.5B). A triple heterozygote would produce 21 bands. In most cases involving multiple genes, however, fewer than the predicted number of bands are actually observed, due to incomplete electrophoretic resolution and co-migration of one or more dissimilar products (Fig. 1.6B). Band number for multimeric enzymes may also be reduced if isozymes specified by different genes reside in different subcellular compartments, in which case intergenic multimers are not formed. The resulting phenotype is then a composite resulting from overlay of essentially independent systems, one for each organelle containing isozyme(s) with appropriate activity. Many routinely examined plant enzymes, for example, have both chloroplast and cytosolic isozymic forms (Schnarrenberger et al., 1975; Simcox and Dennis, 1978; see review by Weeden, 1983; see chapter by Weeden and Wendel in this volume). Because intergenic heteromultimers are not formed, double heterozygotes often display six rather than 10 bands (assuming dimeric enzymes and complete resolution). A more complex example is offered by the major isozymes of maize malate dehydrogenase, which are genetically controlled by five genes (Goodman et al., 1980). The phenotypes observed originate from two overlapping but independent sets of bands, namely, homodimers and heterodimers formed from three mitochondrial isozymes and those resulting from random association among polypeptides of two cytosolic isozymes.

Although it is true that band phenotypes may be simplified by subcellular compartmentalization of isozymes, it should be noted that failure to form intergenic multimers only suggests rather than proves different organellar locations of multigenic isozyme systems. Lack of intergenic hybridization may also be caused by the inability of polypeptides specified by different genetic loci to associate into active enzymes. This phenomenon is most frequently observed with isozymes revealed by nonspecific enzyme stains, such as NADH-dehydrogenase, acid phosphatase, and esterase. For this reason, organelle isolations (methods cited above) are required to substantiate the cellular location of isozymes.

Several additional genetic and non-genetic phenomena may additionally complicate zymogram interpretation. Heterodimers may migrate to positions that are not intermediate to the mobilities of homodimers formed from their constituent polypeptides, or in exceptional cases may even migrate outside of the range of the homodimers (Goodman and Stuber, 1983). "Null" alleles, which are variants that lack enzymatic activity, may confuse interpretation through unexpected band simplification or generation of unexpected phenotypes in crosses (Fig. 1.6). Codominant inheritance, although generally true for allozymes, is not always observed (Cullis, 1979; Weeden and Robinson, 1986; Edwards et al., unpublished data). Additional bands or shifts in migration may arise from post-translational modification of enzymes (Goodman et al., 1980; Endo, 1981; Harry, 1983; Doebley et al., 1986). Artifacts may be introduced during various procedural stages, such as sample preparation, storage, or electrophoresis (reviewed in Harris and Hopkinson, 1976). One of the most common but

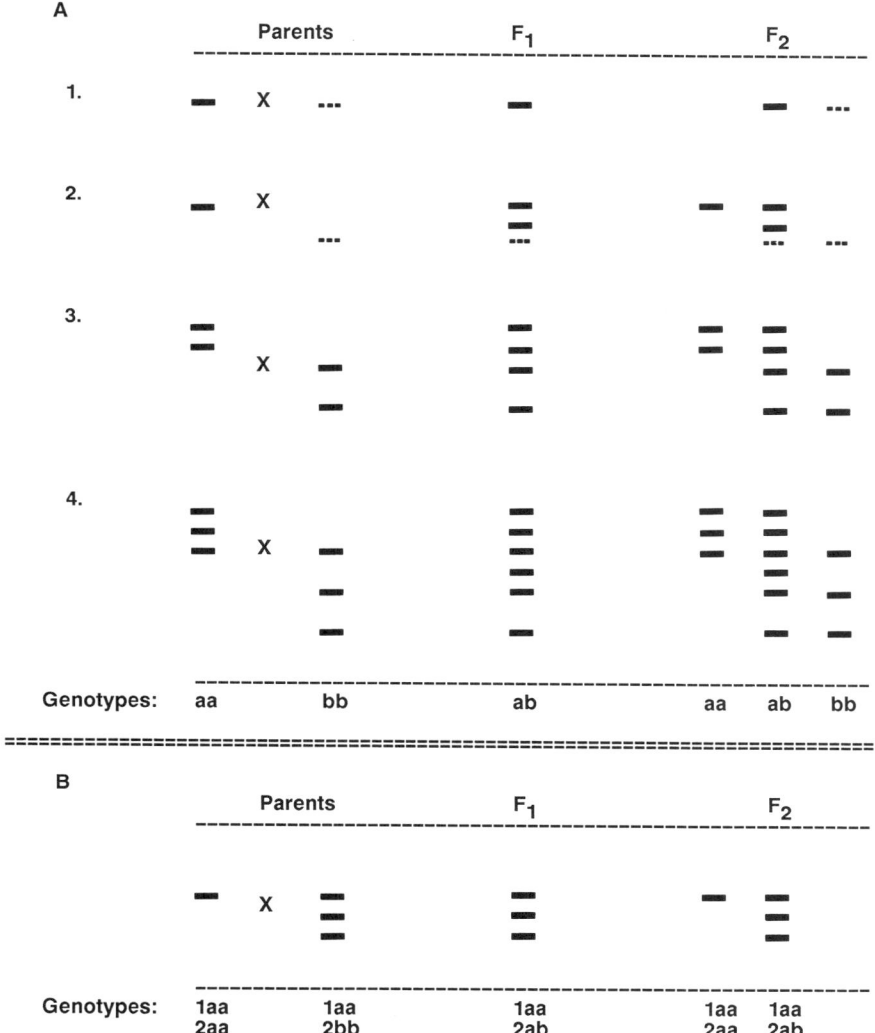

Fig. 1.6. Some genetic and non-genetic phenomena that complicate zymogram interpretation. **A1.** Null alleles (denoted by "- - -") are masked in heterozygous individuals, which segregate only two phenotypic classes (in a 3:1 ratio) upon selfing. **A2.** Some allelic variants are only electrophoretically "null" as homodimers. The ability to form enzymatically active heterodimers with other allelic products demonstrates synthesis of polypeptides specified by the variant allele. **A3, A4.** Many enzymes routinely display "shadow" or "ghost" bands, or multiple-banded phenotypes that segregate as units. Hybrids combine all parental bands (A3), and if the enzyme is dimeric, one or more bands in addition to the expected heterodimer may be observed (A4). These multiple-banded alleles may arise *in vivo* from post-translational modification or *in vitro* from artifacts introduced during sample preparation or electrophoresis. Whatever their origin, multiple-banded phenotypes arising from a single genetic locus mimic phenotypes expected from models involving more than one gene. **B.** Gene duplication may lead to comigration of products specified by different loci. The example given illustrates the phenotypes expected for a two-locus model involving dimeric enzymes, where parents are homozygous for the same allele at locus 1 and differ by a single allele at locus 2. All F_1 individuals are three-banded, and only two phenotypes are recovered in the F_2, in a 1:3 ratio.

underappreciated complications arises from the presence of double-banded or multiple-banded products that genetically behave as single alleles (often referred to as "shadow" bands or "ghost" bands). These are frequently reported for many plant isozymes, including acid phosphatase (Rick and Fobes, 1975; Kahler, 1983), aspartate aminotransferase (Crawford and Wilson, 1977; Guries and Ledig, 1978; Gottlieb, 1981; Neale and Adams, 1981; Harry, 1986), esterase (Tanksley and Rick, 1980), β-glucosidase (Stuber et al., 1977), peroxidase (Quiros and Morgan, 1981; Rick and Tanksley, 1981; Van den Berg and Wijsman, 1982; Millar, 1985), phosphoglucomutase (Neale and Adams, 1981; Wendel and Parks, 1982; Stuber and Goodman, 1983; Wolf et al., 1987), and shikimate dehydrogenase (Tanksley, 1984; Weeden, 1984; Harry, 1986; Jarret and Litz, 1986). It is usually not known whether these multiple-banded allozymes arise *in vivo* through post-translational modification or *in vitro* by chemical or conformational changes, but their frequency of occurrence cautions against unsubstantiated interpretations of "gene duplications" (Fig. 1.6).

Genetic Documentation

As illustrated in Fig. 1.5B, an F_2 resulting from selfing a double heterozygote yields nine progeny classes. Generating the expected progeny class ratios and their phenotypes is a relatively straightforward process and is the standard method used in formal genetic analyses for interpreting allozyme and isozyme variation. The genetic model to test must first be formulated, however. Most commonly, investigators first observe the type of variation displayed in the illustrated progeny arrays in natural populations or breeding materials and consider competing genetic hypotheses for its interpretation. A three-banded phenotype, for example, could result from heterozygosity at a single gene, "fixed heterozygosity" representing two homozygous loci that code for dimeric enzymes with an intergenic heterodimer, homozygosity at three loci encoding monomeric proteins, homozygosity for a single three-banded allozyme, or from several other genetic situations. As a consequence, for some enzyme systems some form of genetic analyses may be necessary.

Although a thorough discussion of the design and implementation of genetic experiments in various plant species is beyond the scope of this chapter, brief consideration of aspects or techniques that are germane to isozyme studies is warranted. Many excellent genetic studies of plant isozymes exist, and the reader is referred to Tanksley and Orton (1983) or Weeden and Wendel (see chapter in this volume) for citations and illustrative examples.

A standard sequence of initial experiments is to screen materials for enzyme variation followed by intercrossing variant phenotypes in several combinations. For many outcrossers the resulting progeny arrays are examined and competing genetic hypotheses may be tested. With self-pollinated plants it is usually necessary to proceed with an additional generation of crosses, as the parental stocks are likely to be homozygous and generate an invariant F_1. These experiments often allow rapid progress toward gel interpretation and are usually sufficient to document codominance and gene number for many enzymes. Additional evidence is often gathered by analyses of pollen samples (see above) or through study of the isozymes from purified organelle preparations (see above). Direct comparisons of pollen extracts (crushed or eluted) with somatic tissues is also of value in determining which bands on a zymogram represent intragenic heteromultimers (Weeden and Gottlieb, 1979). With some enzyme systems, however, it may be necessary to analyze additional crosses from other generations or materials, and for multigene enzymes in particular, proper locus assignment of variants will most likely require additional tests of allelism.

For some applications, it is often useful to understand the linkage relationships among allozyme loci or between allozyme loci and other genetic markers (if they exist). General discussions of linkage theory and the design of linkage experiments may be found in Bailey (1961) and Mather (1951). Presence of linkage is usually detected by significant χ^2 values from contingency table analyses of joint segregation data. Recombination estimates among linked locus-pairs, and their standard errors, are most often calculated using maximum likelihood procedures (Allard, 1956). These analyses may be facilitated by use of various computer programs that are available (e.g., Suiter et al., 1983). Other alternatives for recombination estimation exist, and use of these may be warranted in certain circumstances (Nordheim et al., 1983).

ACKNOWLEDGMENTS

The authors wish to thank D. Farrar and K. Klier for reviewing the manuscript. This work was supported by grants from the National Science Foundation (BSR-8619631) and Pioneer Hi-Bred International of Johnston, Iowa (to JFW), and by the New York State Seed Association and the United States Department of Agriculture Hatch Project #44391 (to NFW).

LITERATURE CITED

Allard, R. W. 1956. Formulas and tables to facilitate the calculation of recombination values in heredity. *Hilgardia* 24: 235–278.

Anderson, J. W. 1968. Extraction of enzymes and subcellular organelles from plant tissues. *Phytochemistry* 7: 1973–1988.

Anderson, J. W., and K. S. Rowan. 1967. Extraction of soluble leaf enzymes with thiols and other reducing agents. *Phytochemistry* 6: 1047–1056.

Arulsekar, S., D. E. Parfitt, W. Beres, and P. E. Hansche. 1986. Genetics of malate dehydrogenase isozymes in the peach. *J. Heredity* 77: 49–51.

Ashton, G. C., and A. W. H. Braden. 1961. Serum β-globulin polymorphism in mice. *Austral. J. Biol. Sci.* 14: 248–254.

Avise, J. C. 1975. Systematic value of electrophoretic data. *Syst. Zool.* 23: 465–481.

Bailey, N. T. J. 1961. *Introduction to the mathematical theory of genetic linkage.* Clarendon Press, Oxford.

Barka, T. 1960. A simple azo-dye method for histochemical demonstration of acid phosphatase. *Nature* 187: 248–249.

Barratt, D. H. P. 1980. Method for the detection of glutamine synthetase activity on starch gels. *Pl. Sci. Lett.* 18: 249–255.

Baum, J. A., and J. G. Scandalios. 1979. Developmental expression and intracellular localization of superoxide dismutases in maize. *Differentiation* 13: 133–140.

Bayer, R. J., and D. J. Crawford. 1986. Allozyme divergence among five diploid species of *Antennaria* (Asteraceae: Inuleae) and their allopolyploid derivatives. *Amer. J. Bot.* 73: 287–296.

Beauchamp, C. O., and I. Fridovich. 1971. Superoxide dismutase: improved assays and assay applicable to acrylamide gels. *Analytical Biochem.* 44: 276–287.

Blackshear, P. J. 1984. Systems for polyacrylamide gel electrophoresis. *Methods in Enzymol.* 104: 237–255.

Brewbaker, J. L. 1971. Pollen enzymes and isoenzymes. *In* J. Heslop-Harrison [ed.], *Pollen development and physiology,* 156–170. Butterworth's, London.

Brewer, G. J., and C. F. Sing. 1970. *An introduction to isozyme techniques.* Academic Press, New York.

Brown, A. H. D., E. Nevo, D. Zohary, and O. Dagan. 1978. Genetic variation in natural populations of wild barley (*Hordeum spontaneum*). *Genetica* 49: 97–108.

Buth, D. G., and R. W. Murphy. 1980. Use of nicotinamide adenine dinucleotide (NAD)-dependent glucose-6-phosphate dehydrogenase in enzyme staining procedures. *Stain Technology* 55: 173–176.

Cardy, B. J., C. W. Stuber, J. F. Wendel, and M. M. Goodman. 1983. Techniques for starch gel electrophoresis of enzymes from maize (*Zea mays* L.). Inst. Stat. Mimeo. No. 1317R, North Carolina State University, Raleigh.

Cerff, R. 1982. Evolutionary divergence of chloroplast and cytosolic glyceraldehyde-3-phosphate dehydrogenases from angiosperms. *Eur. J. Biochem.* 126: 513–515.

Cheliak, W. M., and J. A. Pitel. 1984. Genetic control of allozyme variants in mature tissues of white spruce trees. *J. Heredity* 75: 34–40.

Chrambach, A. 1980. Electrophoresis and electrofocusing on polyacrylamide gels in the study of native macromolecules. *Mol. Cell. Biochem.* 29: 23–46.

_____, and D. Rodbard. 1971 Polyacrylamide gel electrophoresis. *Science* 172: 440–451.
Clayton, J. W., and D. N. Tretiak. 1972. Amine-citrate buffers for pH control in starch gel electrophoresis. *J. Fish. Res. Board Can.* 29: 1169–1172.
Conkle, M. T., P. D. Hodgskiss, L. B. Nunnaly, and S. C. Hunter. 1982. *Starch gel electrophoresis of conifer seeds: a laboratory manual.* General Technical Report PSW-64. U.S.D.A. Forest Service Pacific Southwest Forest and Range Experiment Station, Berkeley.
Coyne, J. 1981. Gel electrophoresis and cryptic protein variation. In M. L. Rattazzi, J. G. Scandalios, G. S. Whitt [eds.], *Isozymes: current topics in biological and medical research,* Vol. 6, 1–32. Alan R. Liss, New York.
Crawford, D. J. 1983. Phylogenetic and systematic inferences from electrophoretic studies. In S. D. Tanksley and T. J. Orton [eds.], *Isozymes in plant genetics and breeding,* Part A, 257–287. Elsevier, Amsterdam.
_____, and H. D. Wilson. 1977. Allozyme variation in *Chenopodium fremontii. Syst. Bot.* 2: 180–190.
Cullis, C. A. 1979. Segregation of the isozymes of flax genotrophs. *Biochem. Genet.* 17: 391–401.
Doebley, J. F., C. W. Morden, and K. F. Schertz. 1986. A gene modifying mitochondrial malate dehydrogenase isozymes in *Sorghum* (Gramineae). *Biochem. Genet.* 24: 813–819.
Ellstrand, N. C., and D. L. Marshall. 1985. The impact of domestication on distribution of allozyme variation within and among cultivars of radish, *Raphanus sativus* L. *Theor. Appl. Genet.* 69: 393–398.
Endo, T. 1981. Developmental modification and hybridization of allelic acid phosphatase isozymes in homo- and heterozygotes for the Acp-1 locus in rice. *Biochem. Genet.* 19: 373–384.
Farinelli, M. P., D. W. Fry, and K. E. Richardson. 1983. Isolation, purification, and partial characterization of formate dehydrogenase from soybean seed. *Pl. Physiol.* 73: 858–859.
Fildes, R. A., and H. Harris. 1966. Genetically determined variation of adenylate kinase in man. *Nature* 209: 262–263.
Gabriel, O. 1971. Locating enzymes on gels. *Methods in Enzymol.* 22: 578–604.
Gastony, G. J., and D. C. Darrow. 1983. Chloroplastic and cytosolic isozymes of the homosporous fern *Athyrium filix-femina* L. *Amer. J. Bot.* 70: 1409–1415.
_____, and L. D. Gottlieb. 1982. Evidence for genetic heterozygosity in a homosporous fern. *Amer. J. Bot.* 69: 634–637.
_____, and _____. 1985. Genetic variation in the homosporous fern *Pellaea andromedifolia. Amer. J. Bot.* 72: 257–267.
Goodman, M. M., and C. W. Stuber. 1983. Maize. In S. D. Tanksley and T. J. Orton [eds.], *Isozymes in plant genetics and breeding,* Part B, 1–33. Elsevier, Amsterdam.
_____, _____, C. N. Lee, and F. M. Johnson. 1980. Genetic control of malate dehydrogenase isozymes in maize. *Genetics.* 94: 153–168.
Gottlieb, L. D. 1977. Electrophoretic evidence and plant systematics. *Ann. Missouri Bot. Gard.* 64: 161–180.
_____. 1979. The origin of phenotype in a recently evolved species In S. Jain, G. B. Johnson, and P. Raven [eds.]. *Topics in plant population biology,* 264–286. Columbia Univ. Press, New York.
_____. 1981. Electrophoretic evidence and plant populations. *Prog. Phytochem.* 7: 1–46.
_____. 1982. Conservation and duplication of isozymes in plants. *Science* 216: 373–380.
_____. 1984. Electrophoretic analysis of the phylogeny of the self-pollinating populations of *Clarkia xantiana. Pl. Syst. Evol.* 147: 91–102.
Guries, R. P., and F. T. Ledig. 1978. Inheritance of some polymorphic isoenzymes in pitch pine (*Pinus rigida* Mill.). *Heredity* 20: 27–32.
Hames, B. D., and D. Rickwood [eds.]. 1981. *Gel electrophoresis of proteins: a practical approach.* IRL Press, Washington, D.C.
Harris, H., and D. A. Hopkinson. 1976. *Handbook of enzyme electophoresis in human genetics.* North Holland, Amsterdam.
Harry, D. E. 1983. Identification of a locus modifying the electrophoretic mobility of malate dehydrogenase isozymes in incense-cedar (*Calocedrus decurrens*), and its implications for

population studies. *Biochem. Genet.* 21: 417–434.

―――. 1986. Inheritance and linkage of isozyme variants in incense-cedar. *J. Heredity* 77: 261–266.

Haufler, C. H., and D. E. Soltis. 1984. Obligate outcrossing in a homosporous fern: field confirmation of a laboratory prediction. *Amer. J. Bot.* 71: 878–881.

Heeb, M. J., and O. Gabriel. 1984. Enzyme localization in gels. *Methods in Enzymol.* 104: 416–439.

Hughes, D. L. 1981. *Identification and translocation of carbohydrates in the cantaloupe (Cucumis melo var. reticulatus) plant and the fate of stachyose during fruit development.* Ph.D. thesis. University of California, Davis.

International Union of Biochemistry Nomenclature Committee. 1984. *Enzyme nomenclature 1984.* Academic Press, New York.

Isola, M. C., and L. Franzoni. 1981. Changes in electrophoretic pattern of ribonucleases during aging of potato tuber slices. *Z. Pflanzenphysiol.* 103: 277–283.

Jaaska, V. 1982. Isoenzymes of superoxide dismutase in wheats and their relatives: alloenzyme variation. *Biochem. Physiol. Pflanzen* 177: 747–755.

Jarret, R. L., and R. E. Litz. 1986. Enzyme polymorphism in *Musa acuminata* Colla. *J. Heredity* 77: 183–188.

Jenkins, C. L. D., and V. J. Russ. 1984. Large scale, rapid preparation of functional mesophyll chloroplasts from *Zea mays* and other C_4 species. *Pl. Sci. Lett.* 35: 19–24.

Johnson, F. M., and H. E. Schaffer. 1974. An inexpensive apparatus for horizontal gel electrophoresis. *Isozyme Bull.* 7: 4–6.

Kahler, A. L. 1983. Inheritance and linkage of acid phosphatase locus Acp4 in maize. *J. Heredity* 74: 239–246.

Kelley, W. A., and R. P. Adams. 1977a. Preparation of extracts from juniper leaves for electrophoresis. *Phytochemistry* 16: 513–516.

―――, and ―――. 1977b. Seasonal variation of isozymes in *Juniperus scopulorum*: systematic significance. *Amer. J. Bot.* 64: 1092–1096.

Kiang, Y. T. 1981. Inheritance and variation of amylase in cultivated and wild soybeans and their wild relatives. *J. Heredity* 72: 382–386.

King, E. E. 1971. Extraction of cotton leaf enzymes with borate. *Phytochemistry* 10: 2337–2341.

Leaback, D. H. 1976. Concentration gradient polyacrylamide gel electrophoresis. *In* I. Smith [ed.], *Chromatographic and electrophoretic techniques,* Vol. II, 250–271. Heinemann Medical Books, London.

Loomis, W. D. 1969. Removal of phenolic compounds during the isolation of plant enzymes. *Methods in Enzymol.* 13: 555–563.

―――. 1974. Overcoming problems of phenolics and quinones in the isolation of plant enzymes and organelles. *Methods in Enzymol.* 31: 528–544.

―――, and J. Battaile. 1966. Plant phenolic compounds and the isolation of plant enzymes. *Phytochemistry* 5: 423–438.

Loukas, M., Y. Vergini, and C. B. Krimbas. 1983. Isozyme variation and heterozygosity in *Pinus halepensis* L. *Biochem. Genet.* 21: 497–509.

Lourenco, E. J., and V. A. Neves. 1984. Partial purification and some properties of shikimate dehydrogenase from tomatoes. *Phytochemistry* 23: 497–499.

Lowrey, T. K., and D. J. Crawford. 1985. Allozyme divergence and evolution in *Tetramolopium* (Compositae: Astereae) in the Hawaiian Islands. *Syst. Bot.* 10: 64–72.

Mcleod, M. J., S. I. Guttman, and W. H. Eshbaugh. 1981. LDH in *Capsicum* (Solanaceae) leaves. *Isozyme Bull.* 14: 76.

Markert, C. L., and L. Faulhaber. 1965. Lactate dehydrogenase isozyme patterns of fish. *J. Exp. Zool.* 159: 319–332.

―――, and F. Moller. 1959. Multiple forms of enzymes: tissue, ontogenetic and species specific patterns. *Proc. Natl. Acad. Sci. USA* 45: 753–763.

Marshall, D. R., and A. H. D. Brown. 1975. The charge state model of protein polymorphism in natural populations. *J. Mol. Evol.* 6: 149–163.

Mather, K. 1951. *The measurement of linkage in heredity,* Second Edition, John Wiley, New York.

Meizel, S., and G. L. Markert. 1967. Malate dehydrogenase isozymes of the marine snail *Ilyanassa obsoleta*. *Arch. Biochem. Biophys.* 122: 753–765.

Micales, J. A., M. R. Bonde, and G. L. Peterson. 1986. The use of isozyme analysis in fungal taxonomy and genetics. *Mycotaxon* 27: 405–449.

Millar, C. I. 1985. Inheritance of allozyme variants in bishop pine (*Pinus muricata* D. Don.). *Biochem. Genet.* 23: 933–946.

Morden, C. W., J. F. Doebley, and K. F. Schertz. 1987. *A manual of techniques for starch gel electrophoresis of Sorghum isozymes.* Texas Agric. Expt. Sta. Misc. Publ. No. 1635. College Station.

Mulcahy, D. L., G. B. Mulcahy, and R. W. Robinson. 1979. Evidence for post-meiotic genetic activity in pollen of *Cucurbita* species. *J. Heredity* 70: 365–368.

Murray, D. R., and S. P. Waters. 1985. Multiple forms of aminopeptidase in representatives of the tribe Vicieae (Leguminosae). *Austral. J. Pl. Physiol.* 12: 39–45.

Neale, D. B., and W. T. Adams. 1981. Inheritance of isozyme variants in seed tissues of balsam fir (*Abies balsamea*). *Canad. J. Bot.* 59: 1285–1291.

———, J. C. Weber, and W. T. Adams. 1984. Inheritance of needle tissue isozymes in douglas-fir. *Canad. J. Genet. Cytol.* 26: 459–468.

Newton, K. J. 1983. Genetics of mitochondrial isozymes. In S. D. Tanksley and T. J. Orton [eds.], *Isozymes in plant genetics and breeding,* Part A, 157–174. Elsevier, Amsterdam.

Nickrent, D. L. 1986. Genetic polymorphism in the morphologically reduced dwarf mistletoes (*Arceuthobium*, Viscaceae): an electrophoretic study. *Amer. J. Bot.* 73: 1492–1502.

Nimmo, H. G., and G. A. Nimmo. 1982. A general method for the localization of enzymes that produce phosphate, pyrophosphate, or CO_2 after polyacrylamide gel electrophoresis. *Anal. Biochem.* 121: 17–22.

Nordheim, E. V., D. M. O'Malley, and R. P. Guries. 1983. Estimation of recombination frequency in genetic linkage studies. *Theor. Appl. Genet.* 66: 313–321.

Odrzykoski, I. J., and L. D. Gottlieb. 1984. Duplications of genes coding 6-phosphogluconate dehydrogenases in *Clarkia* (Onagraceae) and their phylogenetic implications. *Syst. Bot.* 9: 479–489.

O'Malley, D., N. C. Wheeler, and R. P. Guries. 1980. *A manual for starch gel electrophoresis.* Staff Paper Series 11. Dept. Forestry, College of Agric. and Life Sciences, University of Wisconsin, Madison.

Oncelay, C. Y., L. S. Daley, H. M. Vines, G. A. Couvillon, and C. H. Hendershott. 1979. Seasonal fluctuation in malate dehydrogenase, phosphatase, and proteinase activity of dormant peach buds. *Scientia Hort.* 11: 229–239.

Ostrowski, W. 1983. Electrophoretic examination of enzymes. *J. Chromatogr. Libr.* 18B: 287–339.

Ott, L., and J. G. Scandalios. 1978. Genetic control and linkage relationships among aminopeptidases of maize. *Genetics* 89: 137–146.

Pichersky, E., L. D. Gottlieb, and R. C. Higgins. 1984. Hybridization between subunits of triose phosphate isomerase isozymes from different subcellular compartments of higher plants. *Mol. Gen. Genet.* 193: 158–161.

Pierpoint, W. S. 1966. Enzymic oxidation of chlorogenic acid and some reactions of the quinone produced. *Biochem. J.* 98: 567–580.

Pitel, J. A., and W. M. Cheliak. 1984. *Effect of extraction buffers on characterization of isoenzymes from vegetative tissues of five conifer species: a users manual.* Information Report PI-X-34, Petawawa National Forestry Inst., Can. Forest Service.

Poulik, M. D. 1957. Starch gel electrophoresis in a discontinuous system of buffers. *Nature* 180: 1477–1479.

Prakash, S., R. C. Lewontin, and J. L. Hubby. 1969. A molecular approach to the study of genic heterozygosity in natural populations. IV. Patterns of genic variation in central, marginal and isolated populations of *Drosophila pseudoobscura*. *Genetics* 61: 841–858.

Quail, P. H. 1979. Plant cell fractionation. *Ann. Rev. Pl. Physiol.* 30: 425-484.

Quiros, C. F., and K. Morgan. 1981. Peroxidase and leucine-aminopeptidase in diploid *Medicago* species closely related to alfalfa: multiple gene loci, multiple allelism, and linkage. *Theor. Appl. Genet.* 60: 221–228.

Ranker, T. A., and C. R. Werth. 1986. Active enzymes from herbarium specimens: electrophoresis as an afterthought. *Amer. Fern J.* 76: 102–113.

Rick, C. M., and J. F. Fobes. 1975. Allozyme variation in the cultivated tomato and closely related species. *Bull. Torrey Bot. Club* 102: 376–384.

———, and S. D. Tanksley. 1981. Genetic variation in *Solanum pennellii*: comparisons with two other sympatric tomato species. *Pl. Syst. Evol.* 139: 11–45.

Ridgeway, G. J., S. W. Sherburne, and R. D. Lewis. 1970. Polymorphisms in the esterases of Atlantic herring. *Trans. Amer. Fisheries Soc.* 99: 147–151.

Rieseberg, L. H., and D. E. Soltis. 1987. Allozyme differentiation between *Tolmiea menziesii* and *Tellima grandiflora* (Saxifragaceae). *Syst. Bot.* 12: 154–161.

Sadler, R., and M. Shaw. 1978. A caution against the use of polyvinylpyrrolidone in the extraction of plant glutamine synthetase. *Canad. J. Bot.* 56: 1382–1385.

Sari-Gorla, M., C. Frova, G. Binelli, and E. Ottaviano. 1986. The extent of gametophytic-sporophytic gene expression in maize. *Theor. Appl. Genet.* 72: 42–47.

Schaffer, H. E., and F. M. Johnson. 1973. Constant (optimum) power electrophoresis. *Anal. Biochem.* 51: 577–583.

Scheid, H. W., A. Ehmke, and T. Hartmann. 1980. Plant NAD-dependent glutamate dehydrogenase: purification, molecular properties and metal ion activation of the enzymes from *Lemna minor* and *Pisum sativum*. *Z. Naturforsch.* 35C: 213–221.

Schnarrenberger, C., M. Tetour, and M. Herbert. 1975. Development and intracellular distribution of enzymes of the oxidative pentose phosphate cycle in radish cotyledons. *Pl. Physiol.* 56: 836–840.

Schwennesen, J., E. A. Mielke, and W. H. Wolfe. 1982. Identification of seedless table grape cultivars and a bud sport with berry isozymes. *Hort. Sci.* 17: 366–368.

Shannon, L. M. 1968. Plant isoenzymes. *Ann. Rev. Pl. Physiol.* 19: 187–210.

Shumaker, K. M., R. W. Allard, and A. L. Kahler. 1982. Cryptic variability at enzyme loci in three plant species, *Avena barbata*, *Hordeum vulgare*, and *Zea mays*. *J. Heredity*, 73: 86–90.

Shaw, C. R. 1970. How many genes evolve? *Biochem. Genet.* 4: 275–283.

———, and R. Prasad. 1970. Starch gel electrophoresis of enzymes—a compilation of recipes. *Biochem. Genet.* 4: 297–320.

Siciliano, M. J., and C. R. Shaw. 1976. Separation and visualization of enzymes on gels. *In* I. Smith [ed.], *Chromatographic and electrophoretic techniques*, Vol. II, 185–209. Heinemann Medical Books, London.

Simcox, P. D., and D. T. Dennis. 1978. Isoenzymes of the glycolytic and pentose phosphate pathways in proplastids from the developing endosperm of *Ricinus communis*. *Pl. Physiol.* 61: 871–877.

Smith, I. 1976. Starch gel electrophoresis. *In* I. Smith [ed.], *Chromatographic and electrophoretic techniques*, Vol. II, 153–184. Heinemann Medical Books, London.

Smithies, O. 1955. Zone electrophoresis in starch gels: group variation in the serum proteins of normal human adults. *Biochem. J.* 61: 629–641.

Soltis, D. E. 1986. Genetic evidence for diploidy in *Equisetum*. *Amer. J. Bot.* 73: 908–913.

———, C. H. Haufler, D. C. Darrow, and G. J. Gastony. 1983. Starch gel electrophoresis of ferns: a compilation of grinding buffers, gel and electrode buffers, and staining schedules. *Amer. Fern J.* 73: 9–27.

———, C. H. Haufler, and G. J. Gastony. 1980. Detecting enzyme variation in the fern genus *Bommeria*: an analysis of methodology. *Syst. Bot.* 5: 30–38.

———, and P. S. Soltis. 1986. Active enzymes from megaspores of *Marsilea* and *Regnellidium*. *Amer. Fern J.* 76: 17–20.

Stavrakakis, M., and M. Loukas. 1983. The between- and within-grape cultivars genetic variation. *Scientia Hort.* 19: 321–334.

Strauss, S. H., and M. T. Conkle. 1986. Segregation, linkage, and diversity of allozymes in knobcone pine. *Theor. Appl. Genet.* 72: 483–493.

Stuber, C. W., and M. M. Goodman. 1983. Genetic control, intracellular localization, and genetic variation of phosphoglucomutase isozymes in maize (Zea mays L.). Biochem. Genet. 21: 667–689.

_____, and _____. 1984. Inheritance, intracellular localization, and genetic variation of 6-phosphogluconate dehydrogenase isozymes in maize. Maydica 29: 453–471.

_____, _____, and F. M. Johnson. 1977. Genetic control and racial variation of β-glucosidase isozymes in maize (Zea mays L.). Biochem. Genet. 15: 383–394.

Suiter, K. A., J. F. Wendel, and J. S. Case. 1983. Linkage 1: a pascal computer program for the detection and analysis of genetic linkage. J. Heredity 74: 203–204.

Tanksley, S. D. 1980. PGI-1, a single gene in tomato responsible for a variable number of isozymes. Canad. J. Genet. Cytol. 22: 271–278.

_____. 1984. Linkage relationships and chromosomal locations of enzyme-coding genes in pepper, Capsicum annuum. Chromosoma 89: 352–360.

_____, and G. D. Kuehn. 1985. Genetics, subcellular localization, and molecular characterization of 6-phosphogluconate dehydrogenase isozymes in tomato. Biochem. Genet. 23: 441–454.

_____, and T. J. Orton [eds.]. 1983. Isozymes in plant genetics and breeding. Elsevier, Amsterdam.

_____, and C. M. Rick. 1980. Genetics of esterases in species of Lycopersicon. Theor. Appl. Genet. 56: 209–219.

_____, D. Zamir, and C. M. Rick. 1981. Evidence for extensive overlap of sporophytic and gametophytic gene expression in Lycopersicon esculentum. Science 213: 453–455.

Tanner, G. J., L. Copeland, and J. F. Turner. 1983. Subcellular localization of hexose kinases in pea stems: mitochondrial hexokinase. Pl. Physiol. 72: 659–663.

Thorup, O. A., W. B. Strole, and B. S. Leavell. 1961. A method for the localization of catalase on starch gels. J. Lab. Clin. Med. 58: 122–128.

Torres, A. M., U. Diedenhofen, B. O. Bergh, and R. J. Knight. 1978. Enzyme polymorphisms as genetic markers in the avocado. Amer. J. Bot. 65: 134–139.

Turner, V. S., and D. A. Hopkinson. 1979. The use of meldolablue in isozyme stains after electrophoresis. F.E.B.S. Lett. 105: 376–378.

Tyson, H., M. A. Fieldes, C. Cheung, and J. Starubin. 1985. Isozyme relative mobility (R_m) changes related to leaf position: apparently smooth R_m trends and some implications. Biochem. Genet. 23: 641–654.

Vallejos, E. 1983. Enzyme activity staining. In S. D. Tanksley and T. J. Orton [eds.], Isozymes in plant genetics and breeding, Part A, 469–516. Elsevier, Amsterdam.

Van Den Berg, B. M., and H. J. W. Wijsman. 1982. Genetics of the peroxidase isoenzymes in Petunia. Part 3: location and developmental expression of the structural gene prxA. Theor. Appl. Genet. 63: 33–38.

Vodkin, L. O., and J. G. Scandalios. 1981. Genetic control, developmental expression and biochemical properties of plant peptidases. In M. L. Rattazzi, J. G. Scandalios, G. S. Whitt [eds.], Isozymes: current topics in biological and medical research, Vol 5, 1–25. Alan R. Liss, New York.

Wallner, S. J., and J. E. Walker. 1975. Glycosidases in cell wall-degrading extracts of ripening tomato fruits. Pl. Physiol. 55: 94–98.

Warwick, S. I., and L. D. Gottlieb. 1985. Genetic divergence and geographic speciation in Layia (Compositae). Evolution 39: 1236–1241.

Watson, M. A., and G. L. Cook. 1982. Comparison of electrophoretic phenotypes obtained from water hyacinth material prepared in different grinding buffers. Aquat. Bot. 14: 205–210.

Weeden, N. F. 1983. Plastid isozymes. In S. D. Tanksley and T. J. Orton [eds.], Isozymes in plant genetics and breeding, Part A, 139–156. Elsevier, Amsterdam.

_____. 1984. Distinguishing among white seeded bean cultivars by means of allozyme genotypes. Euphytica 33: 199–208.

_____. 1985. Identification and partial characterization of 3 beta-galactosidase isozymes in pea leaves. Pisum Newsl. 17: 76–78.

_____, and L. D. Gottlieb. 1979. Distinguishing allozymes and isozymes of phosphoglucoseisomerases by electrophoretic comparisons of pollen and somatic tissues. *Biochem. Genet.* 17: 287–296.

———, and ———. Isolation of cytoplasmic enzymes from pollen. *Pl. Physiol.* 66: 400–403.

———, and G. A. Marx. 1984. Chromosomal locations of twelve isozyme loci in *Pisum sativum*. *J. Heredity* 75: 365–370.

———, and ———. 1987. Further genetic analysis and linkage relationships of isozyme loci in the pea: confirmation of the diploid nature of the genome. *J. Heredity* 78: 153–159.

———, and R. W. Robinson. 1986. Allozyme segregation ratios in the interspecific cross *Cucurbita maxima* × *C. ecuadorensis* suggest that hybrid breakdown is not caused by minor alterations in chromosome structure. *Genetics* 114: 593–609.

Wendel, J. F., and C. R. Parks. 1982. Genetic control of isozyme variation in *Camellia japonica* L. *J. Heredity* 73: 197–204.

———, M. M. Goodman, C. W. Stuber, and J. B. Beckett. 1988. New isozyme systems for maize (*Zea mays* L.): aconitate hydratase, adenylate kinase, NADH dehydrogenase and shikimate dehydrogenase. *Biochem. Genet.* 26: 421–445.

———, C. W. Stuber, M. D. Edwards, and M. M. Goodman. 1986 Duplicated chromosome segments in maize (*Zea mays* L.): further evidence from hexokinase isozymes. *Theor. Appl. Genet.* 72: 178–185.

Werman, S. D. 1986. Enhancing resolution in horizontal starch gels with cellulose acetate membranes. *Isozyme Bull.* 19: 52.

Werth, C. R. 1985. Implementing an isozyme laboratory at a field station. *Va. J. Sci.* 36: 53–76.

Wheeler, N. C., and R. P. Guries. 1982. Population structure, genic diversity, and morphological variation in *Pinus contorta* Dougl. *Canad. J. For. Res.* 12: 595–606.

———, ———, and D. M. O'Malley. 1983. Biosystematics of the genus *Pinus*, subsection *Contortae*. *Biochem. Syst. Ecol.* 11: 333–340.

Wilson, R. E., and J. F. Hancock. 1978. Comparison of four techniques used in the extraction of plant enzymes for electrophoresis. *Bull. Torrey Bot. Club* 105: 318–320.

Wolf, P. G., C. H. Haufler, and E. Sheffield. 1987. Electrophoretic evidence for genetic diploidy in the bracken fern (*Pteridium aquilinum*). *Science* 236: 947–949.

Womack, J. E., and Y. D. Moll. 1986. Gene map of the cow: conservation of linkage with mouse and man. *J. Heredity* 77: 2–7.

Yazdani, R., and D. Rudin. 1982. Inheritance of fluorescent esterase and β-galactosidase in haploid and diploid tissues of *Pinus sylvestris* L. *Hereditas* 96: 191–194.

Yeh, F. C., and C. Layton. 1979. The organization of genetic variability in central and marginal populations of lodgepole pine, *Pinus contorta* ssp. *latifolia*. *Canad. J. Genet. Cytol.* 21: 487–503.

Zehender, H., D. Trescher, and J. Ullrich. 1983. Activity stain for pyruvate decarboxylase in polyacrylamide gels. *Anal. Biochem.* 135: 16–21.

Zimniak-Przybylska, Z., S. Blixt, and J. Przybylska. 1985. Isoenzyme variation in the genus *Pisum*. IV. Further electrophoretic analysis of amylases from cotyledons of ungerminated seeds. *Genetica Polonica* 26: 303–306.

CHAPTER 2

Genetics Of Plant Isozymes

Norman F. Weeden

Department of Horticultural Sciences
New York State Agricultural Experiment Station
Cornell University
Geneva, New York 14456

Jonathan F. Wendel

Department of Botany
Bessey Hall
Iowa State University
Ames, Iowa 50011

The ability to observe allelic variation at isozyme loci has revolutionized research in the fields of biochemical genetics, population genetics, and evolution. This variation, called allozymic polymorphism, has been used in plants to examine genetic processes at every stage of the life cycle and to ascertain genetic diversity in all major crops as well as many other species. Yet the potential for using allozymes as genetic markers was not immediately predicted when isozyme variability was initially described (Hunter and Markert, 1957; Markert and Moller, 1959). In fact, the extent and prevalence of allozyme polymorphism was a rather disconcerting surprise to evolutionary biologists. Classical evolutionary theory had predicted that the most "efficient" form of an enzyme should, over time, become predominant in isolated populations, with an occasional rare allele produced through mutation. The discovery in many populations of relatively high levels of polymorphism at isozyme loci has forced a major reconsideration of evolutionary theory (Kimura and Crow, 1964; Koehn et al., 1983; Kimura, 1983).

The precise role of selection and mutation in the maintenance of allozyme polymorphism continues to be debated, but such disagreements have not hindered the use of these genetic variants in a myriad of applications. Later chapters in this book discuss how isozyme techniques have been employed in the study of mating systems (Brown et al.), systematics (Crawford), plant breeding (Stuber; Tores), and population biology (Hamrick). Our focus will be the genetic aspects of allozyme variation, including mode of inheritance, gene number, linkage relationships, null alleles, and several other phenomena. We hope that the material presented here, in combination with that given in Chapter 1, will enable the reader to envision beforehand the generalized zymogram for a specific enzyme system in which he or she is interested. The ability to make such predictions not only facilitates the interpretation of the banding patterns observed, but also allows the researcher to identify deviations from expected results. Such deviations may be produced by gene duplications, changes in ploidal level, loss of gene expression, or post-translational modifications, and are often useful indicators of cell physiology (Jones, 1984) or taxonomic groupings (see the chapter by Crawford in this volume).

GENERAL PROPERTIES

Allozymes are more convenient and reliable genetic markers than those previously available due to a suite of attributes. Allozymes generally exhibit Mendelian inheritance, codominant expression, complete penetrance, and absence of pleiotropic and epistatic interactions. The advantages of these properties were emphasized in early papers on enzyme heterozygosity in *Drosophila pseudoobscura* (Hubby and Lewontin, 1966; Lewontin and Hubby, 1966) and have been discussed in detail by several other authors (Peirce and Brewbaker, 1973; Tanksley and Rick, 1980b; Brown and Weir, 1983). Previously, geneticists struggled with traits that showed environmental variation, were under selective pressures, were incompletely penetrant or affected by the genetic background of the organisms, or were under polygenic control. Hubby and Lewontin's demonstration that electrophoretic analysis could reveal widespread and easily interpreted genetic polymorphism quickly led to the establishment of isozymes as indispensable markers in population biology.

Perhaps most significant of all properties of isozymes is the simple genetic basis of most polymorphisms. Because isozymes are proteins, they can directly reflect alterations in the DNA sequence through changes in amino acid composition. Changes in amino acid composition will often alter the charge or, less often, the conformation of the enzyme, thereby producing a change in electrophoretic mobility. Despite the redundancy of the genetic code and the possibility of amino acid substitutions not affecting the overall charge on the protein, changes in the electrophoretic mobility of enzymes provide an extremely useful method of evaluating genetic differences among groups. The more recent findings of introns within structural genes and the existence of multigene families for many proteins have not seriously undermined the ability of isozyme polymorphisms to serve as a direct measure of DNA sequence variation within and among genomes.

Codominance and complete penetrance are properties of allozymes that logically follow from their simple biochemical and genetic basis. An isozyme locus is defined as the structural gene for an enzyme capable of catalyzing a specified biochemical reaction. The observation of codominant expression in nearly all cases examined provides compelling evidence that both alleles are transcribed, and both transcripts are translated. If the gene is expressed, the activity stain provides a direct method of visualizing the products of both alleles.

The freedom from pleiotropic and epistatic interactions again follows from the simple genetic basis of isozymes, but there may be more complicating factors than for codominance and penetrance. The formation of an allozyme is rarely a simple transcription and translation sequence. Most eukaryotic structural genes contain introns, requiring that the primary transcript be processed by tailoring enzymes. Once synthesized, many polypeptides undergo further processing before forming active enzymes. Polypeptides destined for the chloroplast or mitochondrion usually have transport sequences that are cleaved off inside the organelle by a peptidase (Colman and Robinson, 1986). Acid phosphatases usually are glycoproteins (Paul and Williamson, 1987), the carbohydrate moieties added by other enzymes. Whenever enzymatic processing of an RNA sequence or polypeptide occurs, it is possible to introduce new polymorphism into the final active product because the processing enzymes may be heterogeneous. Thus, variation in the primary sequence could be obscured by polymorphism introduced during the processing steps by products of other loci. However, the empirical data indicate that, for the most part, variation in electrophoretic mobility of a protein directly reflects changes in the DNA sequence of the structural gene for that protein and not polymorphism in one or more of the genes involved in the modification steps.

GENE NUMBER CONSERVATION AND SUBCELLULAR LOCALIZATION

Most experimental data on plant isozymes from the last 20 years have not only confirmed that the properties mentioned above are characteristic of nearly all enzyme systems but have also revealed additional traits shared by many enzyme systems. Two of the more important are the stability of locus number for specific enzyme systems and the consistent subcellular compartmentation of the isozymes. This phenomenon was reviewed by Gottlieb (1982), and subsequent findings have, in general, confirmed and extended his conclusions. For enzymes with well-defined biochemical roles, and especially those acting in pathways of primary metabolism (e.g., the Embden-Meyerhof and pentose phosphate pathways, the Krebs and Calvin cycles, and amino acid biosynthesis), a predictable number of loci will be expressed in most diploid plants. Each of the isozymes expressed will have a different subcellular location and a conserved subunit structure. In contrast, enzyme systems such as acid phosphatase, NAD(P)H diaphorase, esterase, peptidase, and peroxidase display considerable variation in total number of isozymes expressed, number of isozymes per subcellular compartment, and number of subunits per enzyme.

To provide a readily available source of information to assist in gel interpretation, data on isozyme number, subcellular location, and subunit structure have been compiled (Table 2.1). The references selected are those best documenting the genetic behavior, or lacking that, the substructure and subcellular localization of the relevant isozymes. Table 2.1 presents isozyme number for an idealized diploid plant and gives only those isozymes generally visualized by routine staining procedures. Isozymes such as the plastid-specific isocitrate dehydrogenase or mitochondrial fumarase are not listed because their activity is usually too faint to be observed under standard conditions.

Even with these stated restrictions, some variation in isozyme number (e.g., aconitate hydratase) and subcellular localization (shikimate dehydrogenase) is evident in Table 2.1. The predictive value of Table 2.1 is, therefore, imperfect. Certainly, quaternary structure is highly conserved in biochemical evolution, enabling one to assume a priori that enzymes such as AAT, GPI, MDH, PGD, and TPI will be dimeric, whereas ACO, PGM, and SKD will act as monomers. Isozyme number and subcellular distribution are more open to variation because specific isozymes may not be strongly expressed in certain species or tissues. In fact, several recent papers have stressed that significant variability in isozyme phenotype exists when tissue specificity and plant age are considered (Jones, 1984; Murray and Ayre, 1987; Pedersen and Simonsen, 1987). This variation may be produced by post-translational modifications of the gene product or by changes in the number of kind or loci expressed. Thus the situation may not be as simple as Table 2.1 implies. In addition, many cases of gene duplications involving isozyme loci have been reported in diploids (see discussion below). Should the number of isozymes observed in a species differ from that given in Table 2.1, it might be appropriate to determine which isozymes are missing or duplicated.

Most of the enzyme systems listed in Table 2.1 have more than one isozyme. An obvious question is why plants require more than one isozyme of any particular enzyme. This redundancy has been attributed to the need for a separate isozyme in each subcellular compartment in which the same reaction occurs (Gottlieb, 1982). Metabolite concentrations and pH are different in the various subcellular compartments, indicating that different isozymes may be necessary for efficient catalysis. The presence of chloroplast isozymes can be postulated to be a natural consequence of this organelle's endosymbiotic origin (Margulis, 1970), presuming that this organelle retained much of its original biochemistry (Weeden, 1981). The endosymbiont hypothesis could also account for mitochondrial and microbody isozymes (Margulis, 1981; Gray and Doolittle, 1982), but the evidence is much less convincing. Organelle-specific isozymes may be present in these latter two subcellular compartments simply because enzymes with properties different from those of cytoplasmic isozymes are required to carry out the appropriate reactions in these two organelles.

The mention of plastid-specific isozymes immediately raises the question of whether the coding gene resides in the nucleus or the organelle. Several studies have demonstrated that plastid-specific isozymes typically exhibit biparental inheritance (e.g., Weeden and Gottlieb, 1980; Lax et al., 1984; Tanksley and Kuehn, 1985). Experiments with plants lacking the ability to synthesize protein within the plastid also confirm that most plastid-specific proteins are synthesized on cytosolic (80S) ribosomes (Boulter et al., 1972; Schaefers and Feierabend, 1976; Feierabend and Mikus, 1977). The one exception to the generalization that isozymes are nuclear encoded is ribulose bisphosphate carboxylase (rubisco). This enzyme consists of two types of subunits. The small subunit is nuclear encoded, whereas the large subunit is coded by a plastid gene (Ellis, 1981). Generally, the variation in this enzyme has been examined by isoelectric focusing after dissociation of the enzyme into its monomeric subunits (Kung et al., 1974). Recently, a method for visualizing the holoenzyme on starch gels has been described (Weeden, 1984), making it possible to use isozyme techniques to search for variation in both the large and small subunits simultaneously. When polymorphism is observed using this latter method, further tests are required (e.g., isoelectric focusing of the subunits or inheritance studies) to determine whether the large or small subunit is varying.

Table 2.1. *Number, subcellular localization, and subunit structure of isozymes for the more commonly assayed enzyme systems*

Enzyme System	Number of Isozymes	Subcellular Distribution[1]	Quarternary Structure	References
ACP	2–4	various	monomer, dimer	de Cherisey, 1985
ACO	1–3	c, mt	monomer	Brouquisse et al., 1987
				Morden et al., 1988
ADK	1–2	c, p	monomer	Kleczkowski and Randall, 1986
				Wendel et al., 1988
ADH	1–3	c	dimer	Freeling and Schwarz, 1973
AMP	2–3	c	monomer	Scandalios and Espiritu, 1969
AAT	4	c, p, mt, mb	dimer	Huang et al., 1976
				Weeden and Marx, 1987
CAT	1	mb	tetramer	Scandalios, 1974
DIA	1–4	c, p, mt	monomer, dimer, tetramer	J. F. Wendel et al., unpubl. data
				Weeden and Lamb, 1987
EST	2–10	c	monomer, dimer	Tanksley and Rick, 1980
				Wehling and Schmidt-Stohn, 1984
FDH	1	unknown	dimer	Farinelli et al., 1983
				Wendel and Parks, 1982
FBA (ALD)	2	c, p	tetramer	Anderson et al., 1975
				Weeden and Gottlieb, 1980
FBP	2	c, p	tetramer	Zimmermann et al., 1978
				Navot and Zamir, 1986
FUM	1	mt	tetramer	Kanarek et al., 1964
				Brouquisse et al., 1987
GAL	1–3	c, p	monomer dimer	Bhalla and Dalling, 1984
				Yazdani and Rudin, 1982
G6PD	2	c, p	dimer	Schnarrenberger et al., 1973
GPI	2	c, p	dimer	Schnarrenberger and Oeser, 1974
				Gottlieb, 1977
GDH	1	c	hexamer	Cammerts and Jacobs, 1983
GLU	1	c	dimer	Stuber et al., 1977
G3PD (NAD, NADP)	3	c, p	tetramer	Cerff and Chambers, 1979
HEX	2–3	c, p, mt	monomer	Miernyk and Dennis, 1983
				Tanner et al., 1983
				Wendel et al., 1986
IDH (NADP)	1	c	dimer	Tanksley, 1984
				Ni et al., 1987
LDH	1	c	tetramer	Mayr et al., 1982
MDH (NAD)	3	c, mt, mb	dimer	Goodman et al., 1980
ME	1	c	tetramer	Navot and Zamir, 1986
				Weeden and Lamb, 1987

[1]Abbreviations for subcellular compartments: c = cytosol, p = plastid, mt = mitochondrion, mb = microbody, cw = cell wall.

Enzyme System	Number of Isozymes	Subcellular Distribution[1]	Quarternary Structure	References
PRX	2–13	c, cw	monomer, dimer	Garcia et al., 1982 Brewbaker et al., 1985
PPO	1	p	monomer	Pryor and Schwartz, 1973 Lax et al., 1984
PGM	2	c, p	monomer	Weeden and Gottlieb, 1980 Navot and Zamir, 1986
PGD	2	c, p	dimer	Weeden and Marx, 1984 Tanksley and Kuehn, 1985
SKD	1–2	p c	monomer monomer	Weeden and Gottlieb, 1980 J. F. Wendel et al., unpubl. data
SOD	3	c, p mt	dimer tetramer	Baum and Scandalios, 1981 Jackson et al., 1978
TPI	2	c, p	dimer	Pichersky and Gottlieb, 1983

A few reports have suggested that certain isozymes occupying different subcellular compartments are coded by the same gene. In pumpkin cotyledons, Chou and Splittstoesser (1972) described seven cytosolic glutamate dehydrogenase isozymes, one of which was active in both the particulate and cytosolic fractions. More recently, Hock (1984) presented evidence indicating that there was a common precursor for glyoxysomal and mitochondrial malate dehydrogenase in watermelon. Beyond these few exceptions, however, analysis of the inheritance of cytosolic, mitochondrial, and plastid isozymes has shown that they are each coded by a different locus (Newton, 1983; Weeden, 1983b).

Two additional points that have been discussed are the comparison of the variability of organelle-specific isozymes relative to their cytosolic counterparts (Gottlieb and Weeden, 1981) and their more anodal mobility (Ireland and Dennis, 1980). Although the observation that plastid-specific glucosephosphate isomerase displays comparatively little polymorphism has been extended to many other species (e.g., Jones, 1984), not all plastid isozymes share this level of conservation. Plastid aspartate aminotransferase shows more variation than the cytosolic isozyme in pea (Weeden and Marx, 1987), and the plastid forms of phosphoglucomutase and 6-phosphogluconate dehydrogenase show significant polymorphism in pea, tomato, lentil, and apple (Weeden and Gottlieb, 1980; Tanksley and Kuehn, 1985; Tadmor et al., 1987; Weeden and Lamb, 1985, 1987). The same generalizations can be extended to polymorphic mitochondrial isozymes such as maize MDH (Goodman et al., 1980; Stuber and Goodman, 1983a) and pea AAT (Weeden and Marx, 1984).

It is also clear that plastid-specific isozymes do not always possess a faster mobility than the cytosolic forms. Although the plastid isozymes of GPI and TPI often migrate faster than the cytosolic, in Phaseolus vulgaris and in certain lycopods cytosolic GPI allozymes migrate anodal to the plastid form. In Cucurbita the cytosolic TPIs all migrate faster than the plastid forms (Weeden and Robinson, 1986). A similar order of isozymes has been reported in Equisetum (Soltis, 1986). Other enzyme systems such as PGM in pea (Weeden and Gottlieb, 1980) and AAT in apple (Weeden and Lamb, 1987) also show this reversal. In maize, mitochondrial MDH is encoded by three loci. The most common allozyme of one of these isozymes migrates faster than the cytosolic forms, whereas the most common allozymes of the other two mitochondrial isozymes migrate more slowly (Goodman et al., 1980). A similar overlapping pattern of isozyme migration is displayed by the five isozymes of maize TPI (J. F. Wendel et al., unpublished data). Thus, neither degree of polymorphism nor relative mobility are reliable parameters for determining subcellular location.

An additional question concerns the reasons for multiplicity of isozymes in esterases, aminopeptidases, and peroxidases. These systems often have three or more isozymes expressed in the same subcellular compartment. Does each isozyme have a unique and important role in cell physiology, or are these merely redundant proteins? Arguments can be presented supporting both hypotheses. In pea, three leucine aminopeptidases can be resolved using the standard substrate, leucyl β-naphthylamide (Scandalios and Espiritu, 1969; Weeden, 1985). However, when different substrates are used in the assay, each isozyme has a distinct preference, although not complete specificity (Scandalios and Espiritu, 1969; Murray and Waters, 1985; Weeden, 1985). The presence of several LAP isozymes may be attributed to the artificial substrate being used and a somewhat broad substrate specificity of the isozymes. The 13 peroxidases identified in maize can each be distinguished by considering substrate and tissue specificity (Brewbaker et al., 1985). Esterase and alcohol dehydrogenase isozymes also have been shown to have different substrate preferences (Tanksley and Rick, 1980a; Gomes et al., 1982; Roose, 1984).

Thus, at least part of the multiplicity of isozymes can be explained by differences in function, but it remains an open question whether or not this explanation holds for tight clusters of duplicated esterase or peroxidase loci such as are found in tomato, rye, and pea (Tanksley and Rick, 1980b; Wehling and Schmidt-Stohn, 1984; Weeden and Marx, 1987). It is known that individual units of small gene families can diverge (Dunsmuir et al., 1983). However, divergence in function has not been demonstrated for compound loci encoding esterase or peroxidase isozymes in plants. Should we dismiss these latter clusters as transient duplicate expression in the lack of evidence to the contrary? If so, why do we observe such clusters primarily in these "non-specific" systems? Certainly, unequal crossing-over must occur around genes coding glucosephosphate isomerase, phosphoglucomutase, aspartate aminotransferase, and other enzyme systems with more stable numbers of isozymes. Recently, 47 "isoperoxidases" were identified in regenerating tobacco callus (Kay and Basile, 1987). Although the number of structural loci responsible for such a plethora of forms was not investigated, it is difficult to envision a distinct function for each of the variants. At present, we simply do not have enough data to explain why clusters of loci exist (or are expressed) in some enzyme systems but not in others. It is probable that studies on the molecular biology of some of these systems will soon provide at least part of the answer.

GENETIC STUDIES

The genetics of isozyme variants have been intensively studied in many important crops as well as in species representing model systems such as conifers and ferns. Table 2.2 presents a summary of the species in which 10 or more polymorphic isozyme loci have been defined using controlled crosses or single parent segregation analysis. Thus, only taxa in which formal genetic studies have been performed are listed. Ten or more isozyme loci have been defined in many other species, using data from surveys of variation in natural populations or phenotypic comparisons across taxa. Designation of isozyme loci in these latter species is accomplished by extrapolation from the studies presented in Table 2.2. Genetic conclusions derived from this indirect approach are subject to several sources of error, as discussed in the previous chapter, and are best considered tentative.

Despite this limitation the list includes a wide range of vascular plant families and provides considerable confidence that the generalizations made in Table 2.1 apply to most vascular plants. On the other hand, even if we included all species subjected to isozyme analysis, there would still be gaps in the coverage. For instance, few monocotyledons other than grasses have been studied, and whole orders of dicots have been ignored. We may have a set of convenient generalizations with which to work, but there will undoubtedly be many additional surprises as new species are investigated.

The status of isozyme research for some of the taxa listed in Table 2.2 was reviewed in the compendium edited by Tanksley and Orton (1983). Other chapters in this book deal with ferns and bryophytes. Thus, it is unnecessary to discuss this same material here. Our intent is to summarize the work in some of the major crops since the 1983 reviews and to describe significant advances in several other taxa.

Table 2.2. Vascular plant genera in which ten or more isozyme loci have been demonstrated by genetic analysis

Genus	Loci	Selected references[1]
		Conifers
Calocedrus	18	Harry, 1986
Picea	20	King and Dancik, 1983; Boyle and Morgenstern, 1985
Pinus	31	Millar, 1985
Pseudotsuga	27	El-Kassaby et al., 1982
		Monocotyledons
Avena	14	Price and Kahler, 1983
Elytrigia	11	Hart and Tuleen, 1983
Hordeum	32	Brown, 1983
Oryza	12	Ranjhan et al., 1988; Sano and Barbier, 1985
Secale	20	Jaaska, 1983; Miller, 1984; Wehling et al., 1985
Setaria	12	de Cherisey et al., 1985
Triticum	78	Hart, 1984; Chenicek and Hart, 1987
Zea	45	Goodman and Stuber, 1983a; Wendel et al., 1986, 1988
		Dicotyledons
Camellia	12	Wendel and Parks, 1982
Capsicum	14	Tanksley, 1984
Citrullus	17	Navot and Zamir, 1986
Citrus	10	Torres et al., 1985
Clarkia	17	Gottlieb, 1986
Cucurbita	20	Kirkpatrick et al., 1985; Weeden and Robinson, 1986
Eucalyptus	20	Moran and Bell, 1983
Glycine	19	Kiang and Gorman, 1985; Griffin and Palmer, 1987
Gossypium	13	Suiter, 1988
Lens	20	Tadmor et al., 1987; F. Muehlbauer et al., 1989
Limnanthes	14	McNeill and Jain, 1983
Lycopersicon	35	Mutschler et al., 1987
Malus	24	Chevreau et al., 1985; Weeden and Lamb, 1987
Oenothera	10	Levy et al., 1975
Persea	12	Torres et al., 1986
Petunia	10	Wijsman, 1983
Phaseolus	10	Weeden, 1986
Pisum	38	Weeden and Marx, 1987; Weeden et al., 1987
Plantago	14	van Dijk, 1985
Stephanomeria	11	Gallez and Gottlieb, 1982
Tolmiea	12	Soltis and Rieseberg, 1986
Turnera	10	Shore and Barrett, 1987
Vitis	10	Loukas et al., 1983

[1]References were selected to provide current yet extensive data. For several species many more references are available.

Genetic analysis of isozyme loci developed rapidly in three crops: wheat, tomato, and maize. Long before genetic studies on isozymes were initiated in most other species, over 20 allozyme polymorphisms had been described in each of these crops, the monogenic nature of the polymorphisms had been demonstrated, and the respective genes had been located on a chromosomal linkage map. These crops are of obvious economic importance; yet their commercial value is only part of the explanation why these crops became preeminent in isozyme studies. For any genetic study one also requires convenient systems and the expertise to take advantage of the systems. Hart grasped the opportunity offered by the availability of the nulli-tetrasomic lines developed by Sears (1966) in the wheat cultivar 'Chinese Spring'. By observing additional bands or increased intensity of already existing bands, Hart and his coworkers have been able to assign more isozyme loci to chromosomes than have been mapped in any other crop (Hart, 1979, 1984). That such progress could be made without the benefit of a pre-existing chromosomal linkage map and with a relatively low level of allozyme variability within the species makes the system and the results even more remarkable.

Both maize and tomato are excellent experimental organisms with which to perform genetic studies and have been the objects of considerable isozyme analysis. Early work was facilitated by the availability of numerous morphological markers and well-developed chromosomal linkage maps. *Zea mays* contains high levels of enzyme polymorphism (e.g., Stuber and Goodman, 1983a; Goodman and Stuber, 1983b; Doebley et al., 1984, 1985), and thus electrophoretic variants were relatively easily incorporated into genetic studies. Analyses of segregation patterns in hundreds of controlled crosses successfully documented the inheritance of numerous isozyme variation patterns and also resulted in an extensive isozyme linkage map (Goodman and Stuber, 1983a). The situation in tomato was not as straightforward because the amount of allozyme polymorphism within *Lycopersicon esculentum* germplasm was relatively low. However, Rick (1975, 1979) identified several interspecific crosses that produced fertile hybrids as well as F_2 and backcross populations. The considerable allozyme variation available through these interspecific crosses was analyzed and mapped using trisomic stocks and the previously mapped mutants (Fobes, 1980; Tanksley and Rick, 1980b).

Since the 1983 reviews, several additional isozyme loci have been mapped in these species. Triosephosphate isomerase, superoxide dismutase, and aconitate hydratase loci have been characterized and located on the wheat homoeologous group 5, 2, and 6, respectively (Pietro and Hart, 1985; Neuman and Hart, 1986; Chenicek and Hart, 1987). Two new peroxidase isozymes were described in wheat extracts by Ainsworth et al. (1984). In tomato 35 isozyme loci have been mapped at least to chromosome and in most cases to a specific region on the chromosome linkage map (Mutschler et al., 1987). The *Est-4—Pgi-1—Pgd-2* linkage group has been assigned to chromosome 12, a chromosome previously deficient in marker loci. Six additional loci (*Est-8, Acon-1,2, Idh-1, Pgd-3,* and *Sod-1*) have been shown to be polymorphic and positioned on the linkage map.

Recent studies in maize have expanded the number of available loci that can be examined by starch gel electrophoresis to approximately 45 with a concomitant increase in knowledge of the linkage relationships of all allozyme markers. Newly mapped allozyme markers include two hexokinase loci (Wendel et al., 1986), five triosephosphate isomerase loci (J. F. Wendel et al., unpublished data), two NADH dehydrogenase loci (Wendel et al., 1988), an additional acid phosphatase locus (Kahler, 1983; Sisco et al., 1987), and a single locus for each of the enzyme systems adenylate kinase (Wendel and Beckett, 1987; Wendel et al., 1988), aconitate hydratase (Wendel et al., 1988), and shikimate dehydrogenase (Wendel et al., 1988). The addition of these new markers has generated a linkage map containing one or more allozyme markers on all 10 chromosomes of maize.

Approximately 40 allozyme-encoding loci have been described in rice (*Oryza sativa*), although few of these have been mapped to chromosomes (Second, 1982, 1985; Sano and Morishima, 1984; Sano and Barbier, 1985; Ranjhan et al., 1987). Progress in linkage mapping has been slowed by the lack of a well developed morphological map and the absence of cytogenetic or morphological tester stocks. The recent development of a complete series of primary trisomics (Khush et al., 1984) offers an important new tool in this regard; these have already been used to map four isozyme loci (Ranjhan et al., 1988).

Another crop that has emerged as an important model for isozyme studies is the garden pea, *Pisum sativum*. Currently, approximately 50 polymorphic isozyme loci have been described within the species, 40 of which have been placed on the chromosomal linkage map (Weeden, 1985, 1988; Weeden and Marx, 1987). As was the case for maize and tomato, pea is a convenient organism with which to perform genetic studies. Many monogenic morphological markers have been described and mapped in this species (Blixt, 1974), and these were especially useful in the mapping of the isozyme loci.

Many other taxa now boast over 20 genetically defined allozyme polymorphisms. Several genera of Triticeae have extensive linkage maps. Progress in *Secale, Hordeum,* and *Agropyron* has been rapid, often facilitated by extrapolation from results in wheat (Hart and Tuleen, 1983; Schmidt et al., 1984; Ainsworth et al., 1984; Wehling et al., 1985; Figueiras et al., 1986). Most of these species are diploid, and it has been possible to produce a sequential arrangement of many of the isozyme loci on the chromosomes. In this way, studies of isozyme loci in barley, rye, and other species related to *Triticum* have often complemented the previous investigations of wheat.

The genetic analysis of isozyme loci in *Capsicum* was also aided by previous knowledge of tomato isozymes. Although the genetic analysis of the observed variation was performed independently in *Capsicum* (Tanksley, 1984), the type of variation was often similar to that in *Lycopersicon* (e.g., a two-banded shikimate dehydrogenase phenotype in homozygous lines). Isozyme studies in two other crops were similarly assisted by the availability of a model system. Sorghum (*Sorghum bicolor*) now has seven isozyme loci identified in three linkage groups, another 19 allozyme polymorphisms having been identified (J. Doebley, personal communication). Previous work on maize isozymes facilitated the development of buffer systems and the interpretation of the zymograms, including the identification of a modifier locus affecting mitochondrial malate dehydrogenase (Doebley et al., 1986). Except for acid phosphatase and esterase, the zymograms of lentil, *Lens culinaris*, were practically identical to those previously described and characterized in *Pisum*. This parallel made identification of homologous loci in the two genera exceptionally easy (Weeden et al., 1987) once the genetic basis of the phenotypes had been defined (Tadmor et al., 1987; F. Muehlbauer et al., 1989).

Several other legumes have also been examined intensively, although with less success. Soybean has proved to be a difficult species to analyze because of low levels of allozyme polymorphism and because of its high chromosome number. Recent studies by Kiang and Gorman (1985) and Griffin and Palmer (1987) have reported several additional polymorphic loci and identified some to linkage groups. The common bean, *Phaseolus vulgaris*, also possesses a relatively low level of allozyme variation, with only 10 polymorphic loci reported (Weeden, 1986a).

Numerous genetic studies of isozyme polymorphism have been performed on members of Asteraceae. Early work on alcohol dehydrogenase was done in the sunflower (*Helianthus annuus*) by Torres (1974). Kahler and Lay (1985) extended to seven the number of genetically defined loci in this species. Progeny tests in several species of *Stephanomeria* have defined 10 polymorphic loci (Gallez and Gottlieb, 1982). Most recently, Kesseli and Michelmore (1986) have examined allozyme polymorphism in *Lactuca*. In the interspecific cross *L. sativa* × *L. saligna*, 11 isozyme loci segregated, and three linkage groups were observed.

Four other crops have been subjected to intensive analysis of isozyme polymorphism. Suiter (1988) studied genetic segregation of isozymes in the Old World diploid cultivated cottons *Gossypium arboreum* and *G. herbaceum*. The codominant inheritance of 13 loci was established, and four linkage groups were recognized. Segregation data were obtained for 15 loci in tea, *Camellia japonica*. Linkage tests for 91 of the 105 possible pair-wise combinations resulted in the identification of five linkage groups (Wendel and Parks, 1982, 1984; Wendel, 1983). In the Cucurbitaceae, significant progress has been made in both *Citrullus* and *Cucurbita*. In watermelon, Navot and Zamir (1986) documented allozyme polymorphism at 19 loci and identified four linkage groups. In *Cucurbita pepo* 12 polymorphic loci were identified (Kirkpatrick et al., 1985; Weeden et al., 1986b). Several interspecific crosses are possible in this genus, and that involving *C. maxima* × *C. ecuadorensis* produced an F_2 with 20 isozyme loci segregating (Weeden and Robinson, 1986). Examination of these populations revealed four multilocus linkage groups.

Although relatively few formal genetic studies have been performed on wild plants not closely related to cultivated species, several recent studies clearly demonstrate the practicality of such endeavors. A series of investigations has defined 17 isozyme loci in *Clarkia*, particularly in the polymorphic species *C. xantiana* (Gottlieb, 1984). Except for a few pairs of loci, linkage analysis has not been performed in this genus. Similar segregation data have been obtained for numerous isozyme loci in *Oenothera* (Levy et al., 1975) and *Limnanthes* (McNeill and Jain, 1983). Both segregation and linkage data were obtained in two disparate taxa, *Plantago lanceolata* and *Turnera ulmifolia*. In *P. lanceolata* 14 isozyme loci as well as the self-incompatibility locus and three male sterility loci could be assembled into five multilocus linkage groups (van Dijk, 1985). Three linkage groups were found in *T. ulmifolia*, one of which included three genes controlling morphological traits in addition to three isozyme loci (Shore and Barrett, 1987). In these species the haploid number of chromosomes is six and five, respectively, suggesting that the isozyme loci span a considerable portion of the linkage map.

The commonly used isozyme assays will visualize the products of about 70 loci in a diploid plant. Of these loci 40 to 50 have been found to be polymorphic in the three most intensively studied species: maize, tomato, and garden pea. In each case, an extensive survey of the germplasm, including cross-compatible wild species when available, was necessary to identify such a large number of variants. Thus, it would appear that, in the absence of any major increase in the number of isozyme assays available, 50 to 60 polymorphic loci may be the maximum available within a large and diverse germplasm sample. Species with a more restricted germplasm base would be expected to exhibit fewer polymorphic loci, although treatment with mutagens such as EMS could generate additional polymorphism (Jones et al., 1986). Interestingly, the mating system does not appear to limit the amount of allozyme polymorphism available. Both maize and tomato are predominantly outcrossing in wild populations, the former wind pollinated and the latter pollinated by various insect species. In contrast, pea is predominantly self-pollinated.

Most evidence indicates that isozyme loci will be widely distributed throughout the genome. Tight clustering of isozyme loci has been observed in relatively few instances, as discussed earlier. There exists little or no evidence for the clustering of loci on the basis of metabolic pathway or subcellular location. This apparent chaos is perhaps a bit frustrating for those attempting to find order to the eukaryotic genome, but the random distribution of these loci makes them especially suitable as genetic markers.

GENE DUPLICATION AND POLYPLOIDY

Duplication of segments of the genome has long been recognized as a potentially important event in the evolution of new characters (Ohno, 1970). Early cytological studies in *Drosophila* (Bridges, 1936) and later work in maize (Gopinath and Burnham, 1956) demonstrated the occurrence of duplications and identified several processes by which duplications could be generated. However, it was not until the development of molecular genetics that the prevalence of duplicated genetic material in the eukaryotic genome was revealed. Not only is the genome interspersed with huge quantities of repetitive DNA, some with unknown functions, but many transcribed and translated sequences such as those coding for leghemoglobin (Lee et al., 1983), actin (Bernatsky and Tanksley, 1986), and the small subunit of rubisco (Dunsmuir et al., 1983) also exist in two or more copies. It is not surprising, therefore, that duplication of isozyme loci also has been observed.

The first report of a duplicated isozyme locus in plants involved alcohol dehydrogenase in *Clarkia* (Gottlieb, 1974). Gottlieb observed that stem extracts from *Clarkia franciscana* displayed an additional set of ADH bands compared to similar extracts from the closely related *C. rubicunda*. Genetic tests confirmed that the additional bands were coded by a separate locus, leading to the conclusion that *C. franciscana* expressed three genes coding ADH subunits, whereas other closely related diploid clarkias displayed only two. *Clarkia franciscana* may have been derived relatively recently from *C. rubicunda*, implying that the additional ADH locus was formed by a duplication event during the speciation process (Gottlieb, 1974). Other duplications of ADH loci have been found in *Stephanomeria exigua* (Roose and Gottlieb, 1980), *Sorghum bicolor* (Ellstrand et al., 1983), *Hordeum spontaneum* (Harberd and Edwards, 1983), *Camellia japonica* (Wendel and Parks, 1984), and *Cynosurus* (Ennos, 1986).

However, all these results must be accepted with some reservation. None of these systems has been examined for gene copy number using DNA probes. There is no doubt that more ADH genes are expressed in certain species when compared to related taxa, but whether the additional isozyme was produced by a gene duplication or merely by induction of a pre-existing but inactive sequence has yet to be clarified. This uncertainty is especially important in systems such as ADH, for not only does the base isozyme number vary from species to species but the genes are generally inducible, especially by anaerobic conditions. If one is considering only isozyme data, the switch from a silenced gene to one inducible under anaerobic conditions could be easily interpreted as a gene duplication event.

Instances of presumed gene duplications also have been indentified for other enzymes. As was the case for alcohol dehydrogenase duplications, none of the examples discussed below has been confirmed by gene hybridization or sequencing studies. However, the appearance of additional isozymes for the enzymes mentioned below can be more reliably interpreted as gene duplications because a stable base number of isozymes has been determined for diploid plants (Table 2.1), and the enzymes are all constitutively expressed.

The gene encoding cytosolic glucosephosphate isomerase has been reported duplicated in many taxa, including *Clarkia* (Gottlieb, 1977; Gottlieb and Weeden, 1979), *Persea* (Goldring et al., 1985; Torres et al., 1986), and *Phaseolus* (Weeden, 1986b). Possible duplications in the plastid-specific GPI have been detected in *Coreopsis* (Crawford, 1983) and *Antennaria* (Bayer and Crawford, 1986). Similarly, several TPI duplications have been identified. Both the gene encoding the cytosolic form and that encoding the plastid-specific form are duplicated in *Clarkia* (Pichersky and Gottlieb, 1983), and the cytosolic TPI duplication is present in other genera of the Onagraceae as well. Although a thorough analysis of the distribution of the cytosolic TPI duplication has not been published, results of preliminary studies (Pichersky and Gottlieb, 1983; N. F. Weeden, unpublished data) did not conform to any reasonable phylogenetic grouping of the genera in the family. Apparently the duplication is an ancient one, with many instances of secondary loss.

Five isozymes of triosephosphate isomerase have been identified in maize, of which two are plastid forms and three are cytosolic. The respective loci have been mapped to four different chromosomes. Several of these loci have been localized to genomic regions that possess other duplicated loci, lending support to the hypothesis that they originated through some sort of duplication process (J. F. Wendel et al., unpublished data).

In certain fern taxa the cytosolic form of triosephosphate isomerase shows a duplicated phenotype (Haufler, 1985; Haufler and Soltis, 1986), and in *Equisetum* the plastid isozymes display a duplicated phenotype (Soltis, 1986). Numerous other duplications of isozyme loci have been reported in diploid plants. Examples include PGM in *Camellia* (Wendel and Parks, 1982), *Clarkia* (Soltis et al., 1987), *Zea* (Stuber and Goodman, 1983b), and *Capsicum* (Tanksley, 1984); PGM and IDH in *Layia* (Gottlieb, 1987); and 6-phosphogluconate dehydrogenase in *Lycopersicon* (Tanksley and Kuehn, 1985), *Zea* (Stuber and Goodman, 1984), and *Clarkia* (Odrzykoski and Gottlieb, 1984).

When only one or two presumptive duplications are identified in a lineage, it is usually interpreted as a duplication of a relatively small section of chromosome. This duplication of chromosomal material can occur by unequal crossing-over, giving rise to tandemly repeated segments or by a series of chromosomal rearrangements, producing homologous segments on different chromosomes. Although this latter process is somewhat complex (Gottlieb, 1977), it has been experimentally documented in maize (Gopinath and Burnham, 1956). In populations already polymorphic for two or more chromosomal rearrangements, duplication of a chromosomal segment could occur relatively easily (Weeden, 1983a).

Should several apparent duplications be present in a species, one begins to suspect polyploidy as the cause. Studies in *Tragopogon* and *Triticum* clearly demonstrated the additive expression of isozyme loci in an allopolyploid (Roose and Gottlieb, 1976; Hart, 1983), and studies of other suspected polyploids also have revealed considerable evidence for duplication of isozyme loci (e.g., Oliver et al., 1983; Murdy and Carter, 1985; Chevreau et al., 1985; Bayer and Crawford, 1986). The interpretation of results becomes much more difficult, however, when only a few of the isozyme loci are duplicated. The most prominent example is maize, which contains several pairs of duplicated isozyme loci. Linkage analyses have demonstrated that several of these sets of duplicate loci occur in parallel linkage groups on different chromosomes (Goodman and Stuber, 1983a; Wendel et al., 1986). This observation has been interpreted to support the hypothesis, first proposed by Rhoades (1951), that the genome of maize has an evolutionary history that includes either extensive chromosome segment duplication or polyploidy. Many maize enzyme systems, however, show the typical diploid level of expression. Two hypotheses can be advanced to account for these data: (1) the genome of maize has an evolutionary history that includes polyploidy, but diploidization has occurred in many genomic regions, or (2) present gene duplications reflect chromosome segment duplication that occurred in an ancestral diploid stock. This matter has recently been re-examined with a much more extensive set of genomic markers and RFLP (restriction fragment length polymorphism) analysis (Helentjaris et al.,

1986, 1987). Using a large number of unique or low-copy-number genomic probes, these investigators mapped the locations of probes that hybridized to a single genomic region and those that hybridized to two or more sequences. Duplications were found on all chromosomes, but genomic regions varied greatly in their degree of fragment duplication. Some chromosome segments contained several duplicated fragments in similar gene orders. In summary, the isozyme and RFLP data are concordant in that duplicate pairs of isozyme loci map to chromosome segments bearing extensive restriction fragment duplication, whereas single-copy isozyme loci apparently do not. The presence of so many genomic regions containing non-duplicated sequences suggests that the present genome of maize is not currently structured as an allopolyploid.

It is often of interest to determine whether a particular species is of auto- or allopolyploid origin. Autopolyploids, at least for some time after their formation, are expected to display polysomic inheritance. Allopolyploids, in contrast, exhibit bigenic disomic inheritance because the homoeologous chromosomes, being derived from two divergent stocks, have already undergone sufficient differentiation to preclude or at least minimize pairing between sets during meiosis. The single-locus segregation ratios expected for the two types of polyploids are quite different. Table 2.3 presents the several alternatives for segregation at a single locus in a tetraploid. In nearly all cases, tetrasomic inheritance generates distinctive segregation ratios and expected phenotypic classes. The one exception is the 1:2:1 ratio expected for both the allotetraploid with a single heterozygous locus and the chromosomal segregation in an unbalanced heterozygous autotetraploid. If the mobility of the B allozyme in the allotetraploid is identical to either of the alleles at the A locus, the predicted phenotypes will also be indistinguishable. However, the phenotypic classes could be distinguished on the basis of relative band intensity (Aa versus AAAa) if the B isozyme possessed a unique mobility.

Isozyme loci are ideal markers for determining whether a polyploid displays disomic or tetrasomic inheritance. Tetrasomic inheritance has been observed for several loci in *Medicago* (Quiros, 1983), *Solanum* (Quiros and McHale, 1985), and in tetraploid *Tolmiea menziesii* (Soltis and Rieseberg, 1986; Soltis and Soltis, 1988). Apple, in contrast, displayed only disomic inheritance at isozyme loci (Chevreau et al., 1985) as did tetraploid wild oats (Hutchinson et al., 1983) and *Cucurbita pepo* (Kirkpatrick et al., 1985). In *Medicago* and *Tolmiea,* the data indicating tetrasomic inheritance appeared to be compatible with chromosomal segregation. Chromatid segregation has yet to be demonstrated unambiguously in tetraploid vascular plants.

Polyploids tend to revert to a diploid level of gene expression (Ferris et al., 1979). Isozymes are often useful for studying this process because the loss of duplicate locus expression is particularly easy to observe in systems containing multimeric proteins. Subunits encoded by duplicate loci will generally hybridize to form active heteromers. The occurrence of an intergenic hybrid molecule can be distinguished from a simple heterozygote by the true-breeding nature of the phenotype (Gottlieb, 1974) or by the presence of the hybrid band in pollen tissue (Weeden and Gottlieb, 1979). A polyploid in the process of diploidization will express these duplicated systems in roughly the same proportion as the percentage of its genome still showing tetraploid expression. Excellent studies on tetraploid fish (Ferris and Whitt, 1977; Bailey et al., 1978; Stoneking et al., 1981) have documented the loss of a significant number of duplicate loci during the evolution of various genera. Loss of duplicate expression appears to be common in polyploid plants as well, although the evidence in most cases is circumstantial (see discussion by Haufler, 1987, and the chapter by Soltis and Soltis in this volume). One of the best documented examples is provided by Wilson et al. (1983), who demonstrated repeated loss of duplicated LAP loci in tetraploid *Chenopodium*.

An intermediate stage in the reversion to diploid expression may be duplicate loci that display different levels of staining activity or divergence in various catalytic properties. In electrophoretic gels, differential staining levels among suspected duplicated isozyme loci are commonly observed (e.g., Goodman et al., 1980; Wendel et al., 1986). In the case of allopolyploids this is not sufficient evidence to warrant an interpretation of structural gene divergence subsequent to the duplication event, as the ancestral loci may have already been divergent at the time of polyploid formation.

The argument is stronger in the case of diploid-level duplication, however. The best characterized of such systems is the cytosolic glucosephosphate isomerases in certain *Clarkia* species (Gottlieb, 1977; Gottlieb and Weeden, 1979). Although these species are diploid, the evolution of the

Table 2.3 Comparison of expected single locus segregation groups and frequencies in allo- and autotetraploids

Parental genotype	Possible gametes	Ratio	Zygotic phenotypes in F_2	Ratio	Frequency of zygotic phenotypes
			Allotetraploid		
AABB	AB	(1)	AABB	(1)	1.0
AaBB	AB	(1)	AABB	(1)	0.25
	aB	(1)	AaBB	(2)	0.50
			aaBB	(1)	0.25
AaBb	AB	(1)	AABB	(1)	0.06
	aB	(2)	AABb	(2)	0.125
	Ab	(2)	AAbb	(1)	0.06
	ab	(1)	AaBB	(2)	0.125
			AaBb	(4)	0.25
			Aabb	(2)	0.125
			aaBB	(1)	0.06
			aaBb	(2)	0.125
			aabb	(1)	0.06
			Autotetraploid (chromosomal segregation)		
AAaa	AA	(1)	AAAA	(1)	0.03
	Aa	(4)	AAAa	(8)	0.22
	aa	(1)	AAaa	(18)	0.50
			Aaaa	(8)	0.22
			aaaa	(1)	0.03
AAAa (or Aaaa)	AA	(1)	AAAA	(1)	0.25
	Aa	(1)	AAAa	(2)	0.50
			AAaa	(1)	0.25
			Autotetraploid (chromatidal segregation)		
AAaa	AA	(3)	AAAA	(9)	0.046
	Aa	(8)	AAAa	(48)	0.245
	aa	(3)	AAaa	(82)	0.418
			Aaaa	(48)	0.245
			aaaa	(9)	0.046
AAAa (or Aaaa)	AA	(15)	AAAA	(225)	0.287
	Aa	(12)	AAAa	(360)	0.459
	aa	(1)	AAaa	(174)	0.222
			Aaaa	(24)	0.031
			aaaa	(1)	0.001

duplicated glucosephosphate isomerase loci appears to follow the same pattern expected in a polyploid with many pairs of duplicate loci. One allele in Clarkia xantiana produces a homodimer that displays a fivefold higher K_m for fructose-6-phosphate than that characteristic of other GPI homodimers (Gottleib and Greve, 1981). This mutation may cause the homodimer to be effectively inactive at physiological concentrations of the substrate. In another study involving the same system, enzyme concentrations in species possessing the duplication were found to be indistinguishable from the concentration in species with a single cytosolic enzyme (Gottlieb and Higgins, 1984). Thus, it appears that the plants with the duplication have compensated for the second dose of gene product by both an overall reduction in gene expression and, in one case at least, a change in the kinetic properties of the gene product.

LINKAGE CONSERVATION

Earlier we mentioned the use of isozyme loci for identifying instances of synteny. Such loci are useful because the conserved nature of many enzyme systems permits one to assume that the locus coding the cytosolic phosphoglucomutase, for example, in one species is homologous to that coding the same enzyme in another species. Similar assumptions cannot be made with morphological or physiological single gene mutants (such as flower color mutants) because the same phenotype can be produced from mutations in a number of different genes (e.g., there are at least four distinct white-flowered mutants in pea). The most conspicuous example in the plant literature of the use of isozymes to demonstrate synteny is the work by Hart and others on the three diploid genomes that are present in hexaploid wheat (Hart, 1979). Although the correlation is not perfect, most enzyme systems show triplicate gene expression, with the three genes located on three homoeologous chromosomes. Many of the same linkage groups can be found in rye and barley (Miller, 1984; Miller and Reader, 1987). This conservation of linkage groups is, perhaps, not as dramatic as the synteny of the X chromosome in mammals (Lalley et al., 1978), or the possible conservation of linkage groups from fish to mammals (Morizot, 1983), but it does permit a synergistic interaction of genetic research among related plant groups.

Conservation of portions of chromosomal linkage groups is suspected in several other cases. Comparison of the *Pisum* and *Lens* linkage maps has revealed that at least five of the 14 chromosomal arms exhibit partial synteny (Weeden et al., 1987). If these groups are conserved between *Pisum* and *Lens* they also should be expected in *Lathyrus, Vicia,* and perhaps even *Cicer.* Of the linkages reported in *Citrullus* and *Cucurbita,* at least two appear to be conserved between the genera. Navot and Zamir (1986) found linkage between the loci coding the plastid and cytosolic isozymes of glucosephosphate isomerase in *Citrullus.* Studies with monosomic alien addition lines in *Cucurbita* placed a homologous pair of loci on a single *C. palmata* chromosome (Weeden et al., 1986a). An *Est—Skdh* linkage was also reported in the two genera (Navot and Zamir, 1986; Weeden et al., 1986b), although in this case the homology is less certain. Two *Skdh* loci were identified in *Citrullus*, whereas only one was defined in *Cucurbita*, and several esterases were observed in both genera.

Conservation of linkage groups also has been observed among conifers. Both *Pinus* and *Abies* display linkage between a locus encoding an aspartate aminotransferase isozyme and one encoding a leucine aminopeptidase, and another linkage between a second aspartate aminotransferase locus and one of the glucosephosphate isomerase loci (Neale and Adams, 1981).

In sorghum, linkage results indicate that chromosome 1 has a group of loci in common with chromosome 6 of maize (J. Doebley et al., personal communication). One comparison that did not identify much synteny was that between *Lycopersicon* and *Capsicum* of the Solanaceae. Despite this close relationship (which in mammals would almost guarantee a high degree of homology) only two sections of the *Lycopersicon* linkage map could be demonstrated in *Capsicum* (Tanksley, 1984). Evidence for synteny is also lacking in comparisons of the current isozyme linkage maps of wheat and maize, garden pea and common bean, and sunflower and lettuce.

Several possible instances of linkage conservation over much larger taxonomic distances have been discussed (Tanksley, 1983). Clusters of peroxidase-encoding genes have been identified in tomato (Tanksley and Rick, 1980b), alfalfa (Quiros, 1983), *Malus* (E. Chevreau, personal communication), and garden pea (Weeden and Marx, 1987). Tight linkage between esterase loci exists in tomato (Rick, 1983), rye (Wehling and Schmidt-Stohn, 1984), and pea (Weeden and Marx, 1987). Linkage between an aminopeptidase locus and a glucosephosphate isomerase locus has been reported in taxa as divergent as *Gossypium* (Suiter, 1988), *Camellia* (Wendel and Parks, 1982), and *Pinus* (Adams and Joly, 1980; Conkle, 1981; Eckert et al., 1981). As was the case for the peroxidases and esterases, it is difficult to verify homology between aminopeptidase loci. Even the homology of the glucosephosphate isomerase loci is questionable in certain cases because subcellular localization was not determined. An alcohol dehydrogenase locus has been found to be linked with a phosphoglucomutase locus in maize (Goodman and Stuber, 1983a), tomato (Rick, 1983), and celery (Arus and Orton, 1984). In some cases the subcellular localization of the PGM isozyme involved has not been determined, leaving the possibility that non-homologous isozyme loci are being compared. Furthermore, plants generally express two or more ADH loci, again presenting the problem of homology. Finally, for both the aminopeptidase—glucosephosphate isomerase and the alcohol dehydrogenase—phosphoglucomutase comparisons there exist genera within the taxonomic unit being examined that lack the respective isozyme linkages. Thus, at present there is no clear evidence for linkage conservation above the familial level.

MODIFIERS

It is generally assumed that simply inherited isozyme polymorphism reflects a DNA sequence change in the structural gene. Yet it is known that many proteins are modified after synthesis either by specific cleavage or by addition of various small molecules such as glucose or phosphate residues (Markert and Whitt, 1968). Loci that produce modification in specific plant isozymes have been described in tomato (Rick et al., 1979), maize (Goodman et al., 1980; Newton and Schwartz, 1980), Sorghum (Doebley et al., 1986), Calocedrus (Harry, 1983), and Pinus (Millar, 1985). Four of these loci altered the mobility of specific malate dehydrogenase isozymes. In both maize and Sorghum it was the mitochondrial isozyme of malate dehydrogenase that was affected. The subcellular location of the altered MDH in Calocedrus or Pinus was not determined. The affected isozyme in tomato was one of the several peroxidases.

Another type of modification observed in plant isozymes is a reversible association with low molecular weight factors in the cytoplasm. The plastid-specific glyceraldehyde-3-phosphate dehydrogenase can exist in different forms depending on the NAD/NADP ratio in the cell at the time of extraction (Cerff, 1978; de Looze and Wagner, 1983). This isozyme system also is peculiar with respect to the presence of two plastid-specific forms: an A_4 homotetramer and an A_2B_2 intergenic tetramer (Cerff and Chambers, 1979). The variation caused by the pyridine nucleotide ratio appears to be independent of that generated by the different subunits. An acid phosphatase from Pinus sylvestris also has been shown to exist in two forms, depending on the presence or absence of a highly charged RNA fragment (Jonsson, 1981). When the low molecular weight RNA associates with the acid phosphatase, the pI of the isozyme is changed by six pH units.

A particularly interesting example of isozyme variation is the alteration of peroxidase, esterase, and acid phosphatase phenotypes observed in flax genotrophs (Cullis, 1981; Fieldes et al., 1981). The term genotroph was developed to describe the heritable change induced in inbred flax genotypes when these were exposed to large concentrations of inorganic nutrients. The progeny from plants so treated often differed significantly in size from progeny of untreated plants, and these differences were maintained in subsequent generations (Durrant, 1962, 1971). These observations represent one of the few known cases of environmental induction of a stable and predictable phenotypic change.

As was the case for the other effects produced by the treatment of flax plants, the modification in isozyme mobility was not observed in the treated plant but only in the succeeding generation. Several isozyme systems were affected as well as certain other glycoproteins. Fieldes and Tyson (1984) suggested that these changes were not caused by alterations in the DNA sequence of the respective structural genes, but were the result of differences in the type or number of carbohydrate moieties added to the proteins post-translationally. Further evidence supporting this hypothesis has recently been presented (Tyson et al., 1986). In this last paper the authors demonstrated that instead of the usual codominant expression of acid phosphatase allozymes, the faster form acted as a dominant allele. Similar results were obtained for the peroxidase variation. In both cases the faster form possessed a lower molecular weight, indicative of limited proteolysis or loss of carbohydrate residues. Furthermore, the segregation pattern for the two enzyme systems was nearly identical, suggesting that the same modifier or two closely linked modifiers were acting on these isozymes. Although the exact nature of this variation has not been determined, it is clearly different than the routine allozyme variation patterns generally expected.

SKEWED SEGREGATION RATIOS

In crosses involving widely divergent taxa, segregation ratios often differ significantly from the 3:1 or 1:2:1 ratios expected for single Mendelian genes (Grant, 1975). Such segregation distortion is occasionally and easily observed at isozyme loci and also has been reported in populations produced from closely related parents (Millar, 1985; Torres et al., 1985, 1986; Wendel et al., 1987). Some instances of segregation distortion have been attributed to gametophytic selection either through pollen competition (Mulcahy, 1974; Pfahler, 1975; Ottaviano et al., 1982) or as a result of linkage to a self-incompatibility locus (Wendel and Parks, 1984; Manganaris and Alston, 1987). In *Lycopersicon,* the cause of distorted ratios in *L. esculentum* × *L. pennelli* F_2 populations appeared not to be caused by a prezygotic mechanism but rather by a selective loss of certain genotypes through zygotic abortion (Gadish and Zamir, 1987). Further investigation of the phenomenon in *Lycopersicon, Capsicum,* and *Lens* demonstrated that nearly half of the isozyme markers examined showed abnormal segregation ratios (Zamir and Tadmor, 1986). The authors concluded that these distorted ratios were generated because the isozyme loci were closely linked to genes or chromosomal segments exposed to strong selection pressures during gametophytic and postzygotic development.

In both maize and *Cucurbita,* abnormal segregation was observed at many isozyme loci distributed over much of the genome (Wendel et al., 1987; Weeden and Robinson, 1986), indicating that many genes are involved in the functioning of gametes or selection of viable zygotes in F_2 and backcross generations. At least for *Cucurbita* this result was unexpected and did not support the cryptic structural rearrangement hypothesis usually invoked to explain hybrid breakdown in this genus. C. Werth (personal communication) has suggested that the abnormal segregation seen in polyploids like *Cucurbita* and *Gossypium* is caused by loss of gene function at homoeologous loci. For example, take two polyploid species both containing duplicated loci coding a specific enzyme. However, locus 1 has been silenced in species A, whereas in species B the second locus is inactive. The interspecific hybrid would be heterozygous for a "null" allele at both loci, and one fourth of the gametes produced by this hybrid would lack the capacity to synthesize the enzyme. Such a deficiency could reduce the fitness of either the gametophyte or a succeeding sporophyte. A single such locus combination would produce distorted segregation ratios within only a small portion of the genome, but one can envisage many such homoeologous loci having this relationship for the two species, thereby causing segregation distortion throughout the genome.

NULL ALLELES

Alleles that are no longer transcribed or that code for defective polypeptides lacking enzymatic activity are generally referred to as "null alleles." The existence of such forms in natural populations was postulated from the occurrence of individuals with recessive syndromes or heritable diseases caused by the lack of a specific enzyme (e.g., phenylketonuria). Null alleles were also predicted in polyploids and other duplicated gene systems on theoretical grounds. Such systems are expected to diverge gradually, with only one of the loci maintaining an enzyme performing the original function (Ohno, 1970). However, null variants often remain undetected because their presence is masked in the heterozygous condition. Consequently, for most plants it is difficult to determine the frequency of null alleles in natural populations, unless it is high enough to generate null homozygotes.

In conifers, the availability of haploid megagametophytic tissue greatly facilitates the study of the frequency and distribution of null alleles. One of the best examples is provided by Allendorf et al. (1982), who reported the frequencies of null alleles at 29 and 27 loci in *Pinus ponderosa* and *P. resinosa*, respectively. Mean frequencies of null alleles were approximately the same for both species (0.003). Moreover, there was little concordance between the two taxa with respect to distribution of null alleles among loci; null alleles were detected at 8 and 4 loci in *P. ponderosa* and *P. resinosa,* respectively, but there were no instances of null alleles being detected at homologous loci in both taxa. These data suggest that there is no differential tendency among loci to generate null variants.

Null alleles can also be recognized by the appearance of a 3:1 (activity present:activity absent) segregation ratio for an enzyme locus. The 3:1 ratio is generated upon self-pollination of a plant heterozygous for a null allele because the heterozygous phenotype often displays an activity band nearly as intense as the homozygous active phenotype. Numerous examples of 3:1 segregation ratios have been reported (e.g., Weeden and Robinson, 1986; Weeden and Lamb, 1987); however, because there are other causes for such a ratio (e.g., protein modification), few have been conclusively shown to be caused by segregation of a non-transcribed sequence or a defective polypeptide.

An alternative method for demonstrating the presence of an inactive polypeptide is available in systems with multimeric enzymes. An example of this approach is described by Jones et al. (1986) in studies on the cytosolic glucosephosphate isomerase duplication in Clarkia. Lines normally expressing a true-breeding three-banded phenotype were treated with EMS, and both single and two-banded mutants were recovered after selfing the progeny. The two-banded phenotype was produced by the loss of activity in one of the homodimers. Apparently the monomeric subunit of this isozyme no longer formed an active homodimer but could still interact with the alternative gene product to form an active heterodimer. The occurrence of such a two-banded phenotype in any dimeric enzyme system is a strong indication of a defective monomer.

An interesting aspect of null alleles is their effect on the viability of an organism. In most cases the heterozygote appears to be as fit as the homozygous active genotype, but the homozygous null can be lethal. In the Clarkia GPI study mentioned above, plants homozygous for null alleles at both loci were never recovered in a progeny from doubly heterozygous individuals (Jones et al., 1986). The authors attributed this absence of a double null phenotype to the requirement for cytosolic GPI activity if normal metabolic processes are to be maintained in the cell.

In maize, two different types of null alleles segregate according to Mendelian expectations for one of the cytosolic TPIs: one that produces active heterodimers with the other cytosolic isozymes and one that does not (J. F. Wendel, unpublished data). It is tempting to speculate that the former is translated but is inactive as a homodimer, whereas the latter allele is either not translated or produces a protein product that remains non-functional as both a homo- and heterodimer. Hybrids synthesized between true-breeding lines for these two null alleles display all possible heteromeric products and generate the expected monogenic segregation ratios and phenotypes in backcross and F_2 progenies.

Maize also possesses two loci encoding cytosolic hexokinase and an additional locus or loci governing mitochondrial hexokinase (Wendel et al., 1986). Fully vigorous and fertile plants were generated that were null for both cytosolic isozymes. Similarly, the major forms of maize malate dehydrogenase are encoded by five nuclear genes, three governing mitochondrial MDH and two encoding cytosolic MDH (Goodman et al., 1980). Goodman et al. (1981) showed that lines lacking cytosolic MDHs are viable but that at least one active allele specifying mitochondrial MDH is required for normal growth.

A null phenotype in an individual may also be explained by causes other than true allelism. Possibilities include denaturation or inactivation of an unstable allozyme during extraction or electrophoresis and epistatic interactions of other genomic factors. A particularly interesting example of the latter phenomenon is offered by the maize β-glucosidase locus, Glu1. Numerous inbred lines have been characterized as having null alleles at this locus (Stuber and Goodman, 1983a). The question of whether or not null phenotypes result from true allelism at Glu1 was investigated by analyses of F_1, F_2, and backcross progeny resulting from intercrossing null lines with each other and with lines with strong β-glucosidase activity (M. D. Edwards, J. F. Wendel, and C. W. Stuber, unpublished data). The data clearly demonstrate that all of the null lines tested possess "hidden" active Glu1 alleles that are masked in inbred lines but revealed in a proportion of hybrids or more advanced generations. By studying joint segregation of Glu1 with numerous additional allozyme markers, it was possible to identify several genomic segments on different chromosomes that influenced intensity of glucosidase staining. Thus, "nullness" for β-glucosidase arises not from allelic variation at the Glu1 locus, but from epistatic effects of genes located in other chromosomal regions.

CONCLUSIONS

Considerable progress has been made in the elucidation of the genetic basis of isozyme polymorphisms, particularly in economically important crops. Chromosome substitution will continue to play an important part in the genetic analysis of isozyme loci in wheat and its relatives, because nearly every chromosomal arm can be independently examined. It appears that within the wheat-barley-rye germplasm nearly all of the approximately 100 loci (per diploid genome) coding enzymes amenable to isozyme analysis will be mapped using electrophoretic methods.

For finer scale linkage mapping and for using isozyme loci as genetic markers, allozyme polymorphisms must be identified. Approximately 70% of the 70 or so loci with products observable by isozyme staining techniques have been determined to be polymorphic in maize and garden pea, and over 50% are polymorphic in *Lycopersicon*. Thus, if considerable effort is devoted to germplasm surveys, we should expect to find a minimum of 30 to 40 polymorphic isozyme loci in species or species-aggregates with a reasonable genetic diversity. We have discussed several examples of genetic studies, particularly linkage analysis, in which one taxon produced information that could be extrapolated to related taxa. To maximize the efficiency of this process, one should proceed from the more simple to the more complex system. In most cases, "simple" translates to a self-compatible diploid species with a short generation time and a large reproductive capacity. A number of "model" genera have already been developed including *Hordeum* and *Zea* (Poaceae), *Lycopersicon* and *Capsicum* (Solanaceae), *Pisum* and *Phaseolus* (Fabaceae), *Lactuca* (Asteraceae), *Clarkia* (Onagraceae), *Citrullus* (Cucurbitaceae), and *Pinus* and *Picea* (Pinaceae). However, most angiosperm families lack such model genera, indicative of the considerable amount of basic work still required before isozyme analysis becomes merely a technical exercise.

Indeed, many stimulating questions regarding isozyme genetics remain unanswered. For example, what are the homologies and gene numbers of the many loci whose products are revealed through the use of non-specific stains, such as "esterase", "NADH dehydrogenase", "peroxidase", and "acid phosphatase"? To what extent is variation in these systems produced by pleiotropic interactions with other "modifier" loci? Do all described duplications represent independent events, or are there silent loci that occasionally become activated? What are the rates of gene silencing for redundant genes (e.g., in polyploids)? Are these rates equal for different genes or genomes? Answers to some of these questions will require information from biochemical and molecular genetic laboratories. However, it has been the detailed surveys and genetic analyses of isozyme polymorphisms that have identified these questions as particularly interesting, and such analyses will certainly play a prominent role in the clarification of these questions.

ACKNOWLEDGMENTS

This work was supported in part by Grant No. US-992-85 from BARD—The United States—Israel Binational Agriculture Research & Development Fund (to NFW) and grants from the National Science Foundation (BSR-8619631) and Pioneer Hi-Bred International of Johnston, Iowa (to JFW).

LITERATURE CITED

Adams, W. T., and R. J. Joly. 1980. Genetics of allozyme variants in loblolly pine. *J. Heredity* 71: 33–40.

Ainsworth, C. C., H. M. Johnson, E. A. Jackson, T. E. Miller, and M. D. Gale. 1984. The chromosomal locations of leaf peroxidase genes in hexaploid wheat, rye and barley. *Theor. Appl. Genet.* 69: 205–210.

Allendorf, F. W., K. L. Knudsen, and G. M. Blake. 1982. Frequencies of null alleles at enzyme loci in natural populations of ponderosa and red pine. *Genetics* 100: 497–504.

Anderson, L. E., R. L. Heinrickson, and C. Noyes. 1975. Chloroplast and cytoplasmic enzymes. Subunit structure of pea leaf aldolases. *Arch. Biochem. Biophys.* 169: 262–268.

Arus, P., and T. J. Orton. 1984. Inheritance patterns and linkage relationships of eight genes of celery. *J. Heredity* 75: 11–14.

Bailey, G. S., R. T. M. Poulter, and P. A. Stockwell. 1978. Gene duplication in tetraploid fish: model for gene silencing at unlinked duplicated loci. *Proc. Natl. Acad. Sci. USA* 75: 5575–5579.

Baum, J. A., and J. G. Scandalios. 1981. Isolation and characterization of the cytosolic and mitochondrial superoxide dismutases of maize. *Arch. Biochem. Biophys.* 206: 249–264.

Bayer, R. J., and D. J. Crawford. 1986. Allozyme divergence among five diploid species of *Antennaria* (Asteraceae: Inuleae) and their allopolyploid derivatives. *Amer. J. Bot.* 73: 287–296.

Bernatzky, R., and S. D. Tanksley. 1986. Genetics of actin-related sequences in tomato. *Theor. Appl. Genet.* 72: 314–321.

Bhalla, P. L., and M. J. Dalling. 1984. Characteristics of a β-galactosidase associated with the stroma of chloroplasts prepared from mesophyll protoplasts of the primary leaf of wheat. *Pl. Physiol.* 76: 92–95.

Blixt, S. 1974. The pea. In R. C. King [ed.], *Handbook of genetics*. Vol. 2, 181–221. Plenum Press, New York.

Boulter, D., R. J. Ellis, and A. Yarwood. 1972. Biochemistry of protein synthesis in plants. *Biol. Rev.* 47: 113–175.

Boyle, T. J. B., and E. K. Morgenstern. 1985. Inheritance and linkage relationships of some isozymes of black spruce in New Brunswick. *Canad. J. For. Res.* 15: 992–996.

Brewbaker, J. L., C. Nagai, and E. H. Liu. 1985. Genetic polymorphisms of 13 maize peroxidases. *J. Heredity* 76: 159–167.

Bridges, C. B. 1936. The "Bar" gene duplication. *Science* 83: 210–211.

Brouquisse, R., M. Nishimura, J. Gaillard, and R. Douce. 1987. Characterization of a cytosolic aconitase in higher plant cells. *Pl. Physiol.* 84: 1402–1407.

Brown, A. H. D. 1983. Barley. In S. D. Tanksley and T. J. Orton [eds.], *Isozymes in plant genetics and breeding*, Part B, 57–78. Elsevier, Amsterdam.

————, and B. S. Weir. 1983. Measuring genetic variability in plant populations. In S. D. Tanksley and T. J. Orton [eds], *Isozymes in plant genetics and breeding*, Part A, 219–239. Elsevier, Amsterdam.

Cammerts, D., and M. Jacobs. 1983. A study of the polymorphism and the genetic control of the glutamate dehydrogenase isozymes in *Arabidopsis thaliana*. *Pl. Sci. Lett.* 31: 65–73.

Cerff, R. 1978. Glyceraldehyde-3-phosphate dehydrogenase (NADP) from *Sinapis alba* L. *Pl. Physiol.* 61: 369–372.

————, and S. E. Chambers. 1979. Subunit structure of higher plant glyceraldehyde-3-phosphate dehydrogenases (EC 1.2.1.12 and EC 1.2.1.13). *J. Biol. Chem.* 254: 6094–6098.

Chenicek, K. J., and G. E. Hart. 1987. Identification and chromosomal locations of aconitase gene loci in Triticeae species. *Theor. Appl. Genet.* 74: 261–268.

Chevreau, E., Y. Lespinasse, and M. Gallet. 1985. Inheritance of pollen enzymes and polyploid origin of apple (*Malus* × *domestica* Borkh.). *Theor. Appl. Genet.* 62: 301–304.

Chou, K.-H., and W. E. Splittstoesser. 1972. Glutamate dehydrogenase from pumpkin cotyledons: characterization and isoenzymes. *Pl. Physiol.* 49: 550–554.

Colman, A., and C. Robinson. 1986. Protein import into organelles: hierarchical targeting signals. *Cell* 46: 321–322.

Conkle, M. T. 1981. Isozyme variation and linkage in six conifer species. In *Proc. Symp. on Isozymes in North Amer. Forest Trees and Forest Insects*. Pacific Southwest Forest and Range Experiment Station. Forest Service, USDA, Berkeley.

Crawford, D. J. 1983. Phylogenetic and systematic inferences from electrophoretic studies. In S. D. Tanksley and T. J. Orton [eds.], *Isozymes in plant genetics and breeding*, Part A, 257–287. Elsevier, Amsterdam.

Cullis, C. A. 1981. Environmental induction of heritable changes in flax: defined environments inducing changes in rDNA and peroxidase isozyme band pattern. *Heredity* 47: 87–94.

de Cherisey, H., M. T. Barreneche, M. Jusuf, C. Ouin, and J. Pernes. 1985. Inheritance of some marker genes in *Setaria italica* (L.) P. Beauv. *Theor. Appl. Genet.* 71: 57–60.

de Looze, S., and E. Wagner. 1983. In vitro and in vivo regulation of chloroplast glyceraldehyde-3-phosphate dehydrogenase isozymes from *Chenopodium rubrum*. II. In vitro modulation of the isozyme pattern. *Physiol. Pl.* 57: 238–242.

Doebley, J. F., M. M. Goodman, and C. W. Stuber. 1984. Isoenzymatic variation in *Zea* (Gramineae). *Syst. Bot.* 9: 203–218.

———, ———, and ———. 1985. Isozyme variation in the races of maize from Mexico. *Amer. J. Bot.* 72: 629–639.

———, C. W. Morden, and K. F. Schertz. 1986. A gene modifying mitochondrial malate dehydrogenase isozymes in *Sorghum* (Gramineae). *Biochem. Genet.* 24: 813–819.

Dunsmuir, P., S. Smith, and J. Bedbrook. 1983. A number of different nuclear genes for the small subunit of RuBPCase are transcribed in petunia. *Nucl. Acids Res.* 11: 4177–4183.

Durrant, A. 1962. The environmental induction of heritable changes in *Linum*. *Heredity* 17: 27–61.

———. 1971. Induction and growth of flax genotrophs. *Heredity* 27: 277–298.

Eckert, R. T., R. J. Joly, and D. B. Neale. 1981. Genetics of isozyme variants and linkage relationships among allozyme loci in 35 eastern white pine clones. *Canad. J. For. Res.* 11: 573–579.

El-Kassaby, Y. A., F. C. Yeh, and O. Sziklai. 1982. Inheritance of allozyme variants in coastal Douglas-fir (*Pseudotsuga menziesii* var. *menziesii*). *Canad. J. Genet. Cytol.* 24: 325–335.

Ellis, R. J. 1981. Chloroplast proteins: synthesis, transport and assembly. *Ann. Rev. Pl. Physiol.* 32: 111–137.

Ellstrand, N. C., J. M. Lee, and K. W. Foster. 1983. Alcohol dehydrogenase isozymes in grain sorghum (*Sorghum bicolor*): evidence for a gene duplication. *Biochem. Genet.* 21: 147–154.

Ennos, R. A. 1986. Allozyme variation, linkage, and duplication in the perennial grass, *Cynosurus cristatus*. *J. Heredity* 77: 61–62

Farinelli, M. P., D. W. Fry, and K. E. Richardson. 1983. Isolation, purification and partial characterization of formate dehydrogenase from soybean seed. *Pl. Physiol.* 73: 858–859.

Feierabend, J., and M. Mikus. 1977. Occurrence of a high temperature sensitivity of chloroplast ribosome formation in several higher plants. *Pl. Physiol.* 59: 863–867.

Ferris, S. D., S. L. Portnoy, and G. S. Whitt. 1979. The roles of speciation and divergence time in the loss of duplicate gene expression. *Theor. Pop. Biol.* 15: 114–139.

———, and G. S. Whitt. 1977. Loss of duplicate gene expression after polyploidization. *Nature* 265: 258–260.

Fieldes, M. A., C. L. Deal, and H. Tyson. 1981. Persistence of differences between peroxidase isoenzymes of flax genotrophs in tissue culture. *Phytochemistry* 20: 403–406.

———, and H. Tyson. 1984. Possible post-translational modification, and its genetic control, in flax genotroph isozymes. *Biochem. Genet.* 22: 99–114.

Figueiras, A. M., M. T. Gonzales-Jaen, and C. Benito. 1986. Biochemical evidence of homoeology between *Triticum aestivum* and *Agropyron intermedium* chromosomes. *Theor. Appl. Genet.* 72: 826–832.

Fobes, J. F. 1980. Trisomic analysis of isozymic loci in tomato species: segregation and dosage effects. *Biochem. Genet.* 18: 401–421.

Freeling, M., and D. Schwartz. 1973. Genetic relationships between the multiple alcohol dehydrogenases of maize. *Biochem. Genet.* 8: 27–36.

Gallez, G. P., and L. D. Gottlieb. 1982. Genetic evidence for the hybrid origin of the diploid plant *Stephanomeria diegensis*. *Evolution* 36: 1158–1167.

Gadish, I., and D. Zamir. 1987. Differential zygotic abortion in an inter-specific *Lycopersicon* cross. *Genome* 29: 156–159.

Garcia, P., P. De La Vega, and C. Benito. 1982. The inheritance of rye seed peroxidases. *Theor. Appl. Genet.* 61: 341–351.

Goldring, A., D. Zamir, and C. Degani. 1985. Duplicated phosphoglucose isomerase genes in avocado. *Theor. Appl. Genet.* 71: 491–494.

Gomes, J., S. Jadric, M. Winterhalter, and S. Brkic. 1982. Alcohol dehydrogenase isoenzymes in chickpea cotyledons. *Phytochemistry* 21: 1219–1224.

Goodman, M. M., K. J. Newton, and C. W. Stuber. 1981. Malate dehydrogenase: viability of cytosolic nulls and lethality of mitochondrial nulls in maize. *Proc. Natl. Acad. Sci. USA* 78: 1783–1785.

———, and C. W. Stuber. 1983a. Maize. In S. D. Tanksley and T. J. Orton [eds.], *Isozymes in plant genetics and breeding*, Part B, 1–33. Elsevier, Amsterdam.

———, and ———. 1983b. Races of maize. VI. Isozyme variation among races of maize in Bolivia. *Maydica* 28: 169–187.

———, ———, C. N. Lee, and F. M. Johnson. 1980. Genetic control of malate dehydrogenase isozymes in maize. *Genetics* 94: 153–168.

Gopinath, D. M., and C. R. Burnham. 1956. A cytogenetic study in maize of deficiency-duplication produced by crossing interchanges involving the same chromosomes. *Genetics* 41: 382–395.

Gottlieb, L. D. 1974. Gene duplication and fixed heterozygosity for alcohol dehydrogenase in the diploid *Clarkia franciscana*. *Proc. Natl. Acad. Sci. USA* 71: 1816–1818.

———. 1977. Evidence for duplication and divergence of the structural gene for phosphoglucoisomerase in diploid species of *Clarkia*. *Genetics* 86: 289–307.

———. 1982. Conservation and duplication of isozymes in plants. *Science* 216: 373–380.

———. 1984. Electrophoretic analysis of the phylogeny of the self-pollinating populations of *Clarkia xantiana*. *Pl. Syst. Evol.* 147: 91–102.

———. 1987. Phosphoglucomutase and isocitrate dehydrogenase gene duplication in *Layia* (Compositae). *Amer. J. Bot.* 74: 9–15.

———, and L. C. Greve. 1981. Biochemical properties of duplicated isozymes of phosphoglucose isomerase in the plant *Clarkia xantiana*. *Biochem. Genet.* 19: 155–172.

———, and R. C. Higgins. 1984. Phosphoglucose isomerase expression in species of *Clarkia* with and without a duplication of the coding gene. *Genetics* 107: 131–140.

———, and N. F. Weeden. 1979. Gene duplication and phylogeny in *Clarkia*. *Evolution* 33: 1024–1039.

———, and ———. 1981. Correlation between subcellular location and phosphoglucose isomerase variability. *Evolution* 35: 1019–1022.

Grant, V. 1975. *Genetics of flowering plants*. Columbia Univ. Press, New York.

Gray, M. W., and W. F. Doolittle. 1982. Has the endosymbiont hypothesis been proven? *Microbiol. Rev.* 46: 1–42.

Griffin, J. D., and R. G. Palmer. 1987. Inheritance and linkage studies with five isozyme loci in soybean. *Crop Sci.* 27: 885–892.

Harberd, N. P., and K. J. R. Edwards. 1983. Further studies on the alcohol dehydrogenases in barley: evidence for a third alcohol dehydrogenase locus, and data on the effect of an alcohol dehydrogenase 1 null mutation in homozygous and in heterozygous condition. *Genet. Res. Camb.* 41: 109–115.

Harry, D. E. 1983. Identification of a locus modifying the electrophoretic mobility of malate dehydrogenase isozymes in incense-cedar (*Calocedrus decurrens*), and its implications for population studies. *Biochem. Genet.* 21: 417–434.

———. 1986. Inheritance and linkage of isozyme variants in incense-cedar. *J. Heredity* 77: 261–266.

Hart, G. E. 1979. Genetical and chromosomal relationships among the wheats and their relatives. *Stadler Symp.* 11: 9–29.

———. 1983. Hexaploid wheat (*Triticum aestivum* L.—Thell). In S. D. Tanksley and T. J. Orton [eds], *Isozymes in plant genetics and breeding*, Part B, 35–56. Elsevier, Amsterdam.

———. 1984. Biochemical loci of hexaploid wheat (*Triticum aestivum*, $2n = 42$, Genomes AABBDD). In S. J. O'Brien [ed.], *Genetic maps*. Vol. 3, 484–490. Cold Spring Harbor Laboratory, New York.

———, and N. A. Tuleen. 1983. Chromosomal locations of eleven *Elytrigia elongata* (=*Agropyron elongatum*) isozyme structural genes. *Genet. Res. Camb.* 41: 181–202.

Haufler, C. H. 1985. Enzyme variability and modes of evolution in *Bommeria* (Pteridaceae). *Syst. Bot.* 10: 92–104.

———. 1987. Electrophoresis is modifying our concepts of evolution in homosporous pteridophytes. *Amer. J. Bot.* 74: 953–966.

———, and D. E. Soltis. 1986. Genetic evidence indicates that homosporous ferns with high chromosome numbers may be diploid. *Proc. Natl. Acad. Sci. USA* 83: 4389–4395.

Helentjaris, T., D. F. Weber, and S. Wright. 1986. Use of monosomics to map cloned DNA fragments in maize. *Proc. Natl. Acad. Sci. USA* 83: 6035–6039.

———, ———, and ———. 1988. Identification of the genomic locations of duplicate nucleotide sequences in maize by analysis of restriction fragment length polymorphisms. *Genetics*: 118: 353–363.

Hock, B. 1984. Processing and organelle import of malate dehydrogenase isoenzymes: is there a common precursor for the glyoxysomal and mitochondrial forms? *Physiol. Veg.* 22: 333–339.

Huang, A. H. C., K. D. F. Liu, and R. J. Youle. 1976. Organelle-specific isozymes of aspartate-α-ketoglutarate transaminase in spinach leaves. *Pl. Physiol.* 58: 110–113.

Hubby, J. L., and R. C. Lewontin. 1966. A molecular approach to the study of genic heterozygosity in natural populations. I. The number of alleles at different loci in *Drosophila pseudoobscura*. *Genetics* 54: 577–594.

Hunter, R. L., and C. L. Markert. 1957. Histochemical demonstration of enzymes separated by zone electrophoresis in starch gels. *Science* 125: 1294–1295.

Hutchinson, E. S., A. Hakim-Elahi, R. D. Miller, and R. W. Allard. 1983. The genetics of the diploidized tetraploid *Avena barbata*. *J. Heredity* 74: 325–330.

Ireland, R. J., and D. T. Dennis. 1980. Isozymes of the glycolytic and pentose-phosphate pathways in storage tissues of different oil seeds. *Planta* 149: 476–479.

Jaaska, V. 1983. Rye and triticale. In S. D. Tanksley and T. J. Orton [eds.], *Isozymes in plant genetics and breeding*, Part B, 79–104. Elsevier, Amsterdam.

Jackson, C., J. Dench, A. L. Moore, B. Halliwell, C. H. Foyer, and D. O. Hall. 1978. Subcellular localization and identification of superoxide dismutase in the leaves of higher plants. *Eur. J. Biochem.* 91: 339–344.

Jones, T. W. A. 1984. Development of phosphogluco-isomerase isozymes in perennial ryegrass (*Lolium perenne*). *Physiol. Pl.* 60: 203–207.

———, E. Pichersky, and L. D. Gottlieb. 1986. Enzyme activity in EMS-induced null mutations of duplicated genes encoding phosphoglucose isomerases in *Clarkia*. *Genetics* 113: 101–114.

Jonsson, I. 1981. Acid phosphatase from needles of *Pinus sylvestris* L. Purification of two interconvertible enzyme forms and characterization of a low-molecular weight factor associated with the enzyme. *Biochim. Biophys. Acta* 660: 204–213.

Kahler, A. L. 1983. Inheritance and linkage of acid phosphatase locus *Acp4* in maize. *J. Heredity* 74: 239–246.

———, and C. L. Lay. 1985. Genetics of electrophoretic variants in the annual sunflower. *J. Heredity* 76: 335–340.

Kanarek, L., E. Marler, R. A. Bradshaw, R. E. Fellows, and R. L. Hill. 1964. The subunits of fumarase. *J. Biol. Chem.* 239: 4207–4211.

Kay, L. E., and D. V. Basile. 1987. Specific peroxidase isoenzymes are correlated with organogenesis. *Pl. Physiol.* 84: 99–105.

Kesseli, R. V., and R. W. Michelmore. 1986. Genetic variation and phylogenies detected from isozyme markers in species of *Lactuca*. *J. Heredity* 77: 324–331.

Khush, G. S., R. J. Singh, S. C. Sur, and A. L. Librojo. 1984. Primary trisomics of rice: origin, morphology, cytology and use in linkage mapping. *Genetics* 107: 141–167.

Kiang, Y. T., and M. B. Gorman. 1985. Inheritance of NADP-active isocitrate dehydrogenase isozymes in soybeans. *J. Heredity* 76: 279–284.

Kimura, M. 1983. The neutral theory of molecular evolution. In M. Nei and R. K. Koehn [eds.], *Evolution of genes and proteins*, 208–233. Sinauer, Sunderland, MA.

———, and J. F. Crow. 1964. The number of alleles that can be maintained in a finite population. *Genetics* 49: 725–738.

King, J. N., and B. P. Dancik. 1983. Inheritance and linkage of isozymes in white spruce (*Picea* glauca). *Canad. J. Genet. Cytol.* 25: 430–436.

Kirkpatrick, K. J., D. S. Decker, and H. D. Wilson. 1985. Allozyme differentiation in the *Cucurbita pepo* complex: *C. pepo* var. *medullosa* vs. *C. texana*. *Econ. Bot.* 39: 289–299.

Kleczkowski, L. A., and D. D. Randall. 1986. Maize leaf adenylate kinase. Purification and partial characterization. *Pl. Physiol.* 81: 1110–1114.

Koehn, R. K., A. J. Zera, and J. G. Hall. 1983. Enzyme polymorphism and natural selection. In M. Nei and R. K. Koehn [eds.], *Evolution of genes and proteins*, 115–136. Sinauer, Sunderland, MA.

Kung, S. D., K. Sakano, and S. G. Wildman. 1974. Multiple peptide composition of the large and small subunits of *Nicotiana tabacum* fraction I protein ascertained by fingerprinting electrofocusing. *Biochim. Biophys. Acta* 262: 138–147.

Lalley, P. A., J. D. Minna, and U. Francke. 1978. Conservation of autosomal gene synteny groups in mouse and man. *Nature* 274: 160–162.

Lax, A. R., K. C. Vaughn, and G. E. Templeton. 1984. Nuclear inheritance of polyphenol oxidase in *Nicotiana*. *J. Heredity* 75: 285–287.

Lee, J. S., G. G. Brown, and D. P. S. Verma. 1983. Chromosomal arrangement of leghemoglobin genes in soybean. *Nucl. Acids Res.* 11: 5541–5553.

Levy, M., E. E. Steiner, and D. A. Levin. 1975. Allozyme genetics in permanent translocation heterozygotes of the *Oenothera biennis* complex. *Biochem. Genet.* 13: 487–500.

Lewontin, R. C., and J. L. Hubby. 1966. A molecular approach to the study of genic heterozygosity. II. Amount of variation and degree of heterozygosity in natural populations of *Drosophila pseudoobscura*. *Genetics* 54: 595–609.

Loukas, M., M. N. Stavrakakis, and C. B. Krimbas. 1983. Inheritance of polymorphic isoenzymes in grape cultivars. *J. Heredity* 74: 181–183.

Manganaris, A. G., and F. H. Alston. 1987. Inheritance and linkage relationships of glutamate oxaloacetate transaminase isoenzymes in apple. 1. The gene GOT-1, a marker for the S incompatibility locus. *Theor. Appl. Genet.* 74: 154–161.

Margulis, L. 1970. *Origin of eukaryotic cells*. Yale University Press, New Haven.

———. 1981. *Symbiosis in cell evolution*. Freeman, San Francisco.

Markert, C. L., and F. Moller. 1959. Multiple forms of enzymes: tissue, ontogenetic, and species specific patterns. *Proc. Natl. Acad. Sci. USA* 45: 753–763.

———, and G. S. Whitt. 1968. Molecular varieties of isozymes. *Experimentia* 24: 977–991.

Mayr, U., R. Hensel, and O. Kandler. 1982. Subunit composition and substrate binding region of potato L-lactate dehydrogenase. *Phytochemistry* 21: 627–631.

McNeill, C. I., and S. K. Jain. 1983. Genetic differentiation studies and phylogenetic inference in the plant genus *Limnanthes* (section Inflexae). *Theor. Appl. Genet.* 66: 257–269.

Miernyk, J. A., and D. T. Dennis. 1983. Mitochondrial, plastid, and cytosolic isozymes of hexokinase in developing endosperm of *Ricinus communis*. *Arch. Biochem. Biophys.* 226: 458–468.

Millar, C. I. 1984. Inheritance of allozyme variants in bishop pine (*Pinus muricuta* D. Don). *Biochem. Genet.* 23: 933–945.

Miller, T. E. 1984. The homoeologous relationship between the chromosomes of rye and wheat. Current status. *Canad. J. Genet. Cytol.* 26: 578–589.

———, and S. M. Reader. 1987. A guide to the homoeology of chromosomes within the Triticeae. *Theor. Appl. Genet.* 74: 214–217.

Moran, G. F., and J. C. Bell. 1983. Eucalyptus. In S. D. Tanksley and T. J. Orton [eds.], *Isozymes in plant genetics and breeding*, Part B, 423–442. Elsevier, Amsterdam.

Morden, C. W., J. F. Doebley, and K. F. Schertz. 1988. Genetic control and subcellular localization of aconitase isozymes in *Sorghum*. *J. Heredity* 79: 294–299.

Morizot, D. C. 1983. Tracing linkage groups from fishes to mammals. *J. Heredity* 74: 413–416.

Mulcahy, D. L. 1974. Adaptive significance of gamete competition. In H. F. Linskens [ed.], *Fertilization of higher plants*, 27–30. North Holland, Amsterdam.

Murdy, W. H., and M. E. B. Carter. 1985. Electrophoretic study of the allopolyploidal origin of *Talinum teretifolium* and the specific status of *T. appalachianum* (Portulacaceae). *Amer. J. Bot.* 72: 1590–1597.

Murray, D. R., and D. J. Ayre. 1987. Isoenzymes from hulls and seeds of developing pea fruits. *J. Pl. Physiol.* 127: 193–201.

———, and S. P. Waters. 1985. Multiple forms of aminopeptidase in representatives of the tribe Vicieae (Leguminosae). *Austral. J. Plant Physiol.* 12: 39–45.

Mutschler, M. A., S. D. Tanksley, and C. M. Rick. 1987. 1987 linkage maps of the tomato (*Lycopersicon esculentum*). *Rpt. Tomato Genet. Coop.* 37: 5–34.

Navot, N., and D. Zamir. 1986. Linkage relationships of 19 protein coding genes in watermelon. *Theor. Appl. Genet.* 72: 274–278.

Neale, D. B., and W. T. Adams. 1981. Inheritance of isozyme variants in seed tissues of balsam fir (*Abies balsamea*). *Canad. J. Bot.* 59: 1285–1291.

———, J. C. Weber, and W. T. Adams. 1984. Inheritance of needle tissue isozymes in Douglas fir. *Canad. J. Genet. Cytol.* 26: 459–468.

Neuman, P. R., and G. E. Hart. 1986. Genetic control of the mitochondrial form of superoxide dismutase in hexaploid wheat. *Biochem. Genet.* 24: 435–446.

Newton, K. J. 1983. Genetics of mitochondrial isozymes. *In* S. D. Tanksley and T. J. Orton [eds.], *Isozymes in plant genetics and breeding*, Part A., 157–174. Elsevier, Amsterdam.

———, and D. Schwartz. 1980. Genetic basis of the major malate dehydrogenase isozymes in maize. *Genetics* 95: 425–442.

Ni, W., E. F. Robertson, and H. C. Reeves. 1987. Purification and characterization of cytosolic NADP specific isocitrate dehydrogenase from *Pisum sativum*. *Pl. Physiol.* 83: 785–788.

Odrzykoski, I. J., and L. D. Gottlieb. 1984. Duplications of genes coding 6-phosphogluconate dehydrogenase in *Clarkia* (Onagraceae) and their phylogenetic implications. *Syst. Bot.* 9: 479–489.

Ohno, S. 1970. *Evolution by gene duplication*. Springer, New York.

Oliver, J. L., J. M. Martinez-Zapater, L. Pascual, A. M. Enriquez, C. Ruiz-Rejon, and M. Ruiz-Rejon. 1983. Different genome amplification mechanisms and duplicate gene expression in Liliaceae. *In* M. C. Rattazzi, J. G. Scandalios and G. S. Whitt [eds.], *Isozymes: current topics in biological and medical research*. Vol. 10, 341–363. Alan R. Liss, Inc., New York.

Ottaviano, E., M. Sari-Gorla, and E. Pe. 1982. Male gametophytic selection in maize. *Theor. Appl. Genet.* 63: 249–254.

Paul, E. M., and V. M. Williamson. 1987. Purification and properties of acid phosphatase-1 from nematode resistant tomato cultivar. *Pl. Physiol.* 84: 399–403.

Pedersen, S., and V. Simonsen. 1987. Tissue specific and developmental expression of isozymes in barley (*Hordeum vulgare* L.). *Hereditas* 106: 59–66.

Peirce, L. C., and J. L. Brewbaker. 1973. Applications of isozyme analysis in horticultural science. *HortSci.* 8: 17–22.

Pfahler, P. L. 1975. Factors affecting male transmission in maize (*Zea mays* L.). *In* D. L. Mulcahy [ed.], *Gamete competition in plants and animals*, 115–124. North Holland, Amsterdam.

Pichersky, E., and L. D. Gottlieb. 1983. Evidence for duplication of the structural genes coding plastid and cytosolic isozymes of triose phosphate isomerase in diploid species of *Clarkia*. *Genetics* 105: 421–436.

Pietro, M. E., and G. E. Hart. 1985. The genetic control of triosephosphate isomerase of hexaploid wheat and other Triticeae species. *Genet. Res. Camb.* 45: 127–142.

Price, S., and A. L. Kahler. 1983. Oats. *In* S. D. Tanksley and T. J. Orton [eds.], *Isozymes in plant genetics and breeding*. Part B, 103–128. Elsevier, Amsterdam.

Pryor, T., and D. Schwartz. 1973. The genetic control and biochemical modification of catechol oxidase in maize. *Genetics* 75: 75–92.

Quiros, C. F. 1983. Alfalfa, luzerne (*Medicago sativa* L.). *In* S. D. Tanksley and T. J. Orton [eds.], *Isozymes in plant genetics and breeding*, Part B, 253–294. Elsevier, Amsterdam.

———, and N. McHale. 1985. Genetic analysis of isozyme variants in diploid and tetraploid potatoes. *Genetics* 111: 131–145.

Ranjhan, S., J. C. Glaszmann, D. A. Ramirez and G. S. Khush. 1988. Chromosomal localization of four isozyme loci by trisomic analysis in rice (*Oryza sativa* L.). *Theor. Appl. Genet.* 75: 541–545.

Rhoades, M. M. 1951. Duplicate genes in maize. *Amer. Naturalist* 85: 105–110.

Rick, C. M. 1975. The tomato. *In* R. C. King [ed.], *Handbook of genetics*. Vol. 2: 247–280. Plenum Press, New York.

———. 1979. Biosystematic studies in *Lycopersicon* and closely related species of *Solanum*. *In* J. G. Hawkes, R. N. Lester, and A. D. Skelding [eds.], *The biology and taxonomy of the Solanaceae*, 667–678. Academic Press, New York.

———. 1983. Tomato. *In* S. D. Tanksley and T. J. Orton [eds.], *Isozymes in plant genetics and breeding*, Part B, 147–166. Elsevier, Amsterdam.

———, S. D. Tanksley, and J. F. Fobes. 1979. A pseudoduplication in *Lycopersicon pimpinellifolium*. *Proc. Natl. Acad. Sci. USA* 76: 3435–3439.

Roose, M. L. 1984. Catalytic properties of alcohol dehydrogenase isozymes specified by duplicate genes in the diploid plant *Stephanomeria exigua*. *Biochem. Genet.* 22: 631–643.

———, and L. D. Gottlieb. 1976. Genetic and biochemical consequences of polyploidy in *Tragopogon*. *Evolution* 30: 818–830.

———, and ———. 1980. Alcohol dehydrogenase in the diploid plant *Stephanomeria exigua* (Compositae): gene duplication, mode of inheritance, and linkage. *Genetics* 95: 171–186.

Sano, R., and H. Morishima. 1984. Linkage relationships among 6 polymorphic isozyme loci in rice cultivars. *Jap. J. Breed.* 34 (suppl. 1): 252–253.

———, and P. Barbier. 1985. Analysis of five isozyme genes and a chromosomal location of *Amp-1*. *Rice Genet. Newsl.* 2: 60–62.

Scandalios, J. G. 1974. Isoenzymes in development and differentiation. *Ann. Rev. Pl. Physiol.* 25: 225–258.

———, and L. G. Espiritu. 1969. Mutant aminopeptidases in *Pisum sativum* I. Developmental genetics and chemical characteristics. *Mol. Gen. Genet.* 105: 101–112.

Schaefers, H.-A., and J. Feierabend. 1976. Ultrastructural differentiation of plastids and other organelles in rye leaves with a high-temperature-induced deficiency of plastid ribosomes. *Cytobiologie* 14: 75–90.

Schmidt, J.-C., P. Seliger, and R. Schlegel. 1984. Isoenzyme als biochemische Markerfaktoren fur Roggenchromosomen. *Biochem. Physiol. Pflanzen* 179: 197–210.

Schnarrenberger, C., and A. Oeser. 1974. Two isoenzymes of glucosephosphate isomerase from spinach leaves and their intracellular compartmentation. *Eur. J. Biochem.* 45: 77–82.

———, and ———, N. E. Tolbert. 1973. Two isoenzymes each of glucose-6-phosphate dehydrogenase and 6-phosphogluconate dehydrogenase in spinach leaves. *Arch. Biochem. Biophys.* 154: 438–448.

Sears, E. R. 1966. Chromosome mapping with the aid of telocentrics. *Proc. 2nd Inter. Wheat Genet. Symp. Hereditas* suppl., 2: 370–381.

Second, G. 1982. Origin of the genic diversity of cultivated rice (*Oryza* ssp.): study of the polymorphism scored at 40 isozyme loci. *Jap. J. Genet.* 57: 25–57.

———. 1985. Evolutionary relationships in the *sativa* group of *Oryza* based on isozyme data. *Genet. Sel. Evol.* 17: 89–114.

Shore, J. S., and S. C. H. Barrett. 1987. Inheritance of floral and isozyme polymorphisms in *Turnera ulmifolia* L. *J. Heredity* 78: 44–48.

Sisco, P. H., J. F. Wendel, and C. W. Stuber. 1987. *Acp4* is the most distal marker on chromosome 1L. *Maize Genet. Coop. News Lett.* 61: 86.

Soltis, D. E. 1986. Genetic evidence for diploidy in *Equisetum*. *Amer. J. Bot.* 73: 908–913.

———, and L. H. Rieseberg. 1986. Autopolyploidy in *Tolmiea menziesii* (Saxifragaceae): genetic insights from enzyme electrophoresis. *Amer. J. Bot.* 73: 310–318.

———, and P. S. Soltis. 1988. Electrophoretic evidence for tetrasomic segregation in *Tolmiea menziesii* (Saxifragaceae). *Heredity:* 60: 375–382.

Soltis P. S., D. E. Soltis, and L. D. Gottlieb. 1987. Phosphoglucomutase gene duplications in *Clarkia* (Onagraceae) and their phylogenetic implications. *Evolution* 41: 667–671.

Stoneking, M., B. May, and J. E. Wright. 1981. Loss of duplicate gene expression in salmonids: evidence for a null allele polymorphism at the duplicate aspartate aminotransferase loci in brook trout (*Salvelinus fontinalis*). *Biochem. Genet.* 19: 1063–1077.

Stuber, C. W., and M. M. Goodman. 1983a. *Allozyme genotypes for popular and historically important inbred lines of corn, Zea mays L.* USDA, ARS, Agric. Res. Results, Southern Series, No. 16.

———, and ———. 1983b. Inheritance, intracellular localization and genetic variation of phosphoglucomutase isozymes in maize (*Zea mays* L.). *Biochem. Genet.* 21: 667–689.

———, and ———. 1984. Inheritance, intracellular localization and genetic variation of 6-phosphogluconate dehydrogenase isozymes in maize. *Maydica* 29: 453–471.

———, ———, and F. M. Johnson. 1977. Genetic control and racial variation of β-glucosidase isozymes in maize (*Zea mays* L.). *Biochem. Genet.* 15: 383–394.

Suiter, K. A. 1988. Genetics of allozyme variation in *Gossypium arboreum* L. and *G. herbaceum* L. (Malvaceae). *Theor. Appl. Genet.* 75: 259–271.

Tadmor, Y., D. Zamir, and G. Ladizinsky. 1987. Genetic mapping of an ancient translocation in the genus *Lens*. *Theor. Appl. Genet.* 73: 883–892.

Tanksley, S. D. 1983. Gene mapping. *In* S. D. Tanksley and T. J. Orton [eds.], *Isozymes in plant genetics and breeding*, Part A, 109–138. Elsevier, Amsterdam.

———. 1984. Linkage relationships and chromosomal locations of enzyme-coding genes in pepper, *Capsicum annuum*. *Chromosoma* 89: 352–360.

———, and G. D. Kuehn. 1985. Genetics, subcellular localization, and molecular characterization of 6-phosphogluconate dehydrogenase isozymes in tomato. *Biochem. Genet.* 23: 441–454.

———, and T. J. Orton [eds.]. 1983. *Isozymes in plant genetics and breeding.* Elsevier, Amsterdam.

———, and C. M. Rick. 1980a. Genetics of esterases in species of *Lycopersicon*. *Theor. Appl. Genet.* 56: 209–219.

———, and ———. 1980b. Isozymic gene linkage map of the tomato: application in genetics and breeding. *Theor. Appl. Genet.* 57: 161–170.

Tanner, G. J., L. Copeland, and J. F. Turner. 1983. Subcellular localization of hexose kinases in pea stems: mitochondrial hexokinase. *Pl. Physiol.* 72: 659–663.

Torres, A. M. 1974. Genetics of sunflower alcohol dehydrogenase: Adh_2, nonlinkage to Adh_1 and Adh_1 early alleles. *Biochem. Genet.* 12: 385–392.

———, T. Mau-Lastovicka, V. Vithanage, and M. Sedgley. 1986. Segregation distortion and linkage analysis of hand pollinated avocados. *J. Heredity* 77: 445–450.

———, ———, T. E. Williams, and R. K. Soost. 1985. Segregation distortion and linkage of *Citrus* and *Poncirus* isozyme genes. *J. Heredity* 76: 289–294.

Tyson, H., M. A. Fieldes, and J. Starobin. 1986. Genetic control of acid phosphatase R_m and its relation to control of peroxidase R_m in flax (*Linum*) genotrophs. *Biochem. Genet.* 24: 369–383.

Van Dijk, H. 1985. Allozyme genetics, self-incompatibility and male sterility in *Plantago lanceolata*. *Heredity* 54: 53–63.

Weeden, N. F. 1981. Genetic and biochemical implications of the endosymbiotic origin of the chloroplast. *J. Mol Evol.* 17: 133–139.

———. 1983a. Evolution of plant isozymes. *In* S. D. Tanksley and T. J. Orton [eds.], *Isozymes in plant genetics and breeding*, Part A, 175–208. Elsevier, Amsterdam.

———. 1983b. Plastid isozymes. *In* S. D. Tanksley and T. J. Orton [eds.], *Isozymes in plant genetics and breeding*, Part A, 139–156. Elsevier, Amsterdam.

———. 1984. Distinguishing among white seeded bean cultivars by means of allozyme genotypes. *Euphytica* 33: 199–208.

———. 1985. An isozyme linkage map for *Pisum sativum*. *In* P. Hebblethwaite, M. C. Heath, and T. Dawkins [eds.], *The pea crop*, 55–66. Butterworths, Guildford.

———. 1986a. Enzyme loci defined in *Phaseolus vulgaris*. *Ann. Rpt. Bean Imp. Coop.* 29: 53.

———. 1986b. Genetic confirmation that the variation in the zymograms of 3 enzyme systems is produced by allelic polymorphism. *Ann. Rpt. Bean Imp. Coop.* 29: 117–118.

———. 1988. Polymorphic isozyme loci identified in *Pisum*. *Pisum Newsl.* 20: 46–48.

———, and L. D. Gottlieb. 1979. Distinguishing allozymes and isozymes of phosphoglucoisomerases by electrophoretic comparisons of pollen and somatic tissues. *Biochem. Genet.* 17: 287–296.

_____, and _____. 1980. The genetics of chloroplast enzymes. *J. Heredity* 71: 392–396.

_____, J. D. Graham, and R. W. Robinson. 1986a. Identification of two linkage groups in *Cucurbita palmata* using alien additions lines. *HortSci.* 21: 1431–1433.

_____, and R. C. Lamb. 1985. Identification of apple cultivars by isozyme phenotypes. *J. Am. Soc. Hort. Sci.* 110: 509–515.

_____, and _____. 1987. Genetics and linkage analysis of 19 isozyme loci in apple. *J. Amer. Soc. Hort. Sci.* 112: 865–872.

_____, and G. A. Marx. 1984. Chromosomal locations of twelve isozyme loci and *Pisum sativum. J. Heredity* 75: 365–370.

_____, and _____. 1987. Further genetic analysis and linkage relationships of isozyme loci in pea. *J. Heredity* 78: 153–159.

_____, and R. W. Robinson. 1986. Allozyme segregation ratios in the interspecific cross *Cucurbita maxima* × *C. ecuadorensis* suggest that hybrid breakdown is not caused by minor alterations in chromosome structure. *Genetics* 114: 593–609.

_____, _____, and J. W. Shail. 1986b. Genetic analysis of isozyme variants in *Cucurbita pepo. Cucurbit Genetics Coop. Rpt.* 9: 104–106.

_____, D. Zamir, and Y. Tadmor. 1987. Applications of isozyme analysis in pulse crops. In R. J. Summerfield [ed.], *Proceedings of the international food legume research conference.* 979–987 Martinus Nijhoff—Dr. W. Junk, Amsterdam.

Wehling, P., and G. Schmidt-Stohn. 1984. Linkage relationships of esterase loci in rye (*Secale cereale* L.). *Theor. Appl. Genet.* 67: 149–153.

_____, _____, and G. Wricke. 1985. Chromosomal location of esterase, peroxidase and phosphoglucomutase isozyme structural genes in cultivated rye (*Secale cereale* L.). *Theor. Appl. Genet.* 70: 377–382.

Wendel, J. F. 1983. *Electrophoretic analysis of genetic variation in wild and cultivated Camellia japonica* L. Ph.D. Dissertation. University of North Carolina, Chapel Hill.

_____, and J. B. Beckett. 1987. A new isozyme marker for the short arm of chromosome 6. *Maize Genet. Coop. News Lett.* 61: 19.

_____, M. D. Edwards, and C. W. Stuber. 1987. Evidence for multi-locus genetic control of preferential fertilisation in maize. *Heredity* 58: 297–301.

_____, M. M. Goodman, C. W. Stuber, and J. B. Beckett. 1988. New isozyme systems for maize (*Zea mays* L.): aconitate hydratase, adenylate kinase, NADH dehydrogenase, and shikimate dehydrogenase. *Biochem. Genet.* 26: 421–445.

_____, and C. R. Parks. 1982. Genetic control of isozyme variation in *Camellia japonica* L. *J. Heredity* 73: 197–204.

_____, and _____. 1984. Distorted segregation and linkage of alcohol dehydrogenase genes in *Camellia japonica. Biochem. Genet.* 22: 739–748.

_____, C. W. Stuber, M. D. Edwards, and M. M. Goodman. 1986. Duplicated chromosome segments in maize (*Zea mays* L.): further evidence from hexokinase isozymes. *Theor. Appl. Genet.* 72: 178–185.

Wijsman, H. J. W. 1983. Petunia. In S. D. Tanksley and T. J. Orton [eds.], *Isozymes in plant genetics and breeding,* Part B, 229–252. Elsevier, Amsterdam.

Wilson, H. D., S. C. Barber, and T. Walters. 1983. Loss of duplicate gene expression in tetraploid *Chenopodium. Biochem. Syst. Ecol.* 11: 7–13.

Yazdani, R., and D. Rudin. 1982. Inheritance of fluorescent esterase and β-galactosidase in haploid and diploid tissues of *Pinus sylvestris* L. *Hereditas* 96: 191–194.

Zamir, D., and Y. Tadmor. 1986. Unequal segregation of nuclear genes in plants. *Bot. Gaz.* 147: 355–358.

Zimmerman, G., G. J. Kelly, and E. Latzko. 1978. Purification and properties of spinach leaf cytoplasmic fructose-1,6-bisphosphatase. *J. Biol. Chem.* 253: 5952–5956.

CHAPTER 3

Isozyme Analysis of Plant Mating Systems

A. H. D. Brown
J. J. Burdon
A. M. Jarosz

CSIRO, Division of Plant Industry, Canberra, Australia

The last decade has seen a marked increase in the number of genetical studies of plant mating systems. Earlier detailed work had largely been limited to economic plants in which morphological markers were available. Fryxell (1957) comprehensively reviewed this period and compiled an extensive list of all plant species studied. Each species was classified into one of seven classes according to its predominant mode of reproduction. Classification was based either on genetical evidence (from the segregation of morphological marker genes) or on studies of reproductive morphology and development. Table 3.1 gives the total number of taxa in each class as listed by Fryxell, and shows the diversity of plant mating systems, even at this simplified level.

The recent upsurge of research on the genetic analysis of plant mating systems has several bases. One reason is the increased awareness of the diversity of plant mating systems and their central position in determining the population biology of plants. As well, isozyme techniques have provided a wide array of genetic markers. Allozyme variants have three distinct advantages over morphological markers for studies of plant mating systems: (1) allozymes are codominantly expressed; (2) many isozyme loci are highly polymorphic in most populations; and (3) allozymes are unlikely themselves to be subject to strong selective forces.

With the growth of studies on plant mating systems, the diversity of approaches has also increased. The objective of this discussion is to give an overview of the major issues currently meriting attention and provide a guide to the necessary techniques and approaches for pursuing these issues. We will distinguish two broad areas of inquiry, namely that of assessing the mating system of individuals, and that of estimating the mating parameters of populations. The former area has focused on using markers to validate observations from natural history (e.g., the visitation patterns of insect pollinators) and to determine the paternity of individuals, with the aim of detecting variation in male and female fertility among adults. The second major area has determined the effect of mating systems on genotype frequencies in populations following the transition from adults to progeny. Although these areas overlap, they differ from a statistical perspective. In studies at the individual level, greater precision is obtained by increased genetic information on each individual (i.e., scoring more loci) even at the expense of sample size. In contrast, more efficient experimental procedures for problems at the populational level usually require the study of a few selected loci on an increased number of individuals (Shaw and Brown, 1982).

Table 3.1 *Summary of Fryxell's (1957) tabulation of modes of reproduction of seed plants*

Predominantly unimodal	n	%	Mixture of modes	n	%
Outcrossing	832	62	Outcrossing + Autogamous	136	10
Autogamous	226	17	Outcrossing + Apomictic	23	2
Apomictic	125	9	Autogamous + Apomictic	4	0.3
			All three modes	2	0.1

n = Number of taxa recorded.
% = Percentage of total determined.

MATING SYSTEM OF INDIVIDUALS

Kinds of Research

The primary aim of studies under this heading is to reveal and characterize the various kinds of mating events at the level of the individual organism. These studies include: (1) whether autogamy is complete or partial on each individual plant; (2) whether apomixis is obligate or facultative, and whether it involves pseudogamy; (3) whether insect visitation in entomophilous species results in complete or partial outcrossing, and whether subsequent visits are effective; (4) whether single fruits trace back to one or more than one source of pollen, i.e., the paternity of fruit (Ellstrand, 1984) and degree of pollen carryover; (5) whether individuals differ in their contribution to the next generation through pollen and through eggs, i.e., the extremes of male and female fertility (Müller-Starck and Ziehe, 1984); (6) whether there are nonrandom spatial patterns of mating for seed within fruit, fruits within inflorescences, and inflorescences within a plant (Brown et al., 1986); (7) whether more than one kind of fruit is formed, e.g., the level of autogamy in both chasmogamous and cleistogamous flowers; and (8) whether gametophytic competition or cryptic self-incompatibility occurs and whether pollen tubes differ in growth rate.

Isozyme markers play a decisive role in answering these questions. Such markers give direct evidence of the kinds of mating and their genetic consequences.

The Basic Model for the Individual Adult

The model for the mating behavior at the individual level has two components. First, we would like to know the relative contribution that a specific individual makes through male and female gametes to the next generation. These are the parameters of gametic fertility. Second, the mating system proper consists of the rules governing how these gametes are united to form zygotes. At the individual level, the first step is to specify what proportion of the embryos on the individual female plant are apomictic, selfed, and outcrossed. Next, considering the outcrossed fraction, we need to specify the degree of *biparental inbreeding* or correlation between allogamous uniting gametes, relative to that if mating were at random in the population. The ideal timing for the census would be at zygosis, but, in practice, surveys may have to be made at seed maturation or dispersal.

Alternatively, it is possible to specify the mating events, based on male function, i.e., specify for all seeds that have the given plant as male parent the proportions that are selfs, crosses, etc. This and other more complicated measures of individual selfing rate (Gregorius et al., 1987) can be generated from knowledge of the male and female fertilities and seed outcrossing rate. Such definitions may be useful for certain specific models, but in general are hard to measure and seem overly complex. They have the disadvantage of confusing the two components of mating, namely fertility and type of mating event.

Thus the major specification (ignoring the inbreeding from causes other than self-fertilization) would require four dimensions. Considering bisexual species, two dimensions are needed for the gametic fertilities. In this plane, each individual is mapped according to its relative contributions to the male and female gametic pool (Müller-Starck and Ziehe, 1984). The remaining two dimensions would consist of an equilateral triangle, in which each individual is specified by the proportion of its seeds that are derived by apomixis (c), self-fertilization (s), or outcrossing (t), ($c+s+t=1$). Fryxell (1957) called this the reproductive triangle and used it to represent species. In contrast, our use here is to map bisexual or female individuals.

This picture immediately points to questions of fundamental research interest. For example, if estimates of fertility and mating system parameters are determined for a number of individuals, what is the pattern of covariation between them? Does male fertility relate positively or negatively to selfing rate?

A check of the topics listed above shows that this basic model is insufficient to address some issues. For example, in the wild relative of soybean, *Glycine argyrea*, several additional specialized parameters are meaningful (Brown et al., 1986): (1) the fraction of seed produced in cleistogamous flowers; (2) the probability that a chasmogamous flower is visited by a pollinator; (3) the probability that two outcrossed seeds taken from the same fruit are full sibs; and (4) the probability that two outcrossed seeds from two separate fruits on the same inflorescence are full sibs.

The Ideal Experimental Situation

Consider a population of mature individuals among which the researcher wishes to determine the patterns of mating. Mating in the current generation is to be studied preferably in separate arrays of progeny from single maternal plants, rather than as a bulk sample. Therefore the particular female parent for each progeny is known directly. Furthermore, we restrict discussion to an experimental design in which the maternal genotype is known for each marker locus. This may be determined by direct assay of maternal tissue, megagametophyte tissue in gymnosperms, or by assaying a sufficiently large sample of progeny for reliable inference.

For a complete determination of the mating parameters at the individual level, the population should possess the following features. Each potential male parent plant (genet) should be uniquely defined genetically (Müller-Starck, 1985), so that any pollen grain produced by that plant can be attributed to it unambiguously. A simple case would be homozygosity of a unique allele, i.e., an allele that occurs in no other adult. In addition, when individuals are homozygous for a unique *combination* of alleles at several loci, their progeny will be unequivocally marked, but only when the female parent carries none of the alleles in the combination. Since each plant should be uniquely specified, the ideal experimental situation will be rare and restricted to small populations.

When considering the plane of gametic fertilities, there is a disparity in the requirements for estimating the male versus the female variation in fertility in the population. To determine individual male fertilities, the entire adult male population must be scored for multilocus genotype. Only then will it be known whether each plant is uniquely marked. In contrast, the frequency distribution of female fertilities in the population can be established from a random sample of adult females. However, a limited sample on the female side will not establish the absolute contribution that any one plant makes to the total seed output of the population, and hence the absolute female fertility of that plant. At least the total seed production from the whole population must be estimated for this. Nor will the extreme values for female fertility be apparent. These limits may not apply on the male side. In the limited sample, suppose sufficient markers are in place to assign paternity either to one of the sampled adults or to a foreign unknown source. The fraction of seed whose male parentage is inexplicable from the limited sample will be known, and hence an individual's relative male fertility can be computed, even when the paternity of some seed remains unknown.

The ideal experimental situation of complete biparental assignment could conceivably be achieved without the highly restrictive requirement of every individual producing only gametes that are uniquely marked. The population needs to be both highly polymorphic and small in size so that the multilocus genotypes of all adults are known. Paternal assignment for each seed follows *deterministically* from a simple *exclusion* algorithm (Ellstrand, 1984). The algorithm takes each locus in turn and excludes any impossible parent diagnosed by that locus. If the maternal allele is known, plants that lack the putative paternal allele are excluded. When the maternal allele is unknown, exclusion is possible only if the plant does not possess either one of the progeny alleles. The process is repeated for each locus. In principle, the exclusion procedure could be used when the maternal origin of a progeny is not known directly from the way the seed were sampled, and has to be determined genetically from markers. Multi-sired fruits, endosperm, gametophytic data, or some other method would be needed to distinguish the direction of the cross.

Individual Mating Systems—Incomplete Specification

Proportions of Kinds of Mating Events on Individual Plants: Variation among the individuals of a population in the proportion of their progeny which are selfed, outcrossed, or apomictic is an important topic with limited information (Hamrick, 1982). Variation in selfing has been documented (e.g., in *Bidens menziesii* by Ritland and Ganders, 1985), yet the extent to which it is genetically determined as opposed to being environmentally induced is usually unknown. Barrett and Shore (1987) found evidence of a genetic component among the plants in a population of *Turnera ulmifolia*, in that the degree of herkogamy (anther-stigma separation) was positively related to outcrossing rate. Study of the kinds of mating events is possible in populations that may be too numerous to allow a complete determination of paternity. The major requirement is sufficient polymorphism in the population at several loci so that as many non-self pollen grains as possible will be genetically distinct. Single-locus studies of variation among individuals in outcrossing rate cannot separate variation due to inherent outcrossing rates from variation in pollen allele frequencies (Brown et al., 1985; Hamrick and Schnabel, 1985). Variation in apomixis among the individuals of a population has rarely been studied.

Paternity Studies: When there is insufficient genetic information (polymorphic loci, or too few alleles) to trace the paternal origin of all zygotes, several options for using partial data are possible. The male source of some of the zygotes may be deduced uniquely by exclusion. For the remainder, only a fraction of the potential paternal sources can be excluded with certainty as inferred from the progeny genotype. In any single instance, the paternal source can then be assigned by likelihood ratio. Ritland (1983) has formulated the appropriate test, as has Meagher (1986). Essentially, the method computes the "segregation likelihood" for all potential male parents and assigns paternity to that plant with the highest score. The overall likelihood depends on the segregation likelihood in the following way.

$$L(g_{jli}) \propto L(g_j|Q_i) \times L(P_{il})$$

where $L(g_{jli}) =$ the total likelihood that a pollen grain with genotype j (g_j) sampled at location l traces to a male parent i (P_i, with genotype Q_i),

$L(g_j|Q_i) =$ the "segregation" likelihood or conditional likelihood that the parental genotype Q_i produces a male gamete with genotype g_j, and

$L(P_{il}) =$ the likelihood that parent P_i contributes to offspring sampled at location l.

The basic proposition behind the segregation likelihood method is that the expected frequency in which each potential adult genotype produces a specific gametic genotype is an index of the credibility that the adult is the source of the male gamete in question.

However, the assignment of diploid parental origin to a single haploid gamete purely on segregation likelihood suffers from two potential problems. First, there is a bias toward homozygous parental origin. Suppose the gamete for which a parental origin (either maternal or paternal) is sought has the rare A_1 allele at the A locus. Exclusion procedures would remove all common homozygotes as impossible parents and leave only the homozygote A_1A_1 and the heterozygotes (A_1A_j; $j=2,3, \ldots$) as potential sources. Then the likelihood procedure would always assign the origin of the A_1 gamete to an A_1A_1 homozygote, provided such exists in the sample from the population. Thus, on a multilocus basis, the procedure would pick the most homozygous source of a specific gamete. This becomes biased when one wishes to sum the paternity estimates of several progeny. That the A_1A_1 homozygote is the most likely source of the single gamete sample A_1 does not mean that it is the sole source of all A_1 gametes in a pollen sample.

At first sight, an apparent remedy to this problem would be to weight the likelihoods by the population frequency of male parent genotypes. If the estimated or presumed frequency of the heterozygote were more than twice that of the homozygote, it would be deemed the more likely parent. But this merely swings the bias the other way, because the *unilateral* attribution would assign all A_1 gametes to the heterozygote.

Second, there are no biological constraints in the segregation algorithm which confine the search in space and time for the "most likely" parent. The search for the most homozygous source based on the segregation likelihood is limited only by the total sample and not by the spatial separation of mates or overlap of their flowering.

These two problems could be major reasons for the apparent contrast between male and female fertility patterns which Meagher (1986) reported in *Chamaelirium luteum*. Since male parentage was commonly assigned by likelihood, whereas female parentage was known, the male fertility variances would be inflated by the concentration of paternal assignments on fewer, more homozygous plants. Further, the variance component of mating distances within paternal half-sibships could be inflated relative to that within maternal half-sibships by virtue of the second problem mentioned.

Some possible remedies for these problems follow. For the first problem: (1) use systems where sets of gametes share the same parental sources, e.g., predominantly singly-sired fruits. Such a system would provide more than one gamete or haploid sample to diagnose a diploid source of the pollen; and (2) attribute paternal parentage fractionally to all possible sources existing in the population in proportion to their segregation likelihoods. This procedure is analogous to that used in outcrossing estimation when maternal genotypes are estimated from a progeny array that is too small to infer the maternal genotype with reasonable certainty (Clegg et al., 1978).

For the second problem, a possible solution might be to introduce spatial (Neale, 1984) and/or temporal (Fripp et al., 1987) restrictions based on independent data. However, problems then arise regarding the choice of probability distribution function for the conditioning variable, and relative weighting for these ecological variables compared to the genetic (segregation, joint inheritance, and parental genotype frequencies) information. We can note that an assumption of the distribution and relative weighting is always implied even if not made explicit. Thus, in the method of segregation likelihood, a uniform distribution and weighting are assumed. Further, having set up a likelihood formulation of parental sources, one cannot use the assignments to determine spatial and temporal distributions of mating events, because this information has gone into the formulation of the parental likelihood. It seems there are many potential pitfalls in using likelihood methods based on partial information, making deterministic procedures preferable.

Substantial yet incomplete deterministic paternal assignments can, however, be sufficient to determine several facets of the mating system. Examples include determining the minimum levels of multiple paternity of fruit (Ellstrand, 1984) and the minimum levels of foreign pollen in circumscribed populations (Ellstrand and Marshall, 1985).

A useful parameter to gauge whether sufficient polymorphism is available for extensive determination of paternity in a very large population is the average exclusion probability. This is the average probability (P_E) that a randomly chosen male will be excluded as a paternal parent of a random offspring from a random maternal array. For a single locus A with k distinct alleles (A_i) and allelic frequencies $\{p_i; i = 1, \ldots, k\}$ in a panmictic population this probability is

$$P_E[A] = 1 - 2(2-h)(1-h) + \sum_i p_i^3 (4-3h-3p_i^2) + 2\sum_i p_i^4$$

(Selvin, 1980) where $h = 1 - \sum p_i^2$ is the gene diversity at that locus.

For n independent loci, the overall exclusion probability is

$$P_E[.] = 1 - \prod_{i=1}^{n} (1 - P_E[i])$$

When the k alleles are equally frequent (i.e., $p_i = k^{-1}$ for all i), $P_E[A]$ is a maximum (Selvin, 1980)

$$P_E[A \mid p_i = k^{-1}] = (k-1)(k^3 - k^2 - 2k + 3)/k$$

The values which $P_E[.]$ assumes for various numbers of such identical loci (n) are shown in Table 3.2. These values show that considerable polymorphism is required for high levels of paternity assignment. They also indicate that a few highly polymorphic loci tend to be more powerful than a large number of weakly polymorphic loci.

Table 3.2 *The average paternity exclusion probability in an infinite population with various numbers (n) of marker loci each of which is polymorphic for k equally frequent alleles and hence have gene diversity value [h = (k−1)/k]*

Number of equally frequent alleles (k)	h	Number of loci (n)						
		1	2	3	4	6	8	10
2	0.50	0.19	0.34	0.46	0.56	0.71	0.81	0.87
3	0.66	0.37	0.60	0.75	0.84	0.94	0.98	0.99
4	0.75	0.50	0.75	0.88	0.94	0.99	1.00	1.00
5	0.80	0.60	0.84	0.93	0.97	1.00		
6	0.83	0.66	0.88	0.96	0.99	1.00		
8	0.88	0.74	0.93	0.98	1.00			
10	0.90	0.80	0.93	0.98	1.00			

MATING SYSTEM OF POPULATIONS

Kinds of Research

A somewhat different set of problems comes into focus when the researcher confronts a population as a unit and wishes to determine the mating system to specify more precisely the future genetic composition of the population. The basic aim of such studies has been to trace the genotype (single-locus or multilocus) frequencies in the population from adult to progeny. This characterizes genetically the input from any one reproductive season into the seed bank of the population. Such a characterization requires a mating model. The model allows the nature of the seed input to be measured for a species or population, as an index relative to the major modes of reproduction, namely vegetative reproduction, apomixis, autogamy and allogamy. However, the flexibility of plant mating systems and their temporal shifts and spatial variation need to be considered. Hence, current research directed at the ecological context of plant mating systems may require more elaborate estimation models.

The Basic Mixed Mating Model

The most common model employed in studying plant populations has been the mixed mating model. The model assumes two types of gametic union. Each zygote is assumed to result from either self-fertilization with fixed probability s, or fertilization with a pollen grain chosen at random from the whole population with probability $t = 1-s$. This model had its origin in the early experimental tests of plant breeders who were looking for contaminating outcrosses in plantings of predominantly self-pollinating crops (Jones, 1916).

Fyfe and Bailey (1951) presented the first formal statistical treatment for estimating the parameters of the model and sampling variances for a recessive marker locus. Since then, the model has been used widely in agricultural and natural populations of inbreeding and outbreeding plants (Brown et al., 1985; Schemske and Lande, 1985; Clegg and Epperson, 1985).

Within the framework of the mixed mating model it is possible to derive estimates of outcrossing rates based solely on the observed frequency of heterozygotes among adults or progeny in the population as a whole. However, the procedure assumes that the mating system is the only force acting on genotype frequencies and that the population is in equilibrium with its mating system. The quantitative estimates are thus not reliable because they are totally dependent on these assumptions. More precise studies require the use of single-plant progeny arrays.

A second experimental choice is whether progeny number per female parent is to be sufficiently large (say exceeding 15) so that the maternal genotype can be reliably inferred. Alternatively, when fewer progeny are assayed the observed genotypes in the progeny can be attributed fractionally to all possible maternal genotypes in proportion to their likelihoods. The former approach is preferable as it is simpler to formulate (Brown et al., 1975).

The simplest case is for a single diallelic locus, with alleles A_1, A_2 (Table 3.3). The observed number of progeny genotypes for each maternal genotype is coded as (O_i; $i=1, \ldots, 6$). The expectations are based on assuming the mating system of mixed selfing and random outcrossing, where the allele frequency of A_1 in the pollen is p. In this simplest (diallelic) case, the expected frequency of A_1A_2 progeny from A_1A_2 maternal plants is independent of the mating parameters (s,p) and is omitted here for convenience. However, this class of progeny must be added for multiple alleles, multiple independent loci (Ritland and Jain, 1981), and for other mixed mating models such as mixed outcrossing, selfing, and apomixis (Marshall and Brown, 1974).

In gymnosperms, the progeny of heterozygous maternal plants yield information on the mating system when the haploid megagametophyte is assayed with each embryonic sporophyte. Complete gametic classification is also theoretically possible for loci expressed in the triploid endosperm of many angiosperms, such as for seed storage protein genes. In these special cases, classes 3 and 4 in Table 3.3 are redefined such that O_3 includes the endosperm-progeny combinations (A_1; A_1A_2) whereas O_4 includes the combinations (A_1; A_1A_2) and (A_2; A_2A_2). The expectations remain as defined in Table 3.3.

The progeny of each maternal genotype provide one degree of freedom, and thus arrays from at

Table 3.3. Basic maternal/offspring matrix for estimation of the parameters of the mating system of mixed self-fertilization and random outcrossing

	Genotypes		Number of Progeny	
	Maternal	**Progeny**	**Observed**	**Expected**
1.	A_1A_1	A_1A_1	O_1	$N_{11}(1-X)$
2.		A_1A_2	O_2	$N_{11}X$
3.	A_1A_2	A_1A_1	O_3	$N_{12}(1-X+Y)/2$
4.		A_2A_2	O_4	$N_{12}(1+X-Y)/2$
5.	A_2A_2	A_1A_2	O_5	$N_{22}Y$
6.		A_2A_2	O_6	$N_{22}(1-Y)$

where $X = tq$
$Y = tp$
$N_{11} = O_1 + O_2$
$N_{12} = O_3 + O_4$
$N_{22} = O_5 + O_6$

least two maternal types are required to estimate both mating parameters. When progenies from all three maternal genotypes are available, a test for goodness-of-fit is possible. Brown et al. (1975) detailed the maximum likelihood estimation procedure for this case (see also Ritland, 1983).

The basic model in Table 3.3 has several features and assumptions (Brown et al., 1985; Hamrick and Schnabel, 1985). The formal statistical and genetic assumptions required for estimation can be summarized as follows: (1) for each maternal genotype, the progeny genotypic classes are independent, identically distributed, multinomial random variables; (2) the expected values of both mating system parameters t and p (and therefore their one-to-one transformations X and Y) are uniform over maternal plants; (3) segregation of the alleles in heterozygous maternal plants is strictly Mendelian in a 1:1 ratio for both pollen and ovule production; and (4) no selection occurs between fertilization and the assay of progeny genotypes.

These fundamental assumptions are sufficient to specify the construction of the model and the estimation of its parameters. However the assumptions themselves or the interpretation of the estimates may require the presence of several other biological features. In particular, the model assumes that inbreeding arises only through self-fertilization. This implies that the pollen involved in each outcrossing event is a random gamete sampled from the entire population. Each outcross is strictly an independent sample from the same uniform population of pollen. The genotype of each outcrossing pollen grain is independent of the maternal genotype, and of that of other outcrossing pollen grains included in the same maternal family. In a similar vein, the probability of outcrossing is assumed to be constant for all maternal plants, unaffected by maternal genotype, and independent of whether any other seed in the sample is an outcross or a self.

The basic mixed mating model was first used extensively in studying crop populations and predominantly inbreeding species. As it was extended to natural populations, particularly those in which outcrossing was frequent, departure from its assumptions became evident. The major problems were the heterogeneity of estimates both from different marker loci and from different individuals, and variation in pollen allele frequencies among subpopulations. These problems led to the use of more elaborate procedures.

More Complex Procedures and Models

Recent methods include the use of multilocus estimates, the estimation of co-ancestry between outcross pollen and maternal plant (inbreeding from mating other than by self-fertilization or *biparental inbreeding*), and the use of progeny arrays when outcrosses share paternity. Here we consider only the first topic and make reference to Ritland (1984) and Schoen and Clegg (1984) for the latter two procedures.

Two multilocus procedures use the concept of detection probability. Each progeny of a maternal array can be classified as to whether or not it is a genetically marked outcross. The probability that an outcross will be detected on maternal genotype i (G_i) depends on whether the pollen grain carries a non-maternal allele at any one of the several loci being scored. When only a small number of marker loci are employed, this probability may vary considerably over maternal genotypes. Green et al. (1980) used a procedure allowing different detection probabilities for each maternal genotype in a *three-locus* analysis of outcrossing in *Lupinus albus*.

The maximum likelihood estimate of outcrossing (t) is obtained as the numerical solution to the equation

$$\sum_i [(N_i - O_i) G_i/(1 - G_i t)] = \sum_i O_i/t$$

where O_i is the observed number of detected outcrosses in a total size of N_i from maternal genotype i. The main problem with the method is that computing the detection probabilities (G_i) requires assumptions about the frequencies of pollen multilocus genotypes.

Shaw et al. (1981) developed and simplified this approach further to cope with many loci and *multilocus* maternal genotypes. Their estimator essentially assumes that the same detection probability applied to all outcrosses irrespective of maternal genotype. Although this assumption is strictly incorrect, the error introduced is negligible when the number of polymorphic loci is large, so that the detection probability is high ($G_i \geq 0.9$). Thus if all $G_i = G$, the estimate is

$$t = \Sigma O_i/(G \Sigma N_i),$$

where G is computed as 1 minus the product over loci of the single-locus probability that the pollen allele will also be present in the maternal plant.

These two multilocus methods are approximate but useful because they can be computed readily. Ritland and Jain (1981) presented a formal complete maximum likelihood estimation procedure based on many independent loci. They demonstrated numerically that the multilocus estimate is less affected by selection and nonrandom outcrossing than are single-locus estimates. Their analysis indicates that in most circumstances three to four loci approach the minimum variance possible.

Variation in Male Fertility

Following the current usage, we have included variation in the fertility or reproductive components of the life cycle as part of the mating system, in the loose sense. Such variation can be assessed among individual plants, or as a fitness component of sets of individuals that share a single-locus or multilocus genotype.

At the individual level, evidence of variation in female fertility is most readily obtained directly from the seed fecundity of individual plants. For male function, however, methods must be indirect. One approach, discussed above, is to use isozyme markers to disentangle the paternity of a sample of seed.

Schoen and Stewart (1986, 1987) have developed another method that estimates the male fertilities of distinct adult genotypes (genets or clones). Their procedure is particularly applicable to gymnosperms where the haploid megagametophyte-embryo combination defines the pollen allele contributed to all zygotes at all loci. Further, the population must comprise a limited number of clones such as in plantations or seed orchards. The elements needed for this procedure are a complete adult census so that the multilocus-segregation matrix (with elements $L(g_j | Q_j)$ as defined above) can be determined, and a census of pollen multilocus genotypes inferred from megagametophyte-embryo combinations.

Schoen and Stewart (1986, 1987) report pronounced variation in male fertility in a plantation of about 30 distinct clones of *Picea glauca*. This variation was partly related to the number of ramets per clone in the population, and their production of male cones. Fertilities also fluctuated between years; yet the extent to which the variation in male fertility is genetically controlled remains unknown. Evidence from cloned genotypes, although suggestive of a genotypic component, may obscure what is happening at the level of the individual plant.

Male fertility variation among individuals in natural populations might be anticipated, just as individual plants show conspicuous variation in seed production. Further, polycarpic species would be expected to vary their male and female fertilities among reproductive cycles. Presumably the genetic component for such variance is under intense selective pressure and could be rapidly exhausted unless opposed by countervailing pressures. In natural populations, such pressures may be coordinated through daylength, for example. In contrast, plantations and seed orchards may consist of individuals of disparate origin and phenology. When grown together the variation in fertility may be extreme and bear little relation to the level of variation found in the original populations.

The linear algebra of Schoen and Stewart's procedure aims to estimate fertilities for all parental clones. Despite this, the estimates may depend numerically on the ratio of expected to observed frequencies in the pollen for multilocus gametes of unique origin. Thus, the 26 distinct adult multi-locus "clones" in the *Picea glauca* plantation can be classified as to whether or not the "clone" can produce a unique and therefore distinctive gamete, and whether or not the clone is classified by the authors as having an exceptionally high or low (as opposed to average) fertility. The contingency table (Table 3.4) indicates that an association between these categories exists, where none is expected. The association implies that the estimated variation in fertility could be either inflated or reduced by variation in sampling errors on each estimate which differ according to the number of sources of pollen multilocus types.

Table 3.4 *The relationship between the estimated male fertility of clones and the ability of clones to produce diagnostic gametes (data from Schoen and Stewart, 1987)*

Clones producing diagnostic gametes	Male fertility	
	Exceptional	Average
Yes	5	1
No	5	15

$\chi^2 = 5.25^*$ [$0.01 < P < 0.05$].

Müller-Starck and Ziehe (1984) used a similar approach in comparing observed and expected gametic frequencies except they used a bulked sample of gymnosperm seeds rather than female parent arrays. They computed male and female fertilities only for clones that carry unique alleles or unique combinations (five clones in a total of 36).

Differential male fertility has also been studied as estimates of fertility attached to single allele or multilocus pollen genotypes (Allard et al., 1977; Clegg, 1983). The basic data in this procedure are genotype frequencies in various phases of the life cycle. Maximum likelihood estimators transform shifts in allele frequency into fertility components of fitness. The main problems in this approach are that relatively intense selection can occur at the diploid level without a noticeable shift in allele frequencies, and estimates may negatively relate to fitness components at other phases of the life cycle in ways which suggest they are very sensitive to sampling variation.

Overall, it appears that variation in male fertility is relatively difficult to measure; yet this component is important to describe because it is crucial to a complete understanding of the mating system of a population.

ECOLOGICAL CONTEXT OF MATING SYSTEMS

The outcrossing rates of individual plants are not immutable. Rather, as several reviewers (e.g., Hamrick, 1982; Schemske and Lande, 1985) have stressed, rates vary at all levels of organization within a species. Markedly different rates can occur between different flowers on the same plant (Brown et al., 1986), different plants in the same population (Humphreys and Gale, 1974), and between different populations of the same species (Allard et al., 1977; Schoen, 1982; Glover and Barrett, 1986).

Several ecological factors affect this variation. The mode of pollination, the architectural complexity of individual flowers and plants, and the size and density of populations interact with the genetic control of the level of self-compatibility to determine the rate of outcrossing. Of these, the mode of pollination has a major impact on the pattern of outcrossing.

Effects of Mode of Pollination

Comparisons of wind- and insect-pollinated plant species have shown the considerable effects that these alternative modes of pollination can have on outcrossing rates (Aide, 1986). Anemophilous species tend to be either predominantly autogamous or highly outcrossed and, as a general rule, display little variability in outcrossing rates at the population level. Such wind-pollinated species are relatively insensitive to short range environmental fluctuations. Their rates of outcrossing are primarily controlled by floral structure (e.g., degree of anther protrusion) and level of self-compatibility.

In contrast, entomophilous species display more variation in outcrossing rates, both within and between species (Brown and Albrecht, 1980). As a group, these species apparently do not show the strong bimodal distribution of rates which is typical of wind-pollinated species (Aide, 1986). Outcrossing among and within populations of an entomophilous species can vary greatly (Schemske and Lande, 1985). Reliance on insects as vectors may confer some advantages (e.g., efficient pollen dispersal for rare species with specialized vectors). However, entomophily appears to carry the cost of greater sensitivity to environmental fluctuations. In adverse climatic conditions the foraging behavior of insect pollinators is reduced (Schmitt, 1983).

At the species level, the degree of self-compatibility and floral structure and development determine how sensitive outcrossing rates are to pollinator activity. Self-incompatible species are at one extreme where pollinator activity affects only the number of viable seed produced, and not the outcrossing rate. In contrast, outcrossing rates in self-compatible species can depend on pollinator activity and behavior, especially in species where selfing will eventually occur in the absence of pollinators.

However, the link between pollinator activity and outcrossing rate can be broken in several ways. When the floral structure makes selfing unlikely in the absence of insects (e.g., some genotypes of *Lycopersicon pimpinellifolium*, Rick et al., 1979), outcrossing may remain high even when pollinator activity is low. Alternatively, outcrossing may remain low, despite pollinator activity, in species where selfing occurs before the flower opens (e.g., species in the *Oenothera biennis* complex, Hoff, 1962). Indeed, in *Eucalyptus* species, the presence of many flowers at various stages within the canopy of one tree allows pollinator activity to effect geitonogamy, despite protandry which discourages selfing within individual flowers.

At the population level, entomophilous, self-compatible species show marked variation in outcrossing rates. Indeed, the six species listed by Schemske and Lande (1985) as showing marked interpopulational variation (*Clarkia exilis*, *C. tembloriensis*, *Collinsia sparsiflora* var. *arvensis*, *Gilia achilleifolia*, *Lupinus nanus*, and *L. succulentus*) are all insect-pollinated. For *Gilia achilleifolia*, controlled variation in outcrossing is associated with genetically controlled factors affecting protandry (Schoen, 1982). However, for *Lupinus nanus*, *L. succulentus*, and *Collinsia sparsiflora* the large variation in outcrossing estimates is apparently due to fluctuations in pollinator activity.

Spatial and Temporal Effects on Outcrossing Rates

The effects of the spatial dynamics of a population on outcrossing rates can for convenience be divided into those caused by changes in population size and those caused by changes in plant density. Very little is known about the effects of population size. However, a sharp decrease in outcrossing has been associated with a decrease in the size of populations of *Clarkia exilis* (Vasek and Harding, 1976).

The effects of plant density on outcrossing rates can be complex. To analyze the effects of plant density, it is important to take into account the potentially confounding influences of differences in the level of self-compatibility, mode of pollen delivery, and the plant's architectural plasticity. The latter feature is perhaps best illustrated by comparing two extreme examples. On the one hand are species like *Echium plantagineum* which are highly plastic in their response to increased density. In situations of low competition and low density, they are multi-stemmed and branched. Although only a few flowers are open on any one cyme, the large number of inflorescences per plant may result in several hundred receptive flowers at any given time. Under these circumstances the average flight distance between flowers on a single plant may be much less than that between flowers on different plants, and geitonogamy is favored. At higher densities, however, individuals are smaller, with far fewer cymes, so that increased outcrossing is likely. In line with this model, outcrossing rates in a low density stand were found to be lower than in high density ones (Burdon et al., 1988).

At the other extreme are species like *Trillium* which produce only one or a few flowers per plant each season. In such species, the actual outcrossing rate may be insensitive to both plant and flower density. However, the apparent outcrossing rate may be inversely related to density as near neighbors are more likely to be closely related at high density (Ennos and Clegg, 1982; Ellstrand and Foster, 1983). In *Helianthus annuus,* apparent outcrossing rate fell with increasing stand density (Ellstrand et al., 1978). As this species is reported to be self-incompatible, the actual selfing rate presumably did not vary.

Floral density is likely to be more important than plant density as an ecological feature affecting pollinator foraging behavior and hence outcrossing rates. As such, it is relevant to studies of within-season temporal variation in outcrossing. Although most plant populations show considerable within-season variation in the density of open flowers, which may affect outcrossing rates (Schmitt, 1983), the effects of such changes have rarely been measured. In a study of the self-incompatible tree *Catalpa speciosa* in which selfed, outcrossed, and non-pollinated flowers could be distinguished visually, Stephenson (1982) found that the frequency of outcrossing increased significantly as the flowering season progressed. On the other hand, outcrossing rates in three populations of *Echium plantagineum* were almost uniform throughout one season (Burdon et al., 1988). Clearly this is an area of research that merits further study. Should temporal variation in outcrossing rates be widespread, then one season's reproductive effort could be partitioned into a series of genetically distinct seed groups. This would enrich the concept of the genetic structure of populations and affect the development of sampling strategies.

It should be noted, however, that not all examples of temporal variation in outcrossing can be ascribed to temporal fluctuations in the environment. Two studies of variation in outcrossing rates over a number of consecutive seasons found that genetic factors were important. In the tree species *Eucalyptus delegatensis* (Moran and Brown, 1980) and *Pinus banksiana* (Cheliak et al., 1985), both of which retain seed crops on the tree for several years after pollination, outcrossing rates were highest in the oldest crops and lowest in the most recent crop. In the case of *E. delegatensis* this might reflect chance variation in incidence of insect pollinators. However, such an explanation cannot apply to *P. banksiana,* a wind-pollinated species. A more likely explanation for the observed temporal variation in both species is the preferential survival of heterozygotes among the progeny of all crops (Hamrick and Schnabel, 1985).

Spatial Array of Mating Events

Quantitative studies of plant mating systems, based on the mixed mating model, assume that the probability of random outcrossing is constant over all flowers among and within individual plants. However, many ecological factors can give rise to varying rates of outcrossing and variable pollen allele frequencies. Among plants, variation in outcrossing may be due to variation in degree of self-compatibility (Hamrick, 1982). For individuals of facultative outcrossers, interactions can occur between pollinator foraging behavior, flower density, and subtle three-dimensional spatial positioning of flowers. These are likely to result in nonrandom spatial arrangement of seed with different degrees of relationship within single plants in a population.

The full complexity of such effects is illustrated by the non-random pattern of paternity found among seed on individual plants in a population of *Glycine argyrea* (Brown et al., 1986). This species has a dual flowering strategy of both chasmogamous and strictly autogamous, cleistogamous flowers. Among the chasmogamous flowers, partial outcrossing was typical ($t = 0.38$), although outcrossing was not randomly distributed across all flowers. About 35% of chasmogamous fruits were entirely produced by selfing, whereas the remaining fruits were of mixed origin. Most of the latter fruits had only a single male parent, whereas the remainder showed evidence of multiple paternity. Fruits borne on the same inflorescence were more likely to share paternity than were fruits from different inflorescences. These differences result in a complex hierarchy of relatedness among the seeds within and between fruits on the same plant. Thus the determination of the overall level of outcrossing in the population requires carefully structured sampling, with some estimate of the effective contribution of various kinds of fruits to the next seed generation. Above all, the existence of such non-random patterns gives scope for the operation of natural selection at any of these hierarchical levels. As well as in facultative outcrossers, analogous patterns could also be set up in facultative apomicts.

CONCLUSIONS

Genetic analysis of plant mating systems is an active and contentious field. In the near future we can look forward to the development of more refined, explicit models that take account of specific reproductive features. With more precise and enriched analysis of plant mating systems will come greater appreciation of their importance in determining population structure and evolutionary potential. This will also require firmer links with the ecological context of mating.

Originally, investigators had to be convinced that genetic analysis was needed as a direct check on inferences from reproductive morphology and anatomy. Now there is a need to recognize that plant mating takes place in an ecological context, and that mating is itself subject to natural selection and demographic factors. The mode of pollination, the dynamics of flower number, density, timing, and arrangement all interact with the mating system such that progeny may be organized in a highly non-random fashion. These patterns will determine the scope for selection and founder events to lead to microevolutionary changes in the mating system.

Finally, the availability of markers and models will increase the scope for experimental and manipulative approaches. Such experiments can now test the potential impact of biological factors such as gametophytic competition, density, and pollen carryover, on the mating system. These factors may help explain the diversity of mating which is such a conspicuous feature of plant populations.

ACKNOWLEDGMENTS

A. H. D. Brown is grateful to Drs. M. T. Clegg, B. Devlin, and N. C. Ellstrand of the University of California, Riverside for sharing their ideas on this subject. We also thank Drs. S. C. H. Barrett and G. F. Moran for helpful comments on the manuscript.

LITERATURE CITED

Allard, R. W., A. L. Kahler, and M. T. Clegg. 1977. Estimation of mating cycle components of selection in plants. In F. B. Christiansen and T. M. Fenchel [eds.], Lecture notes in biomathematics 19. Measuring selection in natural populations, 1–20. Springer-Verlag, New York.

Aide, T. M. 1986. The influence of wind and animal pollination on variation in outcrossing rates. Evolution 40: 434–435.

Barrett, S. C. H., and J. S. Shore. 1987. Variation and evolution of breeding systems in the Turnera ulmifolia L. complex (Turneraceae). Evolution. 41: 340–354.

Brown, A. H. D., and L. Albrecht. 1980. Variable outcrossing and the genetic structure of predominately self-pollinated species. J. Theor. Biol. 82: 591–606.

―――, S. C. H. Barrett, and G. F. Moran. 1985. Mating system estimation in forest trees: models, methods and meanings. In H. R. Gregorius [ed.], Population genetics in forestry, 32–49. Springer-Verlag, Berlin.

―――, J. E. Grant, and R. Pullen. 1986. Outcrossing and paternity in Glycine argyrea by paired fruit analysis. Biol. J. Linn. Soc. 29: 283–294.

―――, A. C. Matheson, and K. G. Eldridge. 1975. Estimation of the mating system of Eucalyptus obliqua L'Herit by using allozyme polymorphisms. Austral. J. Bot. 25: 931–949.

Burdon, J. J., A. M. Jarosz, and A. H. D. Brown. 1988. Temporal patterns of reproduction and outcrossing in weedy populations of Echium plantagineum. Biol. J. Linn. Soc. 34: 81–92.

Cheliak, W. M., B. P. Dancik, K. Morgan, F. C. H. Yeh, and C. Strobeck. 1985. Temporal variation of the mating system in a natural population of jack pine. *Genetics* 109: 569–584.

Clegg, M. T. 1983. Detection and measurement of natural selection. In S. D. Tanksley and T. J. Orton [eds.], *Isozymes in plant genetics and breeding,* Part A, 241–255. Elsevier, Amsterdam.

———, and B. K. Epperson. 1985. Recent developments in population genetics. *Adv. Genetics* 23: 235–269.

———, A. L. Kahler, and R. W. Allard. 1978. Estimation of life cycle components of selection in an experimental plant population. *Genetics* 89: 765–792.

Ellstrand, N. C. 1984. Multiple paternity within the fruits of the wild radish *Raphanus sativus. Amer. Naturalist* 123: 819–828.

———, and K. W. Foster. 1983. Impact of population structure on the apparent outcrossing rate of grain sorghum (*Sorghum bicolor*). *Theor. Appl. Genet.* 66: 323–327.

———, and D. L. Marshall. 1985. Interpopulation gene flow by pollen in wild radish, *Raphanus sativus. Amer. Naturalist* 126: 606–616.

———, A. M. Torres, and D. A. Levin. 1978. Density and the rate of apparent outcrossing in *Helianthus annuus* (Asteraceae). *Syst. Bot.* 3: 403–407.

Ennos, R. A., and M. T. Clegg. 1982. Effect of population substructuring on estimates of outcrossing rate in plant populations. *Heredity* 48: 283–292.

Fripp, Y. T., A. R. Griffin, and G. F. Moran. 1987. Variation in allele frequencies in the outcross pollen pool of *Eucalyptus regnans* F. Muell. throughout a flowering season. *Heredity* 59: 161–171.

Fryxell, P. A. 1957. Mode of reproduction in higher plants. *Bot. Rev.* 23: 135–233.

Fyfe, J. L., and N. T. J. Bailey. 1951. Plant breeding studies in leguminous forage crops. 1. Natural cross-breeding in winter beans. *J. Agric. Sci.* 41: 371–378.

Glover, D. E., and S. C. H. Barrett. 1986. Variation in the mating system of *Eichhornia paniculata* (Spreng.) Solms. (Pontederiaceae). *Evolution* 40: 1122–1131.

Green, A. G., A. H. D. Brown, and R. N. Oram. 1980. Determination of outcrossing in a breeding population of *Lupinus albus* L. *Z. Pflanzenzucht* 84: 181–191.

Gregorius, H. R., M. Ziehe, and M. D. Ross. 1987. Selection caused by self-fertilization. 1. Four measures of self-fertilization and their effects on fitness. *Theor. Pop. Biol.* 31: 91–115.

Hamrick, J. L. 1982. Plant population genetics and evolution. *Amer. J. Bot.* 69: 1685–1693.

———, and A. Schnabel. 1985. Understanding the genetic structure of plant populations: some old problems and a new approach. In H. R. Gregorius [ed.], *Population genetics in forestry,* 50–70. Springer-Verlag, Berlin.

Hoff, V. J. 1962. An analysis of outcrossing in certain complex heterozygous Euoenotheras. I. Frequency of outcrossing. *Amer. J. Bot.* 49: 715–721.

Horovitz, A., and J. Harding. 1972. The concept of male outcrossing in hermaphrodite higher plants. *Heredity* 29: 223–236.

Humphreys, M. O., and J. S. Gale. 1974. Variation in wild populations of *Papaver dubium.* VII. The mating system. *Heredity* 33: 33–42.

Jones, D. F. 1916. Natural cross-pollination in the tomato. *Science* 43: 509–510.

Lande, R., and D. W. Schemske. 1985. The evolution of self-fertilization and inbreeding depression in plants. I. Genetic models. *Evolution* 39: 24–40.

Marshall, D. R., and A. H. D. Brown. 1974. Estimation of the level of apomixis in plant populations. *Heredity* 32: 321–333.

Meagher, T. R. 1986. Analysis of paternity within a natural population of *Chamaelirium luteum.* 1. Identification of most-likely male parents. *Amer. Naturalist* 128: 199–215.

Moran, G. F., and A. H. D. Brown. 1980. Temporal heterogeneity of outcrossing rates in alpine ash. (*Eucalyptus delegatensis* R. T. Bak.). *Theor. Appl. Genet.* 57: 101–105.

Müller-Starck, G. 1985. Reproductive success of genotypes of *Pinus sylvestris* L. in different environments. In H. R. Gregorius [ed.], *Population genetics in forestry,* 118–133. Springer-Verlag, Berlin.

———, and M. Ziehe. 1984. Reproductive systems in conifer seed orchards. 3. Female and male fitness of individual clones realized in seeds of *Pinus sylvestris* L. *Theor. Appl. Genet.* 69: 173–177.

Neale, D. B. 1984. *Population genetic structure of the Douglas-fir Shelterwood regeneration system in Southwest Oregon.* Ph.D. thesis, Oregon State University, Corvallis, OR.

Rick, C. M., M. Holle, and R. W. Thorp. 1979. Rates of cross-pollination in *Lycopersicon pimpinellifolium*: impact of genetic variation in floral characters. *Pl. Syst. Evol.* 129: 31–44.

Ritland, K. 1983. Estimation of mating systems. *In* S. D. Tanksley and T. J. Orton [eds.], *Isozymes in plant genetics and breeding,* Part A, 289–302. Elsevier, Amsterdam.

———. 1984. The effective proportion of self-fertilization with consanguineous matings in inbred populations. *Genetics* 106: 139–152.

———, and F. R. Ganders. 1985. Variation in the mating system of *Bidens menziesii* (Asteraceae) in relation to population substructure. *Heredity* 55: 235–244.

———, and S. K. Jain. 1981. A model for the estimation of outcrossing rate and gene frequencies using n independent loci. *Heredity* 47: 35–52.

Schemske, D. W., and R. Lande. 1985. The evolution of self-fertilization and inbreeding depression in plants. II. Empirical observations. *Evolution* 39: 41–52.

Schmitt, J. 1983. Density-dependent pollinator foraging, flowering phenology, and temporal pollen dispersal patterns in *Linanthus bicolor*. *Evolution* 37: 1247–1257.

Schoen, D. J. 1982. The breeding system of *Gilia achilleifolia*: variation in floral characteristics and outcrossing rate. *Evolution* 36: 352–360.

———, and M. T. Clegg. 1984. Estimation of mating system parameters when outcrossing events are correlated. *Proc. Natl. Acad. Sci. USA* 81: 5258–5262.

———, and S. C. Stewart. 1986. Variation in male reproductive investment and male reproductive success in white spruce. *Evolution* 40: 1109–1120.

———, and S. C. Stewart. 1987. Variation in male fertilities and pairwise mating probabilities in *Picea glauca*. *Genetics* 116: 141–152.

Selvin, S. 1980. Probability of nonpaternity determined by multiple allele codominant systems. *Amer. J. Human Genet.* 32: 276–278.

Shaw, D. V, and A. H. D. Brown. 1982. Optimum number of marker loci for estimating outcrossing in plant populations. *Theor. Appl. Genet.* 61: 321–325.

———, A. L. Kahler, and R. W. Allard. 1981. A multilocus estimator of mating system parameters in plant populations. *Proc. Natl. Acad. Sci. USA* 78: 1298–1302.

Stephenson, A. G. 1982. When does outcrossing occur in a mass-flowering plant? *Evolution* 36: 762–767.

Vasek, F. C., and J. Harding. 1976. Outcrossing in natural populations. V. Analysis of outcrossing, inbreeding, and selection in *Clarkia exilis* and *Clarkia tembloriensis*. *Evolution* 30: 403–411.

CHAPTER 4

Isozymes and the Analysis of Genetic Structure in Plant Populations

J. L. Hamrick

Departments of Botany and Genetics
University of Georgia, Athens, GA 30602

Ecologists and plant evolutionary biologists have long recognized that plants are not distributed at random within communities but, rather, are clustered in distinct patches. Environmental heterogeneity is usually cited as playing a critical role but colonization patterns and stochastic events affecting establishment and mortality are also important. More recently plant evolutionary biologists have demonstrated that genetic variation in plant populations is also distributed nonrandomly (Antonovics, 1971; Allard et al. 1972; Hamrick and Allard, 1972; Turkington and Harper, 1979). Rather, like the plants themselves, genes and genotypes tend to be clumped, with marked genetic differences occurring over short distances. This nonrandom distribution of genetic variation is often referred to as the genetic structure of a population (Loveless and Hamrick, 1984).

Plant biologists were somewhat slow to recognize population genetic structure because spatial genetic variation in characteristics such as height, shape, and size is confounded by environmental influences on the phenotype. Thus, for many traits, particularly those that are most likely to be adaptive, it has been necessary to grow plants from different habitats in "common gardens" or to make reciprocal transplants into different habitats. It is not surprising that such studies involve populations that occur in strikingly different habitats. The classic work of Bradshaw and his colleagues (Antonovics et al., 1971; Bradshaw, 1971) on the heavy metal contaminated soils of old mine spills and the elegant work of Snaydon and Davies (1972, 1976) on the Park-Grass Experiment illustrate research of this type. Studies of more subtle habitat differences, although less abundant in the literature, have also demonstrated significant genetic differences at a local spatial scale (Antonovics, 1971; Warwick and Briggs, 1978; Turkington and Harper, 1979).

Thus, whereas our understanding of the ability of plants to adapt to local environmental conditions has been greatly advanced, for logistic reasons such studies are usually limited to small, herbaceous annuals or perennials. Furthermore, quantitatively inherited traits do not lend themselves to studies of the evolutionary processes that influence the development of genetic structure. For studies of evolutionarily important factors such as gene flow and the breeding system, plant evolutionary biologists prefer to use traits controlled by single Mendelian loci. For many years single-gene morphological traits were used to obtain quantitative estimates of mating systems, gene flow, and occasionally selection. These traits, although providing meaningful data, have at least two practical drawbacks: (1) there are usually very few loci available for any plant species; (2) their expression is often dominant-recessive. As a result, progeny testing is necessary for accurate estimates of genotype frequencies.

Biochemical techniques, most notably starch gel electrophoresis, provided plant biologists with additional single-gene markers with which to study evolutionary processes. Isozyme loci have several advantages over single-gene morphological traits: (1) genetic inheritance of electrophoretically detectable traits can be easily demonstrated; most loci have discrete Mendelian inheritance; (2) most are codominant and allele frequencies can be calculated directly; (3) estimates of levels and distribution of genetic variation can be compared directly between populations or species; (4) an array of enzymatic loci can be assayed using small quantities of material; usually one leaf or a seed will suffice; (5) many loci express at all stages of the life cycle; (6) probably most importantly, isozymes can be resolved for most plant species regardless of habitat, size or longevity.

The sections that follow demonstrate how isozyme analyses have been used to describe levels and distributions of genetic variation. I will also discuss how isozyme loci are used as genetic markers to study the evolutionary mechanisms that produce genetic structuring in populations.

ESTIMATES OF LEVELS OF VARIATION

Isozymes have most often been used to estimate levels of variation within populations. The most commonly used measures of intrapopulational variation are the percent of polymorphic loci, the number of alleles per locus, the effective number of alleles per locus, and the mean proportion of loci heterozygous per individual. This last parameter is the expected mean heterozygosity assuming Hardy-Weinberg equilibrium. Other statistics that are sometimes used are the number of alleles per polymorphic locus, the observed proportion of loci heterozygous per individual, and a measure of genetic diversity equivalent to the Shannon species diversity index.

Levels of intrapopulational allozyme variation have been the subject of several reviews, none of which is recent (Brown, 1979; Hamrick et al., 1979; Gottlieb, 1981). Although each review used different criteria to include studies, the general conclusions are similar; plant species generally maintain relatively high amounts of allozyme variation within populations. In their review of over 100 species Hamrick et al. (1979) found that the average plant species has 37% of its loci polymorphic, 1.69 alleles per locus, and a mean heterozygosity per individual of 0.141.

Hamrick et al. (1979) also noted considerable heterogeneity among species for levels of within-population variation. A significant proportion of this variation was associated with life history and ecological characteristics of the species (Table 4.1). Species that were widespread, long-lived, primarily outcrossed by wind-pollination, had high lifetime fecundities and were characteristic of the later stages of succession maintained higher levels of intrapopulational variation than species with other combinations of traits.

Table 4.1. *Relationships between life history characteristics and electrophoretically detectable genetic variation within plant populations (from Hamrick et al., 1979)*

High levels of variation	Low levels of variation
Gymnosperms	Dicots
Regional distribution	Endemic
Long-lived perennial	Biennial
Primarily outcrossed	Primarily selfed
Wind-pollinated	Selfed
High lifetime fecundity	Low to moderate lifetime fecundity
High chromosome number	Low chromosome number
Late successional species	Weedy or early successional

Since these reviews the number of species analyzed has more than doubled. However, the trends observed in 1979 remain valid. One trend hinted at in the earlier reviews was that woody species maintain more variation within populations than herbaceous species. At the time of the earlier reviews the vast majority of the woody taxa studied was wind-pollinated conifers. Thus, it was unclear whether the higher levels of variation associated with woody plants were real or were due to this bias. Today, although conifer studies are still in numerical dominance, other woody taxa have been studied. Table 4.2 compares levels of allozyme variation found in 16 tropical woody species (Hamrick and Loveless, 1989) with those found in conifers and herbaceous angiosperms (Hamrick et al., 1979; Hamrick et al., 1981). These tropical species have a wide array of pollination and seed dispersal mechanisms yet maintain as much allozyme variation as conifers. This supports the conclusion that woody plants, in general, maintain more variation within their populations than herbaceous species.

Table 4.2. *Within-population allozyme variation in several plant groups*

Group	Number of taxa	Percent Polymorphic loci	Number of alleles	Mean heterozygosity	Source
Tropical tree species	16	60.9	—	0.211	Hamrick and Loveless, 1989
Coniferous trees	20	67.7	2.29	0.207	Hamrick et al., 1981
Dicots	74	31.2	1.46	0.113	Hamrick et al., 1979
All plant species	113	36.8	1.69	0.141	Hamrick et al., 1979

DISTRIBUTION OF VARIATION WITHIN AND AMONG POPULATIONS

Perhaps of more interest to plant evolutionary biologists is the distribution of allozyme variation within and among populations. There are two commonly used measures of population differentiation applied to allozyme data. Wright's (1951) F_{ST} statistic is a measure of variance in allele frequencies among populations relative to the standardized variance based on mean allele frequencies ($F_{ST} = \sigma^2/pq$). It is calculated on each allele at a locus. Nei's (1973) genic diversity statistics also can be used to partition variation within and among populations. His G_{ST} statistic is the genic diversity due to variation among populations (D_{ST}) divided by the total diversity (H_T); $G_{ST} = D_{ST} / H_T$.

The distribution of allozyme variation among populations is the product of interactions among several evolutionary factors. Of primary importance are selection, effective population size, and the ability of the species to disperse pollen and seeds. In general, selection should increase population differentiation, as would genetic drift. On the other hand, species with more pollen or seed movement should have less differentiation than species with restricted gene flow. In support of these predictions Loveless and Hamrick (1984) found that long-lived polycarps common to the later stages of succession had low G_{ST} values (Table 4.3). These species are outcrossed, monoecious or dioecious, wind-pollinated species. Annual monocarps in early successional stages had higher amounts of interpopulational differentiation. They are also predominantly self-fertilized, pollinated by small insects, and have perfect flowers.

Table 4.3. *Relationships between life history characteristics and electrophoretically detectable genetic variation among plant populations (from Loveless and Hamrick, 1984)*

High Differentiation	**Low Differentiation**
Autogamous	Predominantly outcrossed
Hermaphroditic	Monoecious or dioecious
Gravity-dispersed seed	Winged/plumose seed
Annual	Long-lived
Monocarpic	Polycarpic
Early successional stage	Late successional stage

Table 4.4. The influence of plant breeding systems and seed dispersal mechanisms on levels of genetic diversity among populations (from Loveless and Hamrick, 1984)

	Number of Studies	Mean Diversity Among Populations (G_{ST})
Breeding System		
Autogamous	39	0.523
Annual	31	0.560
Perennial	8	0.329
Mixed Mating	48	0.243
Outcrossed	76	0.118
Animal	32	0.187
Wind	44	0.068
Seed Dispersal Mechanism		
Gravity	59	0.446
Animal-Attached	18	0.398
Animal-Ingested	14	0.332
Explosive	24	0.262
Winged/Plumose	48	0.079

Combining the mating system with the pollination mechanism (Table 4.4) gives the best prediction of the distribution of allozyme variation. Annual selfing species have more than 50% of their variation among populations whereas predominantly outcrossed wind-pollinated species (e.g., conifers) have less than 10% of their variation among populations. Seed dispersal mechanisms should have similar effects on the distribution of genetic variation. The patterns observed (Table 4.4), although generally consistent with predictions (i.e., species with gravity-dispersed seeds have the highest and wind-dispersed species have the lowest G_{ST} values), are not as straightforward as those produced by the pollination classification. Animal-dispersed species have high G_{ST} values relative to predictions. This could be due to at least two factors. First, animal-attached or ingested seeds may not move long distances. Many seeds fall directly below the maternal plant, and the passage time for birds and fruit-eating mammals can be quite short. Second, seeds are often deposited in a patchy fashion around roosts. This patchy distribution coupled with the possibility that the seeds deposited within a patch may be from the same maternal plant (Levin, 1983), and thus genetically related, may lead to genetic heterogeneity among patches and the high G_{ST} values observed.

A common and important question is whether patterns of allozyme variation are positively associated with patterns of genetic variation found for other traits. Since isozyme analyses are relatively cheap in terms of time and expenses it would be desirable to predict patterns of variation for polygenic traits from those observed for isozyme loci. There are relatively few studies that directly compare variation at allozyme loci with that at morphometric or other genetically controlled traits. The results of these studies have not produced a consistently positive association among the different types of traits (Table 4.5). Ten of the 20 comparisons presented in Table 4.5 had positive associations between allozyme variation patterns and variation at other sorts of traits, three had mixed results, and in seven cases there was no association. Furthermore, Lewontin (1984) cautioned against such comparisons on theoretical grounds and stated that the results may be misleading or at best uninformative. He demonstrated that for individual loci that influence a quantitative trait, it will be more difficult to detect gene frequency differences at the individual loci than for the quantitative trait when these genes vary among populations in the same direction. He argued then, that randomly chosen loci that have nothing to do with the metric trait (i.e., allozyme loci), but which differ between populations by the same amount as the genes for the quantitative character, will also not be detected as different. Both Rogers (1986) and Felsenstein (1986) have disagreed with elements of Lewontin's (1984) conclusions.

Table 4.5. *Comparisons between levels and patterns of allozyme variation and other types of genetically controlled characteristics.* + = association found, − = no association found between traits

Species	Type of Comparison	Result	Source
Avena barbata	allozyme/morphometric	+	Price et al., 1984
Avena barbata	allozyme/single gene visables/herbicide resistance	+	Price et al., 1985
Avena barbata	allozyme/morphometric	+	Hamrick and Allard, 1975
Avena fatua	allozyme/single gene visables/herbicide resistance	+	Price et al., 1985
Hawaiian *Bidens*	allozyme/morphometric	−	Helenurm and Ganders, 1985
Clarkia williamsonii	allozyme/morphometric	−	Price et al., 1984
Clarkia williamsonii	allozyme/single gene visables/herbicide resistance	−	Price et al., 1985
Hordeum jubatum	allozyme/morphometric	+	Price et al., 1984
Hordeum murinum	allozyme/morphometric	−	Giles, 1984
Hordeum spontaneum	allozyme/morphometric	−	Brown et al., 1978
Hordeum spontaneum	allozyme/morphometric flavonoids	+ −	Bekele, 1984
Hordeum vulgare	allozyme/morphometric	+	Price et al., 1984
Hordeum vulgare	allozyme/DNA	+	Brown and Clegg, 1983
Layia ssp.	allozyme/morphometric	+	Warwick and Gottlieb, 1985
Lisianthius skinneri	allozyme/DNA	+	Sytsma and Schaal, 1985
Phlox drummondii	allozyme/morphometric	+ −	Levin, 1977
Pinus contorta	allozyme/morphometric	−	Wheeler and Guries, 1982
Pseudotsuga menziesii	allozyme/morphometric	+	El-Kassaby and Sziklai, 1982
Silene diclinis	allozyme/morphometric	−	Prentice, 1984
Trifolium hirtum	allozyme/morphometric single gene visables	+	Martins and Jain, 1980

The most convincing study specifically designed to compare variation between allozyme and morphometric traits is that of Price et al. (1984). Comparisons were made between estimates of interpopulational differentiation based on allozyme polymorphisms and measurement characters in three predominantly self-pollinated species, *Avena barbata*, *Hordeum vulgare* and *Hordeum jubatum*, and an outcrossing species, *Clarkia williamsonii*. For each species, genetic diversity was measured at several enzyme loci and for several morphological traits. Morphometric differences among populations were measured by Mahalonobis' distance function, and Hedrick's (1971) measure of genotypic distance was used to estimate allozyme variation among populations. The rank correlation between the two distance measures for *Avena barbata* was positive and highly significant ($r = 1.00$; $P < 0.001$) whereas values for *H. jubatum* ($r = 0.47$; $P = 0.14$) and *H. vulgare* ($r = 0.60$; $P = 0.07$) approached significance. The rank correlation for *C. williamsonii* ($r = -0.07$, $P = 0.64$) was not significant, suggesting that isozyme loci may provide more information about other genes in selfing plants than in outcrossers.

EVOLUTIONARY FACTORS AFFECTING GENETIC STRUCTURE

As several reviews have shown (Brown, 1979; Hamrick et al., 1979; Gottlieb, 1981; Loveless and Hamrick, 1984), characteristics of plant species influence the distribution of genetic variation within and among populations. These traits include the mating system, pollen and seed dispersal, and the adaptive response of species to selective forces. Therefore, to understand how evolutionary forces combine to produce population genetic structure, quantitative estimates of each parameter are required. Allozyme loci provide the genetic markers needed to obtain these estimates.

Plant Mating Systems

Plant mating systems play a crucial role in shaping the genetic composition of populations. The mating system not only determines genotype frequencies in subsequent generations but it also affects population parameters such as neighborhood size, gene flow, and selection. Loveless and Hamrick (1984) found that the mating system plays a central role in determining the distribution of allozyme variation within and among populations (Table 4.4).

Allozyme loci are especially valuable for studies of plant mating systems because the alleles are codominant and are unlikely to affect pollinator behavior. Furthermore, almost every plant species studied electrophoretically has at least one or two polymorphic loci. Given the general availability of allozyme markers and the importance of plant mating systems, surprisingly few quantitative estimates are available. Schemske and Lande (1985) review the results from only 55 species.

The most commonly used estimation procedure is the mixed mating model (Fyfe and Bailey, 1951; Brown and Allard, 1970; see chapter by Brown et al. in this volume). This model assumes (Clegg, 1980): (1) mating events are due to random outcrossing or self-fertilization; (2) pollen allele frequencies are identical over all maternal parents; (3) the rate of outcrossing is independent of maternal genotype; and (4) no selection affecting the markers occurs between mating and the progeny census. Characteristics of natural plant populations are typically inconsistent with these assumptions. The assumptions are, however, easily tested in natural situations, and the tests often provide valuable insights into the biology of the species as well as the genetic organization of populations.

The first assumption states that all inbreeding in the population is due to selfing (Clegg, 1980). This is almost certainly not true for species with limited seed dispersal and matings between near neighbors; a percentage of the inbreeding effects may be due to matings among relatives rather than selfing. In most natural populations it is difficult to distinguish biparental inbreeding from selfing. Ellstrand et al. (1978) estimated rates of outcrossing in natural populations of *Helianthus annuus*. The apparent rate of outcrossing varied from 0.54 in dense populations to 0.91 in low-density populations. They explained their results by arguing that near neighbors in high-density situations were more likely to be related than in the low-density locations. If most matings are among near neighbors apparent selfing rates would be higher due to biparental inbreeding. Sun and Ganders (1988), in a study of gynodioecious Hawaiian *Bidens*, partitioned inbreeding effects due to selfing and biparental matings. Estimates of outcrossing obtained from hermaphroditic individuals in eight populations averaged 0.55 whereas estimates from strictly female plants in the same eight populations were 0.85. This indicates that approximately 15% of the realized selfing in the hermaphroditic individuals was actually due to matings between relatives.

Ellstrand and Foster (1983) used an experimental population of the selfing annual crop *Sorghum bicolor* to study this phenomenon further. They compared the apparent outcrossing rates of experimental populations with seed families spatially arranged in stratified and over-dispersed treatments. Using an alcohol dehydrogenase locus as a genetic marker they found that estimates of outcrossing in the stratified treatment averaged 0.22 over three years and two planting sites whereas the outcrossing rate in the over-dispersed treatments averaged 0.46.

The second assumption of the mixed mating model is also often not realized in natural populations. Specifically, several factors related to the genetic structure of populations may lead to maternal plants receiving pollen loads composed of different allele frequencies (Table 4.6). Any nonrandom distribution of allele frequencies combined with a predominance of matings within the patch will affect the genetic composition of the pollen. In animal-pollinated species this trend is enhanced by the idiosyncratic movement of individual pollinators. There is increasing evidence that individual flowers may be pollinated by only one or a few visitors (Lertzman and Gass, 1983; Smyth and Hamrick, 1987).

Table 4.6. *Plant-to-plant variation in pollen allele frequencies for four plant species. The frequency of the rare allele is given. The data presented were chosen to illustrate the range of heterogeneity seen among maternal plants in each study: mean allele frequencies and χ^2 values are based on the complete data set of each study*

Species: Pollinator: Locus:	*Carduus nutans* Insects PGI	*Eucalyptus obliqua* Animals ADH	*Pinus contorta* Wind ADH	*Tachigali versicolor* Insects IDH
Plant #				
1	0.17	0.30	0.33	0.08
2	0.42	0.40	0.53	0.52
3	0.30	0.05	0.23	0.00
4	0.28	0.25	0.58	0.20
5	0.07	0.10	0.43	0.08
mean allele frequency	0.25	0.30	0.39	0.18
χ^2(df)	28.07 (6)	9.7 (4)	52.81 (29)	221.35 (32)
Source	J. L. Hamrick and C. A. Smyth, unpubl. data	Brown et al., 1975	J. L. Hamrick and C. Smith, unpubl. data	J. L. Hamrick et al., unpubl. data

Of the four species represented in Table 4.6 estimates of pollen allele frequencies vary the least in the wind-pollinated conifer, *Pinus contorta,* and the most in the insect-pollinated tropical tree, *Tachigali versicolor*. The insect-pollinated weed *Carduus nutans* and the animal-pollinated *Eucalyptus obliqua* are intermediate. This result was not unexpected since the flowering population of *T. versicolor* consisted of approximately 80 individuals throughout a 25 km² area on Barro Colorado Island, Republic of Panama. The *P. contorta* progeny were collected from 130 adults located along a 300 m transect placed in a large continuous population of this species.

Experimental studies of the effects of genetic structure on estimates of outcrossing are few but those that exist are consistent with interpretations made from natural populations. Ennos and Clegg (1982) used the insect-pollinated morning glory *Ipomoea purpurea* to study the effect of genetic substructure on estimates of outcrossing. One garden consisted of randomly distributed individuals homozygous for two alleles at an esterase locus. In the second garden, individuals representing the two homozygotes were assigned to opposite halves of rows so that populations consisted of two adjacent blocks of homozygous genotypes. In the structured population, estimates of outcrossing were lower (0.61) than in the treatment with no spatial heterogeneity (0.79).

Observations of heterogeneity among individual plants has led Schoen and Clegg (1984) to propose a different model that assumes that matings are either due to selfing or fertilization by a single outcrossed father. This "single pollen parent model" fits the *Ipomoea* data better than the mixed mating model. For many species, however, this model is probably as unrealistic as the mixed mating model. Whereas matings may be correlated to some degree by the foraging behavior of individual pollinators, seeds within a single flower are usually sired by more than one father (Ellstrand, 1984; Ellstrand and Marshall, 1985, 1986; Marshall and Ellstrand, 1986).

Unlike the two previous assumptions there is little reason to expect that maternal individuals homozygous for different isozyme alleles have different rates of outcrossing. There are, however, few actual tests of this assumption. In the study of Ennos and Clegg (1982) a significant difference in the outcrossing rate estimated for the two homozygous esterase genotypes was noted. Unfortunately, their design did not allow them to determine if maternal plants with different genotypes actually outcross at different rates.

Selection between the time of fertilization and the point of progeny analysis can also affect outcrossing estimates. Several lines of evidence indicate that selection operates against inbred progeny in self-compatible species. Seed set in primarily outcrossed species is often considerably lower when selfing occurs (Hagman and Mikkola, 1963; Sorenson and Miles, 1974; Lindgren, 1975). The selfed seeds presumably have lower rates of survival than do outcrossed seed (Sorenson, 1982). C. Smith and J. L. Hamrick unpublished data) demonstrated that in Pinus contorta selfed strobili produce 80% fewer filled seed than outcrossed strobili. By using allozyme markers and crosses of 1:1 mixtures of selfed and outcrossed pollen they also demonstrated that progeny were approximately 50% more heterozygous (outcrossed) than expectations based on random fertilization.

Selection against selfed progeny can also be inferred from the survival of dormant seed stored in serotinous cones or fruits. The year of pollination can usually be determined by either counting annual rings in the stems or counting whorls of fruit or cones. Although there is considerable year to year variation, the overall trend (Fig. 4.1) is toward an increase in outcrossing rates in older seeds. Selfed seeds apparently have lower survival when stored for long periods.

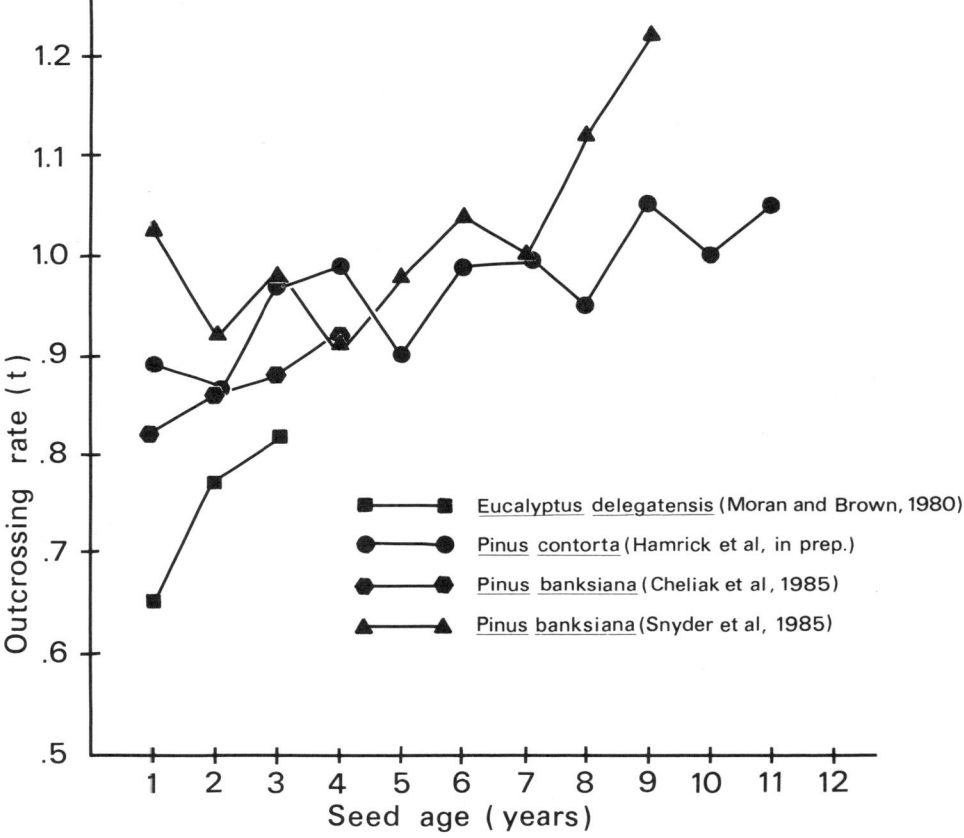

Fig. 4.1. Changes in estimated rates of outcrossing (t) with increasing age of naturally stored dormant seed for three tree species.

We see, therefore, that assumptions of the mixed mating model are often inconsistent with characteristics of natural plant populations. This may be especially serious in animal-pollinated species and may lead to inaccurate estimates of outcrossing rates (Schoen and Clegg, 1984). Recently, a new procedure, paternity exclusion analysis, has been applied to problems dealing with the genetic structure of populations. Because this procedure identifies the father of each seed, the proportions of selfing can be determined directly and are not dependent on assumptions of the mixed mating or correlated mating models. Paternity analysis is dependent on the availability of several polymorphic allozyme loci, and, as a result, this procedure may not be suitable for all species.

Plant mating systems should be just as variable as any other characteristic and, therefore, should vary among populations and years. Several studies (see Hamrick, 1982; Schemske and Lande, 1985, for reviews) have shown that estimates of outcrossing vary greatly among populations. Studies of temporal variation in the plant mating system are less common. Adams and Allard (1982) examined outcrossing in the highly selfing species *Festuca microstachys*. In 12 of 14 populations no outcrosses were observed suggesting that actual outcrossing was less than 1%. In one population 20 outcrosses were observed in 549 progeny producing an estimated outcrossing rate of 0.067. In a subsequent year no outcrosses were found in the 529 progeny examined from this population.

In most mating system studies progeny are obtained from natural populations. As a result little can be said about the genetic basis of the variation observed. In the case of *Festuca microstachys* variation between years is almost certainly due to phenotypic responses to the environment. Schoen (1982a, b) demonstrated that populations of *Gilia achilleifolia* with low rates of outcrossing are genetically different from those with higher outcrossing rates. Furthermore, Kahler et al. (1975) documented that genetic changes across several generations in a cultivated population of *Hordeum vulgare* increased the outcrossing rate by nearly 100%.

To summarize, quantitative estimates of plant mating systems are necessary if we are to understand the forces that affect the genetic organization of natural plant populations. The currently used estimation procedures have demonstrated spatial and temporal variation in rates of outcrossing. Those studies that are available indicate that species can be grouped into two general classes: selfers and predominant outcrossers (Schemske and Lande, 1985). Based on their theoretical analyses Schemske and Lande (1985) argued that species with mixed mating systems should be relatively rare. They cited the results of their review as evidence that this is the case. On the other hand, Waller (1987) stated that this conclusion is based on a small number of species that may not be representative of the actual distribution of plant mating systems. An examination of the studies cited by Schemske and Lande (1985) indicates that annual weedy species and coniferous trees are over-represented in the literature they cite.

Gene Flow

The movement of genes among populations or population subdivisions has a significant influence on the distribution of genetic variation. Species with restricted gene movement should exhibit greater genetic differentiation than species with widely dispersed pollen and seeds. The available evidence, based primarily on the distribution of allozyme loci, supports this prediction (Brown, 1979; Loveless and Hamrick, 1984). There are, however, relatively few direct measures of gene flow in natural plant populations. One complication with seed plants is that genes can be moved in two ways; for an accurate estimate of gene flow both pollen and seed movement must be described and quantified.

Most attempts to measure gene flow involve following the movement of pollen and seed dispersal vectors or the actual movement of pollen or seeds. Studies of dispersal agents have the advantage that movements can be followed from several plants within the population. As a result, pollen or seed dispersal from individuals located in a variety of densities, patch sizes, or positions within patches can be examined. The validity of this approach as a measure of gene flow is, however, based on several assumptions. First, movements of dispersal agents are assumed to represent patterns of pollen or seed deposition. There is increasing evidence that pollinator movements under-represent actual gene movement due to high pollen carryover (Schaal, 1980; Lertzman and Gass, 1983; Levin, 1983). Secondary movement has also been documented for seeds and has the potential to increase gene flow greatly (Beattie and Culver, 1979). Second, this approach provides no information on the mating system. Species with high rates of selfing coupled with long distance movement of released pollen may have less gene flow than predominantly outcrossing species with local pollen movement. Finally, movements of dispersal agents supply no information concerning the success of the pollen grain in fertilization or about successful seed establishment.

Following the actual movement of pollen or seeds rather than their vectors is more difficult but has been accomplished by marking pollen or seed with dyes or radioactive tracers or by measuring pollen or seed deposition around isolated individuals. Since pollen or seed deposition patterns are described, problems caused by pollen carryover or secondary seed movement are circumvented, and some of the problems of estimating rates of selfing/outcrossing are avoided. If it can be assumed that the marked pollen has the same dispersal pattern as unmarked pollen, the major drawback of this procedure is that it provides no information on fertilization or seedling success. Also for many plant species, especially trees, it may be difficult or impossible to mark pollen or seeds.

The availability of allozyme loci has greatly improved estimates of gene flow. Allozyme loci have been used: (1) to follow the dispersal of a unique marker allele; (2) to obtain estimates of gene flow from the distribution of allozyme variation among populations; (3) to identify maternal and paternal individuals using paternity analysis techniques.

Many of the problems of estimating gene flow do not arise if the movement of a unique allele is followed directly. Muller (1977) used a unique LAP allele to measure gene movement in a population of Scots pine, *Pinus sylvestris*. He found the marker allele in progeny of trees located 80 m from the source. Likewise, Schaal (1980) used an allozyme marker to study pollen movement in a population of *Lupinus texensis*. The mean distance of gene movement was 1.8 m, approximately twice the estimate provided by studies of pollinator movement.

There are, however, problems with using single allozyme markers. Since it is often difficult to find individuals with a unique allele, this procedure cannot be applied to all species or populations. Second, if only a few source plants are available, generalizations concerning gene movement in populations or species must be made from the pollen and seed deposition patterns of one or a few individuals. As a result, allozyme markers have frequently been used in artificial populations. A source plant(s) is established and is surrounded with plants homozygous for an alternative allele. Smyth and Hamrick (1987) used two polymorphic allozyme loci to describe pollen movement in four artificial populations of *Carduus nutans*. They found that the marker alleles moved at least 18 m from the source and that the marker alleles were patchily distributed among individuals and flowering heads on the same plants. Some plants, especially those farther from the source, had no heterozygous progeny whereas one-third of the progeny of other, equally distant individuals were heterozygous. Handel (1982, 1983) found similar patterns of pollen deposition in an artificial population of *Cucumis melo*. Although not affecting the overall magnitude of gene flow, such clumped allele distributions should have a marked effect on genetic structure.

It is questionable whether results from studies of gene movement in artificial populations can be used to describe gene movement in natural populations accurately. Often the experimental designs of artificial populations fail to take into consideration dispersion, population shape, or population density, factors that almost certainly have important effects on gene flow in natural populations.

Two indirect methods of estimating gene flow have recently come into use because of the availability of allozyme data. Both methods depend on the distribution of allele frequencies among populations. The first method is based on the relationship $F_{ST} = 1/(4Nm + 1)$ (Wright, 1951). This relationship assumes neutrality for alleles at each locus. Slatkin (1981, 1985; Barton and Slatkin, 1986) has recently developed and refined a second procedure. Frequency distributions of private alleles (i.e., alleles that appear in a single population) are used to estimate the average number of migrants exchanged between local populations (Nm; where N is the population size and m is the fraction of N replaced by immigrants). The logarithm of Nm is approximately linearly related to the log of the mean frequency of private alleles (Slatkin, 1985; Barton and Slatkin, 1986).

These two approaches have great appeal because they can be applied directly to the available allozyme data. It might be argued that the two methods of calculating Nm are redundant since the same data are used in both analyses. However, estimates from the Slatkin procedure are based on low-frequency alleles whereas calculations based on F_{ST} are largely dependent on the distribution of common alleles.

Hamrick and Griswold (1989) calculated Nm values for several groups of plant species using both indirect procedures. The two methods of estimating gene flow gave similar results. Mean estimates of Nm based on the Slatkin procedure ranged from 0.065 for selfing species to 5.38 for outcrossed wind-pollinated species (Table 4.7). Mixed-mating and outcrossed animal-pollinated species had intermediate Nm values. A similar pattern was seen for groups based on seed dispersal mechanisms. Small-seeded, gravity-dispersed species had the lowest Nm values whereas winged- or

Table 4.7. *Estimates of Nm based on two indirect procedures. Slatkin's (1985) method is based on the frequencies of private alleles p̄(1). Wright's (1951) method is based on F_{ST} values (from Hamrick and Griswold, 1988)*

Method	Number of Studies	p̄(1)	Estimates of Nm Based on	
			Slatkin	Wright
Breeding System				
Selfed	19	0.316	0.065	0.265
Mixed-Animal	9	0.200	0.161	0.727
Outcrossed-Animal	16	0.089	0.801	1.154
Outcrossed-Wind	14	0.034	5.380	4.750
Seed Dispersal Mechanisms				
Gravity				
Large	10	0.200	0.161	0.627
Small	15	0.315	0.065	0.380
Animal-Attached	9	0.148	0.292	0.473
Animal-Ingested	4	0.194	0.171	0.621
Winged or Plumose	20	0.038	4.313	2.275

plumose-seeded species had the highest. The range of Nm values was much larger when individual species were compared. Estimates of Nm ranged from 0.050 for *Phlox cuspidata* to 37.8 for *Pinus ponderosa* (Hamrick, 1987).

Comparisons of estimates of gene flow obtained by these two indirect methods with estimates derived from more direct measures are few. Hamrick and Loveless (1989) ranked 16 tropical tree and shrub species based on the species' dispersal potential and found that Nm estimated from private alleles and F_{ST} values were significantly correlated with their rankings. For one of these 16 species, *Tachigali versicolor*, a marker allele was used to estimate rates of gene flow. Estimates of m were 25% over a distance of 500 m or more. Estimates of gene flow into *Pinus taeda* seed orchards (Friedman and Adams, 1985) are also consistent with the low F_{ST} values seen in this species (J. L. Hamrick, unpublished data). In the animal literature, Coyne and Milstead (1987) estimated that *Drosophila melanogaster* can disperse 10 km during a three-month period. Their results are consistent with those of Singh and Rhomberg (1987) who demonstrated that Nm values based on the distribution of private alleles from throughout the world-wide range of this species averaged approximately 2.0. Waples (1987) has also demonstrated for 10 marine fish species that estimates of Nm based on F_{ST} and private alleles were positively associated with predictions of the species' dispersal ability. Thus, the little evidence that is available indicates that these indirect estimates of Nm are consistent with values derived from more direct measures.

The third procedure for estimating gene movement in natural populations, paternity analysis, provides a detailed description of the breeding structure of individual populations. Paternity analyses use polymorphic genetic loci to identify the most probable father (or parents in the case of a seed or seedling) from a set of candidates (Hamrick and Schnabel, 1985; Meagher and Thompson, 1986). The resolution of the procedure is dependent on the number of polymorphic loci, number of alleles per locus, allele frequencies, and the number of potential parents. Once the sire (or parents) of an individual has been identified, patterns of pollen movement within the study area can be determined. Because several maternal plants can be followed simultaneously, pollen dispersal away from paternal individuals can be described (Meagher, 1986), and the proportion of effective pollen originating outside the study area can be determined. By carefully selecting the location of monitored plants or by experimental manipulations, the effects of density, dispersion, and the location of individuals (edge vs. center) on patterns of incoming or outgoing pollen can be described (Ellstrand and Marshall, 1985). This procedure has several advantages over previous approaches, although it requires several polymorphic loci and quite large sample sizes. First, since viable seeds or seedlings are used, patterns of effective pollen or seed movement can be described. Second, since several individuals serve as "pollen sinks" the investigator can prevent the "sample of one" syndrome that haunts studies using a single uniquely marked individual. Third, once the data have been collected they can be used to describe other population parameters such as selfing rates, genetic relatedness of progeny, and male reproductive success. On the negative side, studies using paternity analyses have in common with other approaches the problem of a limited study area. Thus, although the number of progeny produced by immigrant pollen can be calculated, it may not be possible to determine how far the immigrant pollen moved.

The few recent studies that have used multilocus techniques to measure gene flow indicate that gene flow rates may have been underestimated for many plant species. Friedman and Adams (1985) found that as much as 40% of the pollen successfully fertilizing ovules in a loblolly pine (*Pinus taeda*) seed orchard came from wild populations located at least 400 m from the orchard. Studies by Ellstrand and Marshall (1985) of isolated populations of the insect-pollinated wild radish, *Raphanus sativus*, indicate that rates of gene flow of 8–18% may be found between populations separated by as much as a kilometer. Meagher (1986) found a high frequency of matings between individuals separated by as much as 25 m in a population of *Chamaelirium luteum*.

In summary, isozyme loci are the traits of choice for studies of gene flow. They are unlikely to affect pollinator activity or the behavior of seed dispersal agents. They are generally available for all species and can be easily scored. The next few years should see much additional information on gene movement. Recent work has produced higher estimates of gene flow than predicted by earlier authors (Ehrlich and Raven, 1969; Levin and Kerster, 1974). A second observation is that plant species can vary widely in their ability to transport genes. Wind-pollinated conifers have as much gene movement as more mobile animals (Slatkin, 1985), whereas selfing species have very limited gene movement (Hamrick, 1987).

Selection

The discovery of higher than expected levels of allozyme variation led to a major controversy in population genetics concerning whether this variation was maintained by selection or by a combination of mutation and drift (Allard and Kahler, 1972; Kimura and Ohta, 1972). In this discussion I will not summarize the arguments that have been put forth. Rather, I will focus on the use of isozyme loci as genetic markers that facilitate studies of selection. In several of the examples cited, the authors have argued convincingly that selection is acting directly on isozyme loci (e.g., Clegg et al., 1972; Nevo et al., 1982). However, this conclusion is not necessary to illustrate the utility of allozyme markers in studies of selection.

Allozyme traits have proved to be most useful in describing patterns of microhabitat adaptation in several species of selfing plants. Perhaps the best documented case is *Avena barbata* in California. Two multilocus genotypes predominate throughout much of California (Clegg and Allard, 1972). One genotype, termed xeric, is found throughout the warm Mediterranean climate of southern California and the Central Valley of California. The second genotype (mesic) is found in the cool Mediterranean climate surrounding San Francisco Bay (Clegg and Allard, 1972). Three other multilocus genotypes are found in areas of the North Coastal region of California (R. W. Allard, personal communication). In areas such as the Napa Valley, ecotonal habitats occur in which both the mesic and the xeric combinations can be found in specialized microhabitats (Allard et al., 1972; Hamrick and Allard, 1972). In these microhabitats the multilocus association characterizing the mesic genotype predominates in sites with deep, moist soils whereas the xeric genotype predominates on the shallow, drier soils (Hamrick and Holden, 1979). The mosaic patterns of multilocus genotypes found on hillsides in this region can only be explained by selection acting on genotypes marked by the allozyme loci. Supporting evidence that these markers are indicative of differences at a large proportion of the genome is given by Hamrick and Allard (1975) who demonstrated significant differences in several quantitative traits between the mesic and xeric genotypes. Similar patterns of microhabitat heterogeneity were found in *Hordeum spontaneum* (Nevo et al., 1979; Brown et al., 1980) and *Triticum dicoccoides* (Nevo et al., 1982) in Israel. Both of these predominantly selfing species have multilocus combinations of alleles that are predictably found in specific microhabitats.

Isozyme loci have been used to document genetic changes over time in an artificial population of barley (*Hordeum vulgare*) which was constructed by crossing all combinations of several varieties.

The first study documented changes in the genetic composition of two of these composite crosses over several generations. Significant changes in allele frequencies were observed, and a rapid increase in gametic phase disequilibrium was documented (Clegg et al., 1972). Selection has evidently favored the same multilocus combinations in these two independently derived composite crosses. A second study monitored changes in allele and genotype frequencies between different stages of the barley life cycle (Clegg et al., 1978). Multilocus genotype frequencies changed from stage to stage indicating that selection was acting either on the allozyme alleles directly or on the chromosomes that they marked. Clegg and Allard (1973) also documented changes in genotype frequencies at three esterase loci during transitions between seedling and adult and adult and seedling stages in a natural population of *Avena barbata*. Heterozygous genotypes were favored in the viability phase of the life cycle but were not shown to be favored in the fecundity stage. In a third study Burdon et al. (1983) monitored changes in allozyme genotype frequencies between the seedling and adult stages of the life cycle of *Echium plantagineum*. These studies, although providing no direct evidence of selection acting on allozyme loci, illustrate the usefulness of isozyme loci as markers to monitor changes in the genetic composition of populations.

Isozyme markers have also proved useful in studies of the effects of heterozygosity on fitness. There is increasing evidence that more highly heterozygous individuals are favored in plant populations. Since this topic is reviewed elsewhere (Mitton and Grant, 1984; see chapter by Mitton in this volume) suffice it to say that a heterozygote advantage has been reported in species as different as the annual *Avena barbata* (Marshall and Allard, 1970; Hamrick and Allard, 1972; Hamrick and Holden, 1979), the herbaceous perennial *Liatris cylindracea* (Schaal and Levin, 1976), and several species of *Pinus* (*P. ponderosa*, Farris and Mitton, 1984; *P. attenuata*, Strauss, 1986, 1987; *P. rigida*, Ledig et al. 1983; Bush et al., 1987; and *P. contorta*, Knowles and Mitton, 1980).

As mentioned previously, one of the main problems of using isozyme traits to document selection directly is that it is poorly understood how different forms of a locus affect individual fitness. Furthermore, it is difficult in natural populations to insure that the isozyme locus is not linked to some other gene (hitchhiking) at which selection is acting (Lewontin, 1974; Hedrick, 1980, 1982). Whereas it is rather easy to develop adaptive scenarios for morphological and physiological traits, it is often difficult to demonstrate a genetic component to variation in these traits. Much of the difficulty arises from a lack of knowledge about the genetic relationship of plants in the field. Without knowledge of the genetic correlations among individuals one must laboriously establish transplant gardens to demonstrate the genetic basis of phenotypic variation. However, if the parents of individuals are known it should not be difficult to demonstrate that progeny of certain individuals survive, grow, or reproduce better than progeny with a different pedigree. By applying paternity analysis techniques to various stages of the life cycle, significant insights should be gained into the magnitude and direction of the selective processes that shape the genetic structure of populations. Male reproductive success can be monitored directly by identifying individual contributions to the effective pollen pool. Similar analyses can be done on a random sample of dispersed seeds to identify their parents. If the species is dioecious or if haploid female gametophytic tissue is available as it is in many conifer seeds, male and female parents can be identified, and male and female reproductive success can be measured. With bisexual species, success as male or female parents cannot be separated but an individual's contribution to the gamete pool can be calculated. Finally, similar analyses can be applied to established individuals of various ages to determine whether individuals with parents in common survive, grow, or reproduce better than individuals with a different set of parents. In other words, it should be feasible to document whether certain full- or half-sib families demonstrate higher fitness than other sibships. It may also be possible to document whether certain sibships appear to perform better in specific microhabitats than other sibships.

INTERACTIONS AMONG EVOLUTIONARY FACTORS

In the sections above, plant mating systems, gene flow, and selection were treated separately to illustrate how allozyme variation can be used to study evolutionary processes. In actuality these factors interact to produce genetic structure in plant populations. Until recently, studies attempting to unravel the contributions of each factor were difficult because, as we have seen, a knowledge of the genetic relationships of individuals and their parentage is essential. With the application of paternity analyses and genealogical reconstruction techniques the evolutionary factors that affect genetic structure can be quantified, and better predictions of the development and maintenance of genetic structure can be made. With adequate levels of allozyme variation, detailed descriptions of the breeding structure can be made. Seed dispersal distances and patterns can be discerned and the action of natural selection can be documented and quantified.

To my knowledge there are no comprehensive studies at the level of detail that is now possible. T. Meagher (personal communication) has data on a *Chamaelirium luteum* population that includes many but not all of the parameters discussed. Hamrick and Loveless (1986) developed a conceptual model to predict genetic structure based on variation in the breeding structure, seed dispersal, and selection. Breeding structure was divided into random mating and near-neighbor mating. The two patterns of seed dispersal were a leptokurtic pattern characteristic of wind-dispersed seeds and a patchy pattern typical of species dispersed by frugivorous animals. Selection was assumed to occur in one of three ways. First, mortality was assumed to be at random (i.e., no selection). Every individual was equally likely to die regardless of its genotype or spatial position. Second, seedlings produced by different parental combinations were assumed to have different fitnesses; some full-sib or half-sib families were more likely to survive than others. The third selection model assumed that survivorship of sibships was influenced by its genotype, and by microhabitat conditions.

The combination of random mating, leptokurtic seed dispersal, and random mortality was predicted to produce the least genetic structure within populations. As population thinning occurs, genetic structure should decay. At the other extreme the combination of patchy seed dispersal, near-neighbor mating, and the microhabitat selection of particular sibships should produce patches of related individuals. These predictions await empirical verification by detailed multilocus allozyme analyses.

CONCLUSIONS

The availability of allozyme loci to the plant evolutionary biologist has had a significant impact on our ability to study the evolutionary dynamics of natural plant populations. At every level of study, ranging from the relatively straightforward surveys of allozyme variation to the complex selection component analyses or breeding structure descriptions, our ability to obtain quantitative estimates of evolutionary parameters has been greatly enhanced. The greatest advantage of isozyme loci over other sorts of traits is the relative abundance of polymorphic loci which can be used as genetic markers. Probably the greatest impact of the availability of genetic markers has been on studies of the reproductive stage of the plant life cycle. Enough genetic markers are now available to obtain quantitative estimates of the breeding structure, male and female reproductive success, and seed distribution. Until recently, these factors were treated as "black boxes" whose interactions with selection and among themselves to produce genetic structure were the subject of speculation. Today it is realistic to expect to unravel the ways in which the various evolutionary mechanisms interact to generate and maintain genetic structure.

What will be the role of isozyme variation in the future? Will the DNA technology currently available to population geneticists replace isozyme markers in studies of genetic structure? To date, DNA analyses have not been successfully used in studies of the genetic structure of populations. Most of the early attention has been given to the easily extracted chloroplast DNA. Unfortunately, cpDNA has generally proven to be highly conserved with little variation at the population or species level. Chloroplast DNA has proven to be exceptionally useful in studies of higher order systematics, however. Recently, highly variable regions of nuclear DNA have been used to "finger print" varieties of cultivated crops (S. Tanksley, personal communication; G. Kochert, personal communication). Since this technology is reaching the stage where the investigator can expect to analyze several hundred individuals within a reasonable period of time, this procedure may prove to be cost-effective for studies of genetic structure. The possibility of using these highly variable regions in paternity or genealogical reconstruction analyses is truly exciting. But for many other types of studies isozyme analyses will remain the technique of choice. Future research on the genetic structure of plant populations will probably consist of studies that combine and integrate the strengths of isozyme and DNA analyses to examine the dynamic processes that shape the genetic structure of plant populations.

LITERATURE CITED

Adams, W. T., and R. W. Allard. 1982. Mating system variation in *Festuca microstachys*. *Evolution* 36: 591–595.

Allard, R. W., G. R. Babbel, M. T. Clegg, and A. L. Kahler. 1972. Evidence for coadaptation in *Avena barbata*. *Proc. Natl. Acad. Sci. USA* 69: 3043–3048.

———, and A. L. Kahler. 1972. Patterns of molecular variation in plant populations. *In* L. M. Le Cam, J. Neyman, and E. L. Scott [eds.], *Proc. 6th Berkeley Symp. Math. Stat. Prob.*, Vol. 5, *Darwinian, neo-Darwinian, and non-Darwinian evolution*, 237–254. University of California Press, Berkeley.

———, and P. L. Workman. 1963. Population studies in predominantly self-pollinated species. IV. Seasonal fluctuations in estimated values of genetic parameters in lima bean populations. *Evolution* 17: 470–480.

Antonovics, J. 1971. The effects of a heterogeneous environment on the genetics of natural populations. *Amer. Sci.* 59: 593–599.

———, A. D. Bradshaw, and R. G. Turner. 1971. Heavy metal tolerance in plants. *Adv. Ecol. Res.* 7: 1–85.

Barton, N. H., and M. Slatkin. 1986. A quasi-equilibrium theory of the distribution of rare alleles in a subdivided population. *Heredity* 56: 409–415.

Beattie, A. J., and D. C. Culver. 1979. Neighborhood size in *Viola*. *Evolution* 33: 1226–1229.

Bekele, E. 1984. Relationships between morphological variance, gene diversity and flavonoid patterns in the land race populations of Ethiopian barley. *Hereditas* 100: 271–294.

Bradshaw, A. D. 1971. Plant evolution in extreme environments. *In* R. Creed [ed.], *Ecological genetics and evolution*, 20–50. Blackwell Press, Oxford.

Brown. A. H. D. 1979. Enzyme polymorphism in plant populations. *Theor. Pop. Biol.* 15: 1–42.

———, and R. W. Allard. 1970. Estimation of mating systems in open-pollinated maize populations using isozyme polymorphisms. *Genetics* 66: 133–145.

———, and M. T. Clegg. 1983. Isozyme assessment of plant genetic resources. *In* Isozymes: *current topics in biological and medical research*, Vol. 11: *Medical and other applications*, 285–295. A. R. Liss. Inc., New York.

———, M. W. Feldman, and E. Nevo. 1980. Multilocus structure of natural populations of *Hordeum spontaneum*. *Genetics* 96: 523–536.

———, A. C. Matheson, and K. C. Eldridge. 1975. Estimation of the mating system of *Eucalyptus obliqua* L'Herit. by using allozyme polymorphisms. *Austral. J. Bot.* 23: 931–949.

———, E. Nevo, D. Zohary, and O. Dagan. 1978. Genetic variation in natural populations of wild barley (*Hordeum spontaneum*). *Genetica* 49: 97–108.

Burdon, J. J., D. R. Marshall, and A. H. D. Brown. 1983. Demographic and genetic changes in populations of *Echium plantagineum*. *J. Ecol.* 71; 667–679.

Bush, R. M., P. E. Smouse, and F. T. Ledig. 1987. The fitness consequences of multiple-locus heterozygosity: the relationship between heterozygosity and growth rate in pitch pine (*Pinus rigida* Mill). *Evolution* 41: 787–798.

Cheliak, W. M., B. P. Dancik, K. Morgan, F. C. H. Yeh, and C. Strobeck. 1985. Temporal variation and the mating system in a natural population of jack pine. *Genetics* 109: 569–584.

Clegg, M. T. 1980. Measuring plant mating systems. *BioScience* 30: 814–818.

———, and R. W. Allard. 1972. Patterns of genetic differentiation in the slender wild oat species *Avena barbata*. *Proc. Natl. Acad. Sci. USA* 69: 1920–1924.

_____, and _____. 1973. Viability versus fecundity selection in the slender wild oat, *Avena barbata* L. *Science* 181: 667–668.

_____, _____, and A. L. Kahler. 1972. Is the gene the unit of selection? Evidence from two experimental plant populations. *Proc. Natl. Acad. Sci. USA* 69: 2474–2478.

_____, A. L. Kahler, and R. W. Allard. 1978. Estimation of life cycle components of selection in an experimental plant population. *Genetics* 89: 765–792.

Coyne, J. A., and B. Milstead. 1987. Long distance migration of *Drosophila*. 3. Dispersal of *D. melanogaster* alleles from Maryland orchard. *Amer. Naturalist* 130: 70–82.

Ehrlich, P. R., and P. H. Raven. 1969. Differentiation of populations. *Science* 165: 1228–1231.

El-Kassaby, Y. A., and O. Sziklai. 1982. Genetic variation of allozyme and quantitative traits in a selected Douglas-Fir (*Pseudotsuga menziesii* var. *menziesii* (Mirb.) Franco) population. *For. Ecol. Manag.* 4: 115–126.

Ellstrand. N. C. 1984. Multiple paternity within the fruits of the wild radish, *Raphanus sativus*. *Amer. Naturalist* 123: 819–828.

_____, and K. W. Foster. 1983. Impact of population structure on the apparent outcrossing of grain sorghum (*Sorghum bicolor*). *Theor. Appl. Genet.* 66: 323–327.

_____, and D. L. Marshall. 1985. Interpopulation gene flow by pollen in wild radish, *Raphanus sativus*. *Amer. Naturalist* 126: 606–616.

_____, and _____. 1986. Patterns of multiple paternity in populations of *Raphanus sativus*. *Evolution* 40: 837–842.

_____, A. M. Torres, and D. A. Levin. 1978. Density and the rate of apparent outcrossing in *Helianthus annuus* (Asteraceae). *Syst. Bot.* 3: 403–407.

Ennos, R. A., and M. T. Clegg. 1982. Effect of population substructuring on estimates of outcrossing rate in plant populations. *Heredity* 48: 283–292.

Farris, M. A., and J. B. Mitton. 1984. Population density, outcrossing rate and heterozygote superiority in ponderosa pine. *Evolution* 38: 1151–1154.

Felsenstein, J. 1986. Population differences in quantitative characters and gene frequencies: a comment on papers by Lewontin and Rogers. *Amer. Naturalist* 127: 731–732.

Friedman, S. T., and W. T. Adams. 1985. Estimation of gene flow into two seed orchards of loblolly pine (*Pinus taeda* L.). *Theor. Appl. Genet.* 69: 609–615.

Fyfe, J. L., and N. T. J. Bailey. 1951. Plant breeding studies in leguminous forage crops. I. Natural crossbreeding in winter beans. *J. Agric. Sci.* 41: 371–378.

Giles, B. A. 1984. A comparison between quantitative and biochemical variation in the wild barley *Hordeum murinum*. *Evolution* 38: 34–41.

Gottlieb, L. D. 1981. Electrophoretic evidence and plant populations. *Prog. Phytochem.* 7: 1–46.

Hagman, M., and L. Mikkola. 1963. Observations on cross-, self-, and inter-specific pollinations in *Pinus peuce* Griseb. *Silv. Genet.* 12: 73–79.

Hamrick, J. L. 1982. Plant population genetics and evolution. *Amer. J. Bot.* 69: 1685–1693.

_____. 1987. Gene flow and distribution of genetic variation in plant populations. In K. Urbanska [ed.], *Differentiation patterns in higher plants*, 53–67. Academic Press, New York.

_____, and R. W. Allard. 1972. Microgeographical variation in allozyme frequencies in *Avena barbata*. *Proc. Natl. Acad. Sci. USA* 69: 2100–2104.

_____, and _____. 1975. Correlations between quantitative characters and enzyme genotypes in *Avena barbata*. *Evolution* 29: 438–442.

_____, and G. B. Griswold. 1989. Association between Slatkin's measure of gene flow and the dispersal ability of plant species. *Amer. Naturalist*. In press.

_____, and L. R. Holden. 1979. Influence of microhabitat heterogeneity on gene frequency distribution and gametic phase disequilibrium in *Avena barbata*. *Evolution* 33: 521–533.

_____, Y. B. Linhart, and J. B. Mitton. 1979. Relationships between life history characteristics and electrophoretically-detectable genetic variation in plants. *Ann. Rev. Ecol. Syst.* 10: 173–200.

_____, and M. D. Loveless. 1986. The influence of seed dispersal mechanisms on the genetic structure of plant populations. In A. Estrada and T. H. Fleming [eds.], *Frugivores and seed dispersal*, 211–223. Dr. W. Junk Publ., The Hague, Netherlands.

———, and ———. 1989. Associations between the breeding system and the genetic structure of tropical tree populations. In J. Bock and Y. B. Linhart [eds], *Evolutionary ecology of plants*. Westview Press, Boulder, CO. In press.

———, J. B. Mitton, and Y. B. Linhart. 1981. Levels of genetic variation in trees: influence of life history characteristics. In M. T. Conkle, [ed.], *Isozymes of north-american forest trees and forest insects*, 35–41. Pacific SW For. Range Expt. Sta. Tech. Report #48.

———, and A. Schnabel. 1985. Understanding the genetic structure of plant populations. Some old problems and a new approach. In H.-R. Gregorius [ed.], *Population genetics in forestry*, 50–70. Springer-Verlag, Berlin.

Handel, S. N. 1982. Dynamics of gene flow in an experimental population of *Cucumis melo* (Cucurbitaceae). *Amer. J. Bot.* 69: 1538–1546.

———. 1983. Pollination ecology, plant population structure, and gene flow. In L. Real [ed.], *Pollination biology*, 163–211. Academic Press, New York.

Hedrick, P. W. 1971. A new approach to measuring genetic similarity. *Evolution* 25: 276–280.

———. 1980. Hitchhiking: a comparison of linkage and partial selfing. *Genetics* 94: 791–808.

———. 1982. Genetic hitchhiking: a new factor in evolution? *BioScience* 32: 845–853.

Helenurm, K., and F. R. Ganders. 1985. Adaptive radiation and genetic differentiation in Hawaiian *Bidens*. *Evolution* 39: 753–765.

Kahler, A. L., M. T. Clegg, and R. W. Allard. 1975. Evolutionary changes in the mating system of an experimental population of barley (*Hordeum vulgare* L.). *Proc. Natl. Acad. Sci. USA* 72: 943–946.

Kimura, M., and T. Ohta. 1972. Population genetics, molecular biometry and evolution. In L. M. Le Cam, J. Neyman, and E. L. Scott [eds.] *Proc. 6th Berkeley Symp. Math. Stat. Prob.*, Vol. 5: *Darwinian, neo-Darwinian and non-Darwinian evolution*, 43–68. University of California Press, Berkeley.

Knowles, P., and J. B. Mitton. 1980. Genetic heterozygosity and radial growth variability in *Pinus contorta*. *Silv. Genet.* 29: 114–118.

Ledig, F. T., R. P. Guries, and B. A. Bonefeld. 1983. The relation of growth to heterozygosity in pitch pine. *Evolution* 37: 1227–1238.

Lertzman, K. P., and C. L. Gass, 1983. Alternative models of pollen transfer. In C. E. Jones and R. J. Little, Jr. [eds.], *Handbook of experimental pollination biology*, 474–489. Van Nostrand Reinhold, New York.

Levin, D. A. 1977. The organization of genetic variability in *Phlox drummondii*. *Evolution* 31: 477–494.

———. 1983. Plant parentage: an alternate view of the breeding structure of populations. In C. E. King and P. S. Dawson [eds.], *Population biology: retrospect and prospect*, 171–188. Columbia University Press, New York.

———, and H. W. Kerster. 1974. Gene flow in seed plants. *Evol. Biol.* 7: 139–220.

Lewontin, R. C. 1974. *The genetic basis of evolutionary change*. Columbia University Press, New York.

———. 1984. Detecting population differences in quantitative characters as opposed to gene frequencies. *Amer. Naturalist* 123: 115–124.

Lindgren, D. 1975. The relationship between self-fertilized, empty seeds and seeds originating from selfing as a consequence of polyembryony. *Studia Forestalia Suecica* Nr. 126.

Loveless, M. D., and J. L. Hamrick. 1984. Ecological determinants of genetic structure in plant populations. *Ann. Rev. Ecol. Syst.* 15: 65–95.

Marshall, D. L., and N. C. Ellstrand. 1985. Proximal causes of multiple paternity in wild radish, *Raphanus sativus*. *Amer. Naturalist* 126: 596–605.

———, and ———. 1986. Sexual selection in *Raphanus sativus*: Experimental data on nonrandom fertilization, maternal choice, and consequences of multiple paternity. *Amer. Naturalist* 127: 446–461.

Marshall, D. R., and R. W. Allard. 1970. Maintenance of isozyme polymorphisms in natural populations of *Avena barbata*. *Genetics* 66: 393–399.

Martins, P. S., and S. K. Jain. 1980. Interpopulation variation in rose clover: a recently introduced species in California rangelands. *J. Heredity* 71: 29–32.

Meagher, T. R. 1986. Analysis of paternity within a natural population of *Chamaelirium luteum* I. Identification of most likely male parents. *Amer. Naturalist* 128: 199–215.

———, and E. Thompson. 1986. The relationship between single parent and parent pair genetic likelihoods in genealogy reconstruction. *Theor. Pop. Biol.* 29: 87–106.

Mitton, J. B., and M. C. Grant. 1984. Associations among protein heterozygosity, growth rate and developmental homeostasis. *Ann. Rev. Ecol. Syst.* 15: 479–500.

Moran, G. F., and A. H. D. Brown. 1980. Temporal heterogeneity of outcrossing rates in alpine ash (*Eucalyptus delegatensis* R. T. Bak.). *Theor. Appl. Genet.* 57: 101–105.

Muller, G. 1977. Cross-fertilization in a conifer stand inferred from enzyme gene-markers in seeds. *Silv. Genet.* 26: 223–226.

Nei, M. 1972. Genetic distance between populations. *Amer. Naturalist* 106: 283–292.

———. 1973. Analysis of gene diversity in subdivided populations. *Proc. Natl. Acad. Sci. USA* 70: 3321–3323.

Nevo, E., E. Golenberg, A. Beiles, A. H. D. Brown, and S. Zohary. 1982. Genetic diversity and environmental associations of wild Wheat, *Triticum dicoccoides*, in Israel. *Theor. Appl. Genet.* 62: 241–254.

———, D. Zohary, A. H. D Brown, and N. Haber. 1979. Genetic diversity and environmental associations of wild barley, *Hordeum spontaneum*, in Israel. *Evolution* 33: 815–833.

Prentice, H. C. 1984. Enzyme polymorphism, morphometric variation and population structure in a restricted endemic, *Silene diclinis* (Caryophyllaceae). *Biol. J. Linn. Soc.* 22: 125–143.

Price, S. C., R. W. Allard, J. E. Hill, and J. Naylor. 1985. Associations between discrete genetic loci and genetic variability for herbicide reaction in plant populations. *Weed Sci.* 33: 650–653.

———, K. N. Schumaker, A. L. Kahler, R. W. Allard, and J. E. Hill. 1984. Estimates of population differentiation obtained from enzyme polymorphisms and quantitative characters. *J. Heredity* 75: 141–142.

Rogers, A. R. 1986. Population differences in quantitative characters as opposed to gene frequencies. *Amer. Naturalist* 127: 729–730.

Schaal, B. A. 1980. Measurement of gene flow in *Lupinus texensis*. *Nature* 284: 450–451.

———, and D. A. Levin. 1976. The demographic genetics of *Liatris cylindracea* Michx. (Compositae). *Amer. Naturalist* 110: 191–206.

Schemske, D. W., and R. Lande. 1985. The evolution of self-fertilization and inbreeding depression in plants. II. Empirical observations. *Evolution* 39: 41–52.

Schoen, D. J. 1982a. Male reproductive effort and breeding system in a hermaphroditic plant. *Oecologia* 53: 255–257.

———. 1982b. Genetic variation and the breeding system of *Gilia achilleifolia*. *Evolution* 36: 361–370.

———, and M. T. Clegg. 1984. Estimation of mating system parameters when events are correlated. *Proc. Natl. Acad. Sci. USA* 81: 5258–5262.

Singh, R. S., and L. R. Rhomberg. 1987. A comprehensive study of genic variation in natural populations of *Drosophila melanogaster*. I. Estimates of gene flow from rare alleles. *Genetics* 115: 313–322.

Slatkin, M. 1981. Estimating levels of gene flow in natural populations. *Genetics* 99: 323–335.

———. 1985. Rare alleles as indicators of gene flow. *Evolution* 39: 53–65.

Smyth, C. A., and J. L. Hamrick. 1987. Realized gene flow via pollen in artificial populations of musk thistle, *Carduus nutans*. *Evolution* 41: 613–619.

Snaydon, R. W., and M. S. Davies. 1972. Rapid population differentiation in a mosaic environment. II. Morphological variation in *Anthoxanthum odoratum*. *Evolution* 26: 390–405.

———, and ———. 1976. Rapid population differentiation in a mosaic environment. IV. Populations of *Anthoxanthum odoratum* at sharp boundaries. *Heredity* 37: 9–25.

Snyder, T. P., D. A. Steward, and A. F. Strickler. 1985. Temporal analysis of breeding structure in jack pine (*Pinus banksiana* Lamb.). *Canad. J. For. Res.* 15: 1159–1166.

Sorenson, F. C. 1982. The roles of polyembryology and embryo viability in the genetic system of conifers. *Evolution* 36: 725–733.

_____, and R. S. Miles. 1974. Self-pollination effects on Douglas-fir and ponderosa pine seeds and seedlings. *Silv. Genet.* 23: 135–138.

Strauss, S. H. 1986. Heterosis at allozyme loci under inbreeding and crossbreeding in *Pinus attenuata*. *Genetics* 113: 115–134.

_____. 1987. Heterozygosity and developmental stability under inbreeding and crossbreeding in *Pinus attenuata*. *Evolution* 41: 331–339.

Sun, M., and F. R. Ganders. 1988. Mixed mating systems in Hawaiian *Bidens* (Asteraceae). *Evolution*: 42: 516–527.

Sytsma, K. J., and B. A. Schaal. 1985. Genetic variation, differentiation, and evolution in a species complex of tropical shrubs based on isozymic data. *Evolution* 39: 582–593.

Turkington, R., and J. L. Harper. 1979. The growth, distribution, and neighbor relationships of *Trifolium repens* in a permanent pasture. IV. Fine-scale biotic differentiation. *J. Ecol.* 67: 245–254.

Waller, D. M. 1987. Is there disruptive selection for self-fertilization? *Amer. Naturalist* 128: 421–426.

Waples, R. S. 1987. A multispecies approach to the analysis of gene flow in marine shore fishes. *Evolution* 41: 385–400.

Warwick, S. I., and D. Briggs. 1978. The genecology of lawn weeds. I. Population differentiation in *Poa annua* L. in a mosaic environment of bowling green lawns and flower beds. *New Phytol.* 81: 711–723.

_____, and L. D. Gottlieb. 1985. Genetic divergence and geographic speciation in *Layia* (Compositae). *Evolution* 39: 1236–1241.

Wheeler, N. C., and R. P. Guries. 1982. Population structure, genic diversity, and morphological variation in *Pinus contorta* Dougl. *Canad. J. For. Res.* 12: 595–606.

Wright, S. 1951. The genetical structure of populations. *Ann. Eugen.* 15: 323–354.

CHAPTER 5

Isozyme Variation in Colonizing Plants

Spencer C. H. Barrett

Department of Botany, University of Toronto, Toronto, Ontario, Canada M5S 1A1

Joel S. Shore

Department of Biology, University of New Brunswick, Fredericton, New Brunswick, Canada E3B 6E1

The colonization of new environments is an integral feature of the biology of most plant species. However, a relatively small number of plants, commonly referred to as colonizers, weeds, or invading species, possess attributes that enable them to establish populations continuously in areas or habitats that they have not previously occupied. In comparison with other plant life forms, and considering their number, colonizing species have received disproportionate attention from biologists. This interest is probably because some are of economic importance in agriculture and, because, from an evolutionary perspective, others provide excellent experimental systems for microevolutionary studies. The rapid life cycles of many colonizers and their ability to invade diverse environments, often in a short period, provide population biologists with a series of evolutionary experiments that are not available in most other plant groups.

Prior to the advent of enzyme electrophoretic techniques for measuring the levels of genetic variation in plant populations, much of the focus on colonizing species involved the identification of ecological attributes responsible for colonizing success. Generalizations arising from the symposium on "The Genetics of Colonizing Species" (Baker and Stebbins, 1965) stimulated comparative work on a variety of plant species to determine the adaptive significance of variation in traits such as dormancy, rate of development, phenotypic plasticity, fecundity, reproductive effort, and seed dispersal (reviewed in Baker, 1974; Harper, 1977; Grime, 1979; Jain, 1979). More recently, studies of colonizing species have compared levels of genetic diversity in related species and examined the consequences of colonizing episodes on genetic variation and opportunities for evolutionary response in novel environments. Although it is generally recognized that colonizing species employ a diverse array of ecogenetic strategies for invading unoccupied territory, several recurrent patterns have emerged from electrophoretic studies conducted during the past decade. Polyploidy, uniparental reproductive systems, depauperate levels of genetic variation, marked interpopulation differentiation, and a high degree of multilocus association have commonly been reported in colonizing species (reviewed in Brown and Marshall, 1981; Rice and Jain, 1985; Barrett and Richardson, 1986).

In this chapter we review electrophoretic evidence from isozyme surveys to evaluate whether generalizations can be made as to the processes influencing population genetic structure and evolutionary response in colonizing plants. We also assess the problems associated with determining the relative importance of historical factors, reproductive traits, and environmental heterogeneity in affecting population genetic structure. Because many successful colonizers are of polyploid origin, part of our review deals with isozyme variation in polyploid colonizers and considers how polyploidy might contribute to the evolution of colonizing ability.

MEASUREMENT OF POPULATION GENETIC STRUCTURE

The measurement of plant population genetic structure using enzyme electrophoresis has received considerable attention during the past decade (see the chapters by Hamrick and Brown et al. in this volume). Brown and Weir (1983) considered the assumptions, advantages, and pitfalls of various measures as well as formulae for obtaining estimates of the relevant parameters and their variances. Here we briefly consider the parameters commonly estimated in studies of population genetic structure with emphasis on their use in colonizing species. In addition, we briefly consider the application of the methods to populations of polyploid plants.

Estimates of within-population genetic diversity are summary measures made from a data set composed of genotype frequencies for several isozyme loci, obtained by assaying a sample of individuals from a population. As a result of codominant allelic expression at isozyme loci, the data are easily translated into allele frequencies for the isozyme loci. Brown (1975) and Brown and Weir (1983) have considered a range of experimental designs for the estimation of these genetic parameters.

Six statistics are commonly estimated from isozyme data: the percentage of loci that are polymorphic, the average number of alleles per locus, Nei's (1973) index of gene diversity, the Shannon-Weaver information index, the observed proportion of heterozygosity, and Wright's fixation index. The first four statistics are indices of genetic diversity that express, to a greater or lesser extent, the amount of allelic richness in the population and the evenness in distribution of allele frequencies (Brown and Weir, 1983). The final two statistics provide summary measures of the distribution of genotype frequencies in the population. Observed heterozygosity is the proportion of heterozygotes at each locus averaged over all loci. Wright's fixation index measures the deviation from a panmictic genotype distribution. It may be adjusted for mixed mating systems to measure the deviation from neutral inbreeding equilibrium when estimates of mating system parameters are available (Brown and Weir, 1983).

Multilocus measures of variation are also available but are less commonly employed. Brown et al. (1980) suggested a composite measure of multilocus organization, the standardized variance of the distribution of the number of heterozygous loci in two randomly chosen gametes. This parameter provides a measure of multilocus heterozygosity. For inbreeding or clonal species with little isozyme variation it may also be possible to enumerate the number and frequency of multilocus genotypes that occur in a population because relatively few may occur.

In colonizing species it is of interest to assess the manner in which genetic diversity is partitioned among different colonizing populations or among founding and source populations. Statistics commonly employed to assess subpopulation structure include the F-statistics developed by Wright (1951) and recently re-examined by Weir and Cockerham (1984), and the diversity analysis of Nei (1973). More recently, the spatial structure of genetic variation has been investigated using spatial autocorrelation analysis (Sokal and Wartenberg, 1983). This method is particularly useful in the study of colonizing populations and has been applied to the study of spatial pattern of flower color morphs in weedy *Ipomoea purpurea* (Epperson and Clegg, 1986).

Polyploid Populations

The types of polyploidy were originally defined using cytogenetic criteria (Kihara and Ono, 1926) but some controversy still surrounds the classification of polyploids (Stebbins, 1984; Jackson, 1984). For our purposes, we are only concerned with the influence of meiotic pairing on genetic transmission and the partitioning of genetic variation within individuals, populations, and species. We simplify a complex situation that in reality is a continuum determined by both chromosomal and genetic factors. Indeed, the chromosomal system itself evolves and cannot be treated as a static entity (Darlington, 1958). We refer to polyploid populations that exhibit preferential pairing among homologous chromosomes and show only disomic inheritance as allopolyploids. We consider autopolyploids to be euploids that possess more than two copies of each homologous chromosome, show random pairing among homologues at meiosis, and exhibit polysomic inheritance (see also Stebbins, 1950).

All the statistics considered above can be applied to polyploid populations. In addition, two

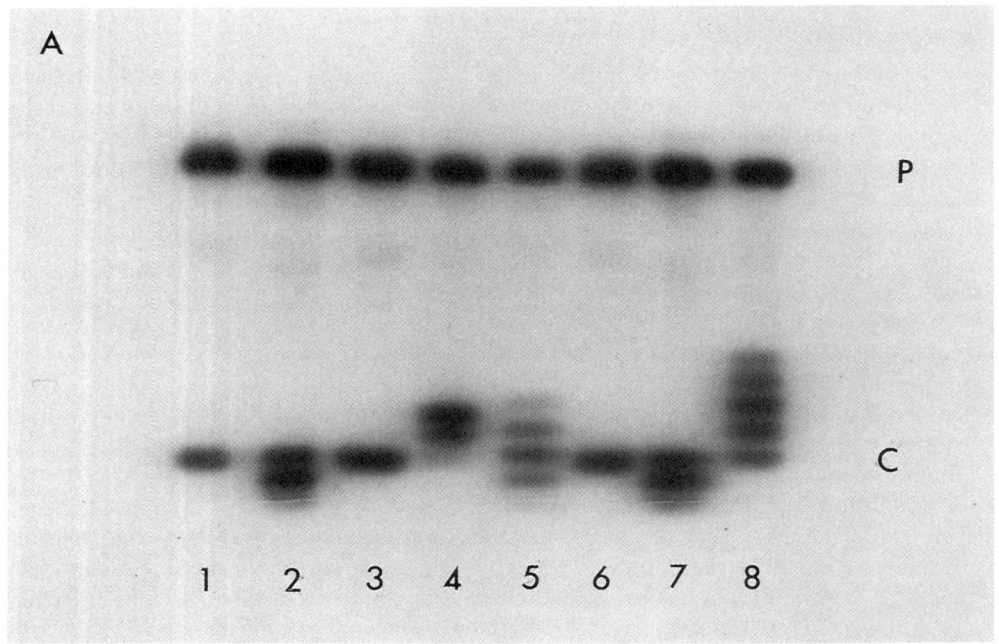

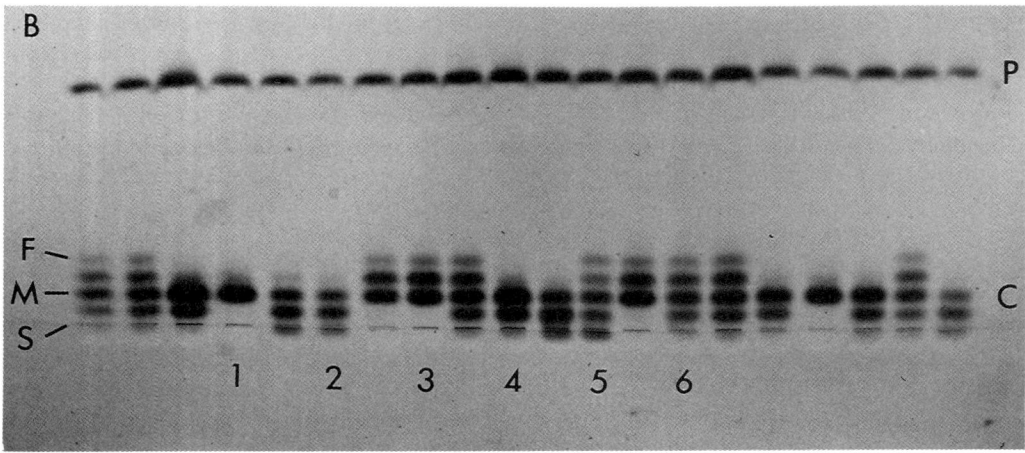

Fig. 5.1. Starch gels of *Turnera ulmifolia* assayed for glucosephosphate isomerase (GPI) activity. GPI is dimeric and two isozyme loci occur in diploid plants, one localized in the plastids (P) and the other in the cytosol (C). **A** For the cytosolic locus, lanes 1 and 2 are diploid homozygote and heterozygote respectively, lanes 3, 4 and 5 are three different genotypes from autotetraploid var. *elegans*, and lanes 6, 7 and 8 are three plants from different allohexaploid varieties of *T. ulmifolia*. The three genotypes in lanes 6, 7 and 8 are homozygous; no segregation is seen among selfed progeny. Genotypes in lanes 7 and 8 are "fixed heterozygotes". The number of loci coding for cytosolic GPI in the allohexaploids is not known with certainty. Lane 6 could result from one, two or three loci (having co-migrating gene products), lane 7 could be accounted for by two or three loci, and the banding pattern of lane 8 is consistent with three loci. Similarly, it is not clear how many loci code for the isozyme of GPI localized in the plastids of hexaploids; one, two or three loci are possible. **B** Segregation for three alleles of cytosolic GPI in autotetraploid *T. ulmifolia* var. *elegans*. The three alleles (F, M, S) code for polypeptides which, upon formation of homodimers, migrate to the positions indicated on the gel as F (fast allozyme), M (intermediate), and S (slow). Other bands are heterodimers. Six different genotypes occur on the gel for this tetrasomic locus. The lane marked 1 is a homozygous genotype, the remainder (2–6) are different heterozygotes. The genotypes are: 1-*MMMM*, 2-*MMSS*, 3-*FMMM*, 4-*MMMS*, 5-*FMSS*, 6-*FMMS*.

other measures might be useful when allopolyploids are studied. The first is the proportion of duplicate loci that exhibit "fixed heterozygosity" (e.g. Fig. 5.1A); that is, are monomorphic for different alleles. A second measure that can be employed in allopolyploids involves partitioning genetic variation into components that occur within genomes versus between homoeologous genomes (Brown and Marshall, 1981).

Few isozyme analyses have been undertaken on autopolyploid species (see below); however, all the statistics considered above can be applied to autopolyploid populations. Nei's index of gene diversity is applicable but cannot be regarded as a measure of panmictic heterozygosity, due to polysomic segregation. The extension of Wright's fixation index to autopolyploids has been considered by Wright (1969), neutral inbreeding equilibrium has been treated by Bennett (1968) and McConnell and Fyfe (1975), and mating system estimation by Bennett (1968) and Barrett and Shore (1987). Observed levels of heterozygosity may be determined by enumerating the number of heterozygotes versus homozygotes at a given gene locus.

FACTORS INFLUENCING POPULATION GENETIC STRUCTURE

The spatial and temporal organization of genetic variation within plant populations results from the joint action of mutation, migration, selection, and genetic drift. In colonizing species certain historical, life history, and ecological factors play a prominent role in determining the patterns of genetic diversity within and among populations. Because repeated colonizing episodes are a feature of invading species, genetic bottlenecks, founder effects, and drift are likely to play a more significant role in influencing genetic diversity than in many other plant life forms. Bottlenecks associated with long-distance founding events, the absence of repeated migration as a source of genetic enrichment, and the possibility of novel selection pressures in new environments are all likely to lead to a loss of genetic variation during colonizing events, particularly those that involve long-distance migration (Clegg and Brown, 1983; Barrett and Husband, 1989). In addition, demographic, life history, and reproductive features play an important role in influencing levels of genetic variation, as does the environment that a population occupies (Hamrick et al., 1979; Barrett, 1982; Loveless and Hamrick, 1984; Brown and Burdon, 1987). The complex forces that act upon genetic variation in populations of colonizing plants make it difficult to disentangle the relative importance of individual factors. Below we review several isozyme studies that have attempted to do so, and in addition, examine both large scale surveys of isozyme variation and studies of individual taxa to see whether recurrent patterns of genetic organization emerge.

Historical Factors

Migration and population establishment in colonizing species often involve a small number of immigrants. In theory this can lead to a loss of genetic variation through sampling effects. A small sample of individuals is unlikely to contain all of the variation present in the source population. Nei et al. (1975) studied the theoretical consequences of this process on genetic diversity and concluded that the level of variability declines, but that this depends not only on the number of immigrants involved, but also on the frequency of bottlenecks following establishment. Much of the literature on bottleneck effects relates to outbreeding organisms and considers the influence of small population size on levels of heterozygosity and the loss of rare alleles. However, in self-fertilizing plants with little heterozygosity, of more significance is the number of different populations in the source region that have supplied immigrants to the newly colonized area and the likelihood of cross-fertilization between them. Since many successful plant colonizers are primarily self-fertilizing, and single propagules are capable of founding populations following long-distance dispersal, genetic bottlenecks should be a standard expectation, particularly in cosmopolitan weeds.

Intercontinental migration of the annual tetraploid barnyard grass *Echinochloa microstachya* from North America to Australia provides evidence of the importance of genetic bottlenecks in selfing species. A survey of genetic variation in 20 North American populations of the species revealed a high degree of interpopulation genetic differentiation for isozymes and quantitative life history traits (S. C. H. Barrett and A. H. D. Brown, unpublished data). Each population was largely composed of a unique multilocus isozyme genotype. This pattern of genetic differentiation is anticipated in wide-ranging, selfing species. Echinochloa microstachya was introduced during this century to rice fields in New South Wales, probably as a contaminant of imported rice stocks from California (McIntyre and Barrett, 1985; Barrett, 1988a). Since introduction, E. microstachya has spread throughout the rice-growing area as a weed of rice. A survey of isozyme variation in Australian populations revealed a very different pattern from that observed in the native North American range. Of the 20 populations assayed, 18 were genetically uniform and composed of the same multilocus genotype. Two variant loci (HK, Lap) were evident in the remaining populations. Clearly, migration of E. microstachya to Australia from North America has been associated with a major genetic bottleneck, because Australian populations are almost entirely devoid of isozyme variation. Of particular interest was the finding that the predominant Australian genotype could be identified from among the North American sample of populations. The genotype occurred in a population from northern California close to Biggs Rice Experiment Station, the major exit point of cultivated rice varieties shipped to Australia in the 1920s. This illustrates the utility of isozymes as genetic markers for testing historical hypotheses concerned with migration and founder effects.

Colonization events may be associated with mating system shifts, and can complicate attempts to determine the influence of historical factors on the patterns of genetic variation in source and colonial populations (Brown and Marshall, 1981). This difficulty was encountered in a study of continental and island populations of Eichhornia paniculata, a short-lived annual or perennial, diploid emergent aquatic that colonizes seasonal pools, ditches, and rice fields in northeastern Brazil, Cuba, and Jamaica (Barrett, 1985; Glover and Barrett, 1987). A survey of isozyme variation in 11 populations from northeastern Brazil and Jamaica indicated the importance of both founder effects and increased levels of self-fertilization on genetic diversity. Populations from Jamaica were genetically depauperate with respect to isozyme variation, containing a significantly lower number of polymorphic loci, number of alleles per locus, mean observed heterozygosity, and genetic diversity than Brazilian populations (Table 5.1). These measures were significantly correlated with the outcrossing rate (t) of populations. Brazilian populations contained significantly more variation within populations than between populations, whereas the reverse situation occurred among Jamaican populations. This pattern is largely the result of differences in the mating systems of populations in the two regions (Glover and Barrett, 1986). The relatively low levels of isozyme variation in Jamaican populations of E. paniculata, compared to those in northeastern Brazil, probably result from a limited number of long-distance dispersal events to the island. However, following establishment, genetic drift and the primarily selfing habit of Jamaican plants probably contributed to reduced levels of genetic variation within populations.

Table 5.1. *Comparisons of genetic variation in continental (Brazil) and island (Jamaica) populations of* Eichhornia paniculata *based on a survey of 21 isozyme loci. After Glover and Barrett (1987)*

Parameter	Symbol	Brazil	Jamaica
% loci polymorphic per population	P	23.8	7.6
Observed % loci heterozygous	H_O	7.8	2.0
Total gene diversity	H_T	0.15	0.06
Gene diversity within populations	H_S	0.091	0.027
Gene diversity between populations	G_{ST}	0.40	0.57
Inter- to intra- population ratio	R_{ST}	0.81	1.7
Average genetic distance	d	0.085	0.049

Brazil $N = 6$ populations, Jamaica $N = 5$ populations

Recent studies of multilocus organization within and among Jamaican populations of E. paniculata provide evidence of the role of genetic drift in structuring genetic diversity in colonizing species (B. C. Husband and S. C. H. Barrett, unpublished data). Among five populations examined, 38 multilocus genotypes were identified with different genotypes predominating within each population

(Figure 5.2). The relatively small number of multilocus genotypes occurring in the five populations and their uneven distribution on the island suggest that populations have been founded by a small number of individuals, and restricted gene flow and inbreeding have preserved the specific allelic combinations found within each population.

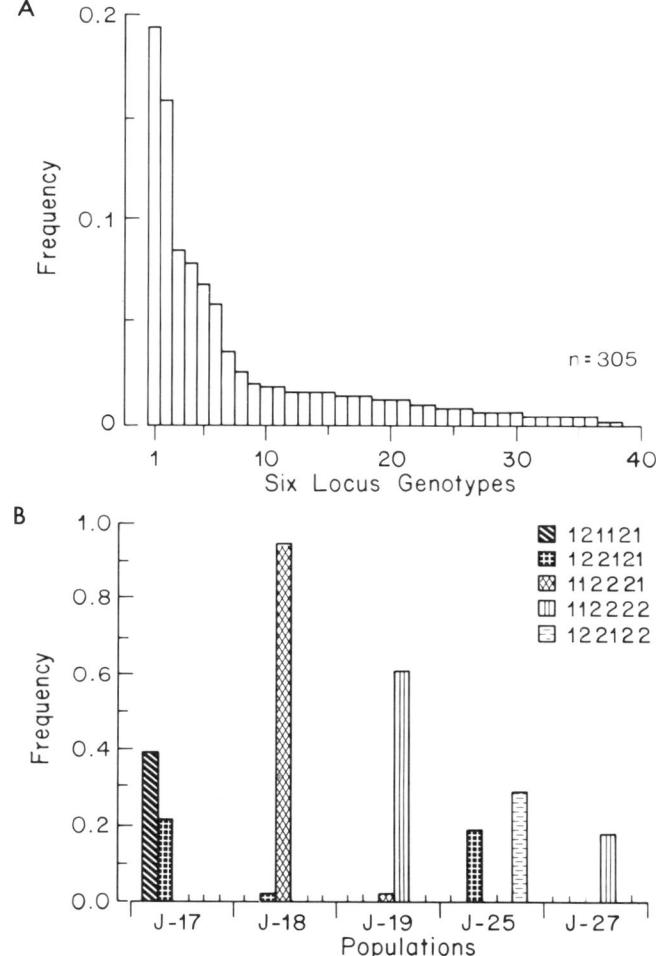

Fig. 5.2. Frequencies of multilocus genotypes in Jamaican populations of *Eichhornia paniculata*. **A** Frequency distribution of 38 six locus isozyme genotypes in five populations (n = 305 plants). **B** Frequencies of the five most common six locus genotypes in each of five Jamaican populations. Sample sizes for each population are 56, 43, 66, 68, 62 for J-17, J-18, J-19, J-25, and J-27 respectively. Polymorphic loci are *Got-1, 2, 3; Pgm-1, GPI-2; Per-1*. The five populations are different from those surveyed in Table 1.

To distinguish between the relative importance of historical factors and effects of the mating system on multilocus organization, examination of the degree of linkage disequilibrium among loci is required. This approach has been employed in both inbreeding and outcrossing colonists (reviewed in Brown, 1984). Isozyme surveys of inbreeders have generally revealed a limited number of multilocus associations with certain genotypes occurring more frequently than would be expected if alleles at different loci occurred at random with respect to one another (Allard et al., 1972; Brown et al., 1980; Warwick, 1989). In contrast, in *Echium plantagineum*, an insect-pollinated, outcrossing, annual weed introduced to Australia, only small, nonsignificant multilocus associations were detected (Brown and Burdon, 1983). In this species it appears that cross-pollination encourages sufficient recombination to remove any transient disequilibrium that might arise from bottlenecks in population size during early founding events.

Not all long-distance colonizing events result in reduced levels of genetic diversity in colonial populations in comparison with source populations. In outcrossing weeds that expand rapidly following establishment, differences in population genetic structure between the native and introduced range may be less marked. Warwick et al. (1987a) found little difference in the levels and patterns of genetic diversity among native European and introduced Canadian populations of the self-incompatible, annual Apera spica-venti (silky bentgrass) (Table 5.2).

A difficulty in comparing native and introduced populations of colonizing plants arises out of the choice of material sampled from the presumed source region. When the species in question has a widespread distribution, as in many cosmopolitan weeds of Eurasian origin, problems of sampling inevitably occur unless historical information on the source of immigrants is available. Ideally, chronological genetic studies of introduced populations into new habitats or regions should be conducted so that the dynamics of evolutionary change under colonization can be studied more precisely.

Table 5.2. *Comparisons of genetic variation in native European and introduced Canadian populations of* Apera spica-venti *(silky bentgrass) based on a survey of 12 isozyme loci. After Warwick et al. (1987)*

Region	N (pops)	P	A	H_O	H_T	H_S	G_{ST}
Europe	6	0.62	2.5	0.23	0.21	0.20	0.01
Canada	9	0.57	2.5	0.23	0.21	0.21	0.02

Reproductive Systems

Flowering plants display a wide range of reproductive systems, with colonizing species forming a diverse but distinctive sample. Most annual weeds are predominantly self-fertilizing and are incapable of clonal propagation. Perennial species display a diversity of reproductive systems, including selfing, outcrossing, and apomixis. Many successful perennial colonizing species possess well-developed clonal propagation and, particularly in aquatic weeds, sexual reproduction occurs only rarely (Barrett, 1982; Wain et al., 1985). As is true of all plants, the variety of reproductive modes in colonizing species is paralleled by differences in population genetic structure. This is because the mating system governs the character of genetic transmission in populations, and the occurrence of sexual reproduction provides opportunities for recombination and genetic experimentation (see the chapters by Brown et al. and Hamrick in this volume).

Large-scale surveys of isozyme variation in plants with differing reproductive systems have provided evidence of the important role of the mating system in influencing the amount of genetic variation and its organization within and among populations (Loveless and Hamrick, 1984, see chapter by Hamrick in this volume). Inbreeding species exhibit lower polymorphism, lower heterozygosity, and more pronounced interpopulation differentiation than outcrossers. In clonal and apomictic plants a range of population genetic structures is evident. In a review of 27 studies, Ellstrand and Roose (1987) found that multiclonal populations of intermediate diversity and evenness were most commonly found. Among several of the colonizing species in their sample (e.g., Taraxacum spp., Cyperus spp.), however, populations containing little or no genetic diversity were encountered (Van Oostrum et al., 1985; Horak and Holt, 1986).

Investigation of the reproductive systems of colonizing species is critical to understanding population genetic structure. A recent study (Wolf et al., 1988) of Bracken Fern, *Pteridium aquilinum*, one of the world's most widespread plant species, indicates how earlier misinterpretation of its reproductive biology led to a false picture of the clonal structure of populations. Although clonal propagation is a prominent feature of bracken populations, a survey of genetic variation at 13 isozyme loci in four geographically distinct populations (two in the U.K. and two in the U.S.A.) points to the importance of sexual reproduction. The survey revealed a high level of genetic diversity with 61% of loci polymorphic and an average of 3.5 alleles per locus. The results caution against the assumption that plants with prolific clonal propagation, such as bracken, will necessarily form vast areas of genetic monotony.

Isozyme studies have also challenged the assumption that reduced levels of genetic diversity are likely to occur in asexual plants (Richards, 1986). A particularly intriguing example concerns the sterile triploid grass, *Puccinellia* × *phryganodes,* a widely distributed colonizer of Arctic coastal wetlands. Although plants undergo vigorous clonal growth, flowering is infrequent, and seed set has never been reported in North American populations. Jefferies and Gottlieb (1983) therefore predicted that low levels of genetic variability would occur both within and between populations. Unexpectedly, an examination of isozyme patterns in clones from three widely separated populations in Arctic Canada indicated high levels of genetic diversity with many clones genetically unique. Jefferies and Gottlieb (1983) suggest that, despite triploidy, some residual fertility and occasional seed production may occur in Arctic populations of *P. phryganodes*. It is also possible that sexual tetraploids reported from northern Europe occur in North America and have contributed to the patterns of genetic variation observed. Regardless of the origins of genetic diversity in *P. phryganodes,* it appears that local triploid populations in the Hudson Bay Lowlands may contain sufficient genetic variation for adaptive responses to the impact of geese grazing (Sadul, 1987).

A significant number of isozyme surveys of primarily selfing colonizing species has revealed populations that are either genetically uniform or contain very low levels of polymorphism and heterozygosity (Table 5.3; see also Brown and Marshall, 1981; Barrett, 1982; Barrett and Richardson, 1986). Most studies were conducted in the introduced rather than the native range of colonizers, and

Table 5.3. *Taxa of colonizing plants displaying extensive monomorphism at isozyme loci both within and among populations*

Taxon	Life history	Ploidal level	Region sampled	Major habitat	Source
Abutilon theophrasti	A	P	E. Canada	agrestal	Warwick, 1988
Avena barbata	A	P	S. California	ruderal	Allard et al., 1978
Bromus tectorum	A	D	N.W. U.S.A.	rangelands	S. Novak and R. Mack (unpubl. data)
Chenopodium spp.	A	D	W. U.S.A.	ruderal	Crawford and Wilson, 1979
Datura stramonium	A	D	E. Canada	agrestal	Warwick, 1988
Echinochloa spp.	A	P	Australia	agrestal	S. C. H. Barrett and A. H. D. Brown (unpubl. data)
Emex spinosa	A	D	Australia	agrestal	Marshall and Weiss, 1982
Erodium spp.	A	P	California	ruderal	S. Novak (unpubl. data)
Hordeum murinum	A	D	Europe	ruderal	Giles, 1983
Hydrocharis morsus-ranae	P	D	E. Canada	aquatic	Scribailo et al., 1984
Panicum miliaceum	A	P	S. Ontario	agrestal	Warwick, 1988
Polygonum lapathifolium	A	P	E. Canada	ruderal	L. Consaul (unpubl. data)
Salicornia spp.	A	D/P	N. temperate	salt marsh	Wolff and Jefferies, 1987a, b
Senecio viscosus	A	P	N. Europe	ruderal	Koniuszek and Verkeij, 1982
Setaria faberi	A	P	E. Canada	agrestal	Warwick, 1988
Sorghum halepense	A	P	E. Canada	agrestal	Warwick, 1988
Taraxacum obliquum	A	P	N. Europe	sand dunes	Van Oostrum et al., 1985
Typha spp.	P	P?	E. U.S.A.	wetlands	Sharitz et al., 1980
Xanthium spp.	A	P	Australia	ruderal	Moran and Marshall, 1978

Life history—annual (A) or perennial (P); Ploidy—diploid (D) or polyploid (P)

the majority of species occurred in either agrestral or ruderal habitats.

Agricultural weeds are especially susceptible to the processes that lead to a loss of genetic variation in populations. Introduction of crop weeds most commonly results from contamination of seed lots with weed seeds. Often few propagules are introduced, and where long distance shipments are involved, the likelihood of recurrent invasion is low. The selfing habit of many weeds enables establishment after long distance dispersal, resulting in the build-up of highly homozygous populations in the introduced range. In addition, the transient nature of many weed populations, particularly those subject to eradication by weed control practices, can prevent the build-up of large, stable population systems capable of maintaining large stores of genetic variation. Some of the factors influencing the population genetics and evolution of agricultural weeds are reviewed by Barrett (1988a).

A recent survey (Warwick, 1989) of isozyme variation in five agricultural weeds in Canada clearly illustrates that high levels of genetic variation are not a prerequisite for colonizing success. *Abutilon theophrasti*, *Datura stramonium*, *Panicum miliaceum*, *Setaria faberi*, and *Sorghum halepense* are weeds of maize and soybean fields in the corn belt region of the U.S.A. During the past 15–20 years they have extended their ranges northward into southern Ontario, in association with the expansion of maize and soybean production. All species are self-pollinating annuals that display a high degree of phenotypic plasticity, and four of the five are polyploid. Electrophoretic surveys of populations revealed a striking lack of genetic polymorphism at isozyme loci in all five species. Most populations were composed of one, or more rarely, several multilocus genotypes (Table 5.4).

Table 5.4. Comparisons of isozyme variation in five annual, self-fertilizing weeds of corn and soybean from southern Ontario, Canada. After Warwick (1989)

	Abutilon theophrasti	Panicum miliaceum	Setaria faberi	Sorghum halepense	Datura stramonium
Chromosome number	$2n = 24$	$2n = 36$	$2n = 36$	$2n = 40$	$2n = 24$
Number of populations	39	39	8	13	9
Number of loci	27	19	24	21	22
Number (%) of loci monomorphic	25 (93%)	18 (95%)	21 (88%)	18 (86%)	22 (100%)
Number (%) of loci polymorphic	2 (7%)	1 (5%)	3 (12%)	3 (14%)	0
Number (%) of duplicated loci with enzyme multiplicity	14 (52%)	8 (42%)	13 (54%)	3 (14%)	2 (9%)
Number of multilocus genotypes	4	2	9	10	1

Not all cases of isozyme uniformity in colonizing species involve weeds of agricultural or ruderal habitats. Several aquatic and wetland species with depauperate levels of isozyme variation have been reported, including *Typha* spp. (Mashburn et al., 1978) and *Hydrocharis morsus-ranae* (Scribailo et al. 1984). The general topic has recently been reviewed by Les (1988). Although *Typha* and *Hydrocharis* are both capable of extensive clonal propagation, the observed genetic uniformity is unlikely to be due entirely to an absence of sexual reproduction, because seedlings derived from populations are also largely devoid of isozyme variability.

One of the most striking examples of genetic uniformity in wetland species involves the annual halophyte *Salicornia*, which colonizes bare sediments in coastal salt marshes, brackish pools, and inland saline lakes throughout the north temperate zone. Jefferies and Gottlieb (1982) and Wolff and Jefferies (1987a, b) surveyed electrophoretic variation in a large number of European and North American populations of diploid and tetraploid taxa within the *Salicornia europaea* species complex. They found an almost complete absence of isozyme variation within taxa, although different species were fixed for different alleles at several loci. Although *Salicornia* is largely inbreeding through the production of cleistogamous flowers, this characteristic alone is unlikely to account for the extensive regional monomorphism of each species. Wolff and Jefferies (1987a) suggested that the electrophoretically uniform populations observed in northeastern North America may have arisen by a severe genetic bottleneck associated with the destruction of coastal habitats following glacial advance during the Pleistocene. They also speculated that the highly specialized salt marsh environment, with its high predictability and low biotic diversity, may account, in part, for the paucity of isozyme variation. The idea that open habitats with restricted biotic diversity select for low genetic polymorphism is discussed further below.

Although populations of predominantly selfing species, such as *Salicornia,* are often genetically uniform at isozyme loci, this does not necessarily imply that they are devoid of variation at other gene loci. In studies of quantitative life history traits, considerable inter- and intrapopulation genetic variation has been revealed in colonizing species with little isozyme variation, (Moran and Marshall, 1978; Moran et al., 1981; Giles, 1983; Warwick and Black, 1986.; Warwick et al., 1987b; S. C. H. Barrett and A. H. D. Brown, unpublished data). The difficulties of comparing patterns of genetic variation at isozyme loci and quantitative traits have been reviewed by Lewontin (1984), Price et al. (1984), Brown and Burdon (1987) and are discussed further in the chapter by Hamrick (this volume).

Environmental Factors

Whereas historical factors and the specific life history and reproductive characteristics of colonizers play important roles in determining the patterns of genetic diversity within and among populations, the nature of the physical and biotic environment that a colonist occupies will ultimately determine the maintenance and adaptive significance of this variation. Environmental factors that potentially affect the relative fitness of genotypes include physical factors such as temperature, moisture, and soil type; and biotic factors such as interspecific competitors, pests, and diseases.

It has been suggested that the open early-successional habitats in which most colonizing species appear are primarily influenced by physical factors and tend to be relatively homogeneous environments. In contrast, in later successional stages biotic complexity results in a high degree of environmental heterogeneity (Hamrick et al., 1979). Theoretical studies predict a positive relationship between the levels of genetic variation in populations and the degree of environmental heterogeneity (Levins, 1968; Antonovics, 1971; Hedrick et al., 1976). Isozyme surveys of species classified by successional stage indicate that species inhabiting early successional stages exhibit significantly lower levels of genetic diversity and more pronounced interpopulation differentiation than those from mature climax communities (Hamrick et al., 1979; Loveless and Hamrick, 1984; see the chapter by Hamrick in this volume). Although these patterns are consistent with theoretical expectations, the problem of intercorrelation among life history traits and successional stage makes it exceedingly difficult to resolve fully the direction of causality between environmental factors and genetic variation.

One way to reduce some of the difficulties of interpretation posed by large surveys encompassing species of diverse phylogenetic histories is to examine population genetic structure in closely related taxa. Although this approach controls many variables, it still may not clearly distinguish between the relative importance of the different factors governing genetic variation.

An example of an attempt to examine the relationship between environmental heterogeneity and genetic variation in colonizing species involves a comparative study of isozyme variation in Californian populations of the annual, selfing, hexaploid barnyard grasses *Echinochloa crus-galli* and *E. oryzoides* (S. C. H. Barrett and A. H. D. Brown, unpublished data). The former can be viewed as a generalist weed; it is cosmopolitan in distribution and invades a wide range of disturbed environments (Maun and Barrett, 1986). In contrast, *E. oryzoides* is a highly specialized rice mimic that has invaded most of the world's rice-growing regions as a seed contaminant of imported rice stocks; it is largely confined in distribution to flooded rice fields (Barrett, 1983). Like most annual monocultures, the Californian rice agroecosystem is deliberately made uniform by the grower and is highly predictable in space and time. In contrast, ruderal habitats such as paths, ditches, and open waste ground are more complex both in terms of physical and biotic diversity. As a result of the differences in environmental heterogeneity that occur between agricultural and ruderal habitats, it was predicted that populations of the two species would display different amounts of genetic diversity within populations (Barrett, 1982).

At each of 10 rice field sites in the Central Valley of California, open-pollinated families of the two species were collected. Populations of *E. crus-galli* were sampled from the mosaic of disturbed habitats around the edges of rice fields, whereas those of E. oryzoides were obtained from the flooded interior of rice fields. Both species were relatively low in genetic diversity with virtually no heterozygosity at polymorphic loci and a high degree of genetic differentiation among populations. This finding is consistent with data from other annual colonizing species that are primarily self-fertilizing. At most of the sites that were sampled, populations of the generalist E. crus-galli were more variable both in terms of the proportion of isozyme loci that were polymorphic and the number of alleles present within populations (Table 5.5). Differences in genetic polymorphism for loci controlling the enzyme alcohol dehydrogenase (Adh) were particularly striking in the two species (Figure 5.3). In E. crus-galli, 12 homozygous multilocus genotypes were evident in the sample as a result of polymorphism at one to three loci. Populations contained different numbers of Adh genotypes, ranging from one to six. In contrast, all but one population of E. oryzoides contained the same multilocus Adh genotype. The exceptional population was fixed for a variant allele at a single locus.

Table 5.5. Comparison of genetic variation in Californian rice-field populations of the generalist weed Echinochloa crus-galli and the specialist, crop mimic, Echinochloa oryzoides based on 31 and 32 isozyme loci, respectively

	Echinochloa crus-galli			Echinochloa oryzoides		
Site	Number of polymorphic loci	Number of alleles at polymorphic loci	Diversity	Number of polymorphic loci	Number of alleles at polymorphic loci	Diversity
A	8	26	1.899	2	4	0.569
B	6	14	2.788	1	2	0.391
D	8	18	2.543	3	6	0.375
E	7	15	1.992	2	4	0.356
F	3	7	1.647	2	4	0.615
G	7	16	1.854	2	4	0.605
H	3	7	0.816	3	6	0.996
I	9	18	2.327	2	4	0.250
J	3	6	0.800	2	4	0.836
K	3	14	2.077	1	2	0.231
L	6	12	0.897	2	4	0.977
Mean	5.7	13.0	1.785	2.0	4.0	0.564

What factors might account for the striking differences in the amounts and kinds of enzyme polymorphism in Californian populations of the two barnyard grass species? Although the data are in accord with the suggestion that differences in environmental heterogeneity may play a role, this is difficult to prove in selfing species with little recombination. The difference in levels of enzyme polymorphism in E. crus-galli and E. oryzoides may result from their differing histories of introduction into California. Multiple introductions from a wide geographical range are likely to occur in a generalist weed such as E. crus-galli. In contrast, a relatively small number of contaminated rice seed lots, probably from Asia (Barrett and Seaman, 1980), may have been responsible for the introduction of E. oryzoides. If this is true the contrasting patterns may simply reflect the joint action of repeated founder effects, finite population size, and genetic drift.

Few studies of colonizing species have attempted to relate the degree of heterozygosity at isozyme loci to habitat disturbance. Bosbach and Hurka (1981), in a survey of 17 loci in 81 European populations of the annual Capsella bursa-pastoris, reported higher levels of heterozygosity in populations that were judged to be in more disturbed sites. However, several problems are apparent in the study, including small sample sizes, lack of statistical analysis, and questionable genetic interpretation of isozyme data (see their Table 1). The authors suggest that high levels of disturbance result in periodic germination of genetically diverse seeds from the seed bank. In less disturbed sites they suggest that strong biotic selection pressures imposed by "successional stress" operate on individuals and lead to populations with less genetic diversity. This interpretation contrasts with the suggestion that increased levels of biotic complexity, mediated through competitors and pest and disease

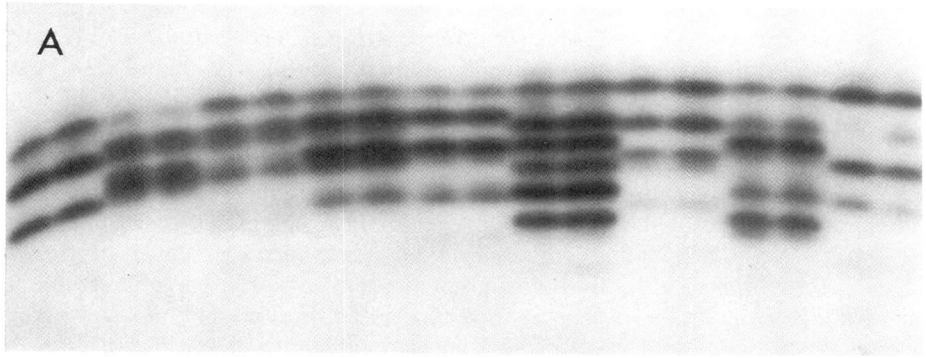
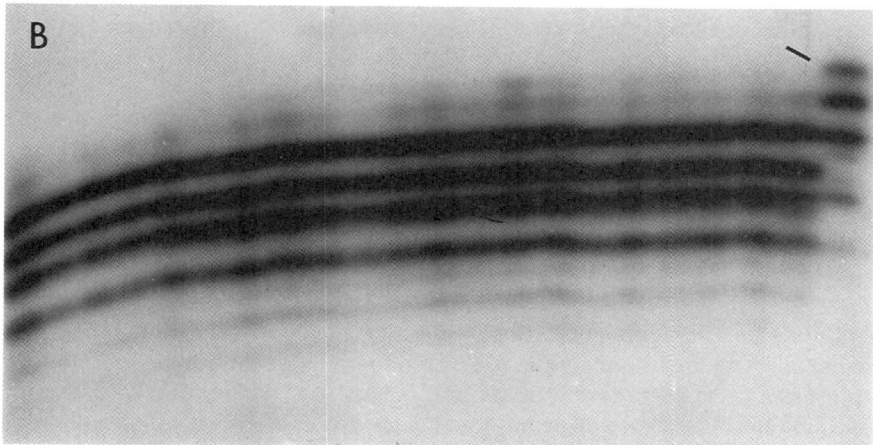

Fig. 5.3. Starch gel of Californian rice-field populations of hexaploid *Echinochloa crus-galli* (A) and *Echinochloa oryzoides* (B) assayed for alcohol dehydrogenase (ADH) activity. ADH is dimeric with three polymorphic loci readily visible in *E. crus-galli* and four predominantly monomorphic loci in *E. oryzoides*. A single variant genotype is visible in *E. oryzoides* and results from polymorphism at a single locus. All genotypes illustrated are multiple homozygotes as a result of the predominantly selfed mating system of the two barnyard grass species.

pressure, will place a selective premium on genetic variation (Levin, 1975; Clarke 1976; Burdon, 1985). Because few genetic studies of plant populations have examined temporal changes in diversity associated with successional change, it is difficult at present to evaluate the importance of biotic selection pressures in regulating patterns of genetic variation.

POLYPLOIDY AND COLONIZING ABILITY

Is there an association between polyploidy and colonizing ability, and if so, how are they causally related? Can enhanced colonizing ability be attributed to the polyploid condition, and how might polyploidy bring this about? These questions have often been posed (e.g., Baker, 1965; Stebbins, 1965; Ehrendorfer, 1965; Mulligan, 1965; Brown and Marshall, 1981; Barrett and Richardson, 1986), but the answers remain unclear. The most recent evidence for such an association was noted by Brown and Marshall (1981) and Clegg and Brown (1983); all 18 of the world's worst weeds (Holm et al., 1977) are polyploid.

An approach employed in the search for an association between polyploidy and colonizing ability has been the use of survey data. Heiser (1950), Stebbins (1965), and Mulligan (1965) demonstrated the absence of correlation between weediness and polyploidy in the weed floras of Indiana, California, and Canada, respectively. However, attributes associated with polyploidy complicate attempts to examine such correlations. For example, Müntzing (1936) demonstrated that polyploidy is less frequent among annuals than perennials within the same genera of herbs. Heiser (1950) showed that in the weed flora of Indiana low levels of polyploidy are correlated with a high frequency of the annual habit, and Gustafsson (1948) also pointed out that weed floras composed chiefly of annuals have a lower percentage of polyploid species.

Breeding systems impose constraints on whether or not polyploidy will evolve in populations and confound simple correlative analysis. It has been frequently argued (e.g., Grant, 1971; Stebbins, 1971) that there is a greater likelihood of the establishment of polyploidy in selfing species. This is because self-fertilization of a polyploid variant would result in fertile polyploid progeny whereas in an outcrossing species backcrossing to diploids would result in sterile progeny. In many groups of related annual and biennial species, the selfing species contain allopolyploids whereas the outcrossers are strictly diploid (Grant, 1971). The kind of polyploidy also varies with breeding system (Stebbins, 1957). Generally, allopolyploid species tend to be selfing whereas autopolyploids are outcrossing (Stebbins, 1957; Bingham, 1980). It is clear that if survey data are used to draw inferences concerning the causal relationship between colonizing success and polyploidy, careful statistical approaches that control for several variables must be undertaken.

Stebbins (1965) has cautioned against the use of survey data and suggests that a more fruitful approach is intense investigation within taxonomic groups containing diploids and related polyploids. He suggests that whereas colonizing ability does not require polyploidy, where diploids and related polyploids both occur, the polyploids have a greater likelihood of becoming weeds. A correlation between weediness (or colonizing ability) and polyploidy is apparent in many groups, e.g., *Knautia* spp. (Ehrendorfer, 1965), *Claytonia perfoliata*, *Amsinckia* spp. (Stebbins, 1965), *Ageratum* spp. (Baker, 1965), *Deschampsia cespitosa* (Rothera and Davy, 1986), and *Eichhornia* (Barrett, 1988b). However, in other cases, there is a correlation between diploidy and weediness, or diploids and polyploids may be equally weedy. For example, diploids appear to be better weeds than polyploids in the *Eupatorium microstemon* species aggregate (Baker, 1965, 1967), and in *Oxalis pes-caprae* weediness has clearly preceded the occurrence of pentaploidy since diploids and tetraploids are more successful as weeds in the native South African range than the sterile pentaploid (Ornduff, 1987). Historical factors have probably led to the introduction of only the pentaploid to other parts of the world where it is now a noxious, widespread weed. Grant (1967) found no apparent relationship between chromosome number and geographic distribution of weedy species of the genus *Amaranthus*. Further, he found no association between polyploidy and weediness in the genus *Celosia* (Amaranthaceae) and concluded that polyploidy does not appear to be a cause of weediness. Thus, although there appears to be some association between polyploidy and colonizing success in many groups, it is by no means a universal association.

The intercorrelation of several factors makes it difficult to sort out what attributes are under a selective premium in colonizing plants. Is polyploidy, breeding system, habit, or some other factor critical, or is it an interaction of two or more of these attributes? Perhaps in addition to survey data and comparative studies, a population biology approach to addressing these questions is required, much as Stebbins (1976, 1985) has undertaken for species of *Erharta*, *Stipa*, *Elymus*, and *Phalaris*. Long-term demographic studies of natural and synthesized polyploids (both auto- and allopolyploids) and their diploid progenitors under a wide range of natural field conditions are most likely to provide the experimental data necessary to resolve these issues. Unfortunately, such studies require a major research commitment of both time and effort.

Release of Variation in Colonizing Polyploid Populations

Emphasis on the causal connection between polyploidy and colonizing ability has focused on the greater biochemical diversity of polyploids potentially providing individuals increased homeostasis and therefore the capacity to respond to new environmental challenges upon colonization (reviewed by Levin, 1983). Here we consider whether the polyploid condition might, in addition, provide enhanced levels of genetic variation upon which selection can act allowing adaptation to novel environments.

In colonizing species it is important to know whether the potential for evolutionary response is possible in founding populations. Autopolyploids can maintain large stores of variability that can be released through segregation and recombination with polysomic inheritance. Polysomic inheritance has a conservative effect on genetic variation and reduces the stochastic loss of variation (Haldane, 1930; Mayo, 1971). This may be important in colonizers that experience repeated bottlenecks in population size. In allopolyploids, particularly those that are primarily inbreeding, considerable biochemical diversity may be tied up within individuals as "fixed heterozygosity" (e.g., Roose and Gottlieb, 1976). However, it is not clear whether this potential source of variation can be released. In theory, mechanisms such as gene silencing, intergenomic recombination, and the regulatory divergence of duplicate loci could alter variation, but the extent to which these processes occur in allopolyploid plant populations is unknown.

That allopolyploids can undergo adaptive evolution is best exemplified by studies of island colonization and evolutionary divergence in Hawaiian *Bidens* (Helenurm and Ganders, 1985). Allohexaploid *Bidens* species on different Hawaiian islands exhibit dramatic morphological and ecological differentiation but show little isozyme divergence or intersterility. The possible role of released homoeologous genetic variation in the divergence of these allopolyploids is not known. However, if homoeologous genetic variation has been released, it has not altered the expression of isozyme variation but instead has been important in generating variability in characters of morphological and ecological significance.

Genetic Variation in Polyploid Colonizers

The levels of genetic variation have been assessed in a number of allopolyploid colonizing species using gel electrophoresis. These include *Chenopodium* ssp. (Wilson and Heiser, 1979; Wilson, 1981; Al Mouemar and Gasquez, 1983); *Avena* spp. (e.g., Jain and Singh, 1979; Kahler et al., 1980); *Hordeum jubatum* (Babbel and Wain, 1977); *Oryza* spp. (Second, 1982); and *Echinochloa* spp. (S. C. H. Barrett and A. H. D. Brown, unpublished data). The recently derived allotetraploids *Tragopogon mirus* and *T. miscellus* exhibit 43% and 33% duplicate loci with "fixed heterozygosity", respectively (Roose and Gottlieb, 1976). Much of the genetic diversity within allopolyploid *Tragopogon* spp. is intergenomic; very little intragenomic diversity occurs (Brown and Marshall, 1981). Using isozyme markers, Rieseberg and Warner (1987) recently documented hybridization between *T. mirus* and *T. miscellus*. Although the evolutionary significance of this hybridization is unclear, its occurrence demonstrates that levels of genetic variation can potentially increase through hybridization among related polyploids of independent origin. Intersterility among the progenitors of allopolyploid species usually precludes gene exchange at the diploid level.

Populations of an older allopolyploid, *Bromus mollis*, display more intragenomic diversity than occurs in *Tragopogon* spp., 22.2% versus 2.7%, respectively (Brown and Marshall, 1981). This demonstrates that allopolyploidy does not preclude the maintenance of intragenomic variability, and this variation may provide the basis for evolutionary response during colonization. Warwick (1989) estimated the percentage of duplicate loci exhibiting "fixed heterozygosity" in four polyploid weed species (Table 5.4). The values range from 14% for *Sorghum halpense* to 54% for *Setaria faberi*. Little intragenomic variation occurs in populations of these species. Thus, allopolyploids may maintain high levels of heterozygosity, resulting from genetic differences between each of their component genomes despite being uniform intragenomically, or they may benefit from both sources of genetic diversity. Weedy allopolyploid populations may possess an enhanced capacity to respond to new selection pressures upon colonization, if they can liberate the potential variation that is tied up intergenomically. In addition, it is possible that the "fixed heterozygous" condition is of direct adaptive value under colonizing situations.

Studies of isozyme variation in autopolyploids in general are few (Soltis and Rieseberg, 1986) and studies of colonizing or weedy autopolyploids are even rarer. Perhaps this is because most colonizers are selfing and autopolyploidy is almost always found in outcrossing species (Stebbins, 1957; Bingham, 1980). Isozyme studies of weedy autopolyploids have been carried out in *Haplopappus spinulosus* (Hauber, 1986) and *Turnera ulmifolia* (Shore, 1986; J. S. Shore and S. C. H. Barrett, unpublished data). Studies of isozyme variation in nonweedy autopolyploids include *Veronica peregrina* (Keeler, 1978); *Medicago* spp. (Quiros, 1982); *Solanum tuberosum* (Martinez-

Zapater and Oliver, 1984); *Galax urceolata* (Epes and Soltis, 1984); *Coreopsis grandiflora* var. *longipes* (Crawford and Smith, 1987); *Heuchera grossulariifolia* (Wolf et al., 1987); *H. micrantha* (Ness et al., 1986), and *Tolmiea menziesii* (Soltis and Rieseberg, 1986).

Turnera ulmifolia is a weedy, perennial, polyploid complex native to the New World Tropics and composed of diploid, tetraploid, and hexaploid varieties (Barrett, 1978; Barrett and Shore, 1987). Although two varieties (tetraploid var. *elegans* and hexaploid var. *angustifolia*) have been introduced to the Old World where they are roadside weeds, our observations of New World populations have not revealed any association between ploidal level and degree of weediness. Diploid and tetraploid varieties in the complex are distylous, self-incompatible, and outcrossing, whereas hexaploid varieties are homostylous, self-compatible, and inbreeding. Isozyme studies indicate that tetraploids in the complex are autopolyploids whereas hexaploids display considerable "fixed heterozygosity" at isozyme loci and are allopolyploids (Figure 5.1).

We examined levels of isozyme variation in six diploid and 16 autotetraploid populations of *Turnera ulmifolia* var. *intermedia* and seven populations of autotetraploid *T. ulmifolia* var. *elegans*. Diploid populations of var. *intermedia* were collected from South and Central America, tetraploid var. *intermedia* from Puerto Rico and the Dominican Republic, and tetraploid var. *elegans* was collected from Brazil. Fourteen isozyme loci were examined. Island populations of autotetraploid var. *intermedia* show the lowest levels of genetic variation, perhaps as a result of bottlenecks associated with island colonization (Table 5.6). Autotetraploid populations of var. *elegans* from Brazil exhibit the highest levels of genetic variation for all parameters that were estimated. This suggests the occurrence of hybridization during or subsequent to the origin of this mainland variety. The high levels of observed heterozygosity in var. *elegans* likely result from the joint effects of its breeding system, tetrasomic inheritance, and high genetic diversity in the species. Bingham (1980) noted that in contrast to allopolyploidy there are no examples among crop plants of successful polysomic polyploid species which are self-pollinated. He suggested that the biochemical and physiological advantages conferred by heterozygosity are important components of autopolyploid vigor. The data for weedy *T. ulmifolia* are consistent with these observations from crop plants. Autopolyploidy is associated with self-incompatibility and outcrossing, whereas allopolyploid members of the complex self-pollinate to varying degrees.

Table 5.6. *Comparisons of genetic variation in diploid and autotetraploid varieties of* Turnera ulmifolia *(x = 5) based on a survey of 14 isozyme loci. Values are the means of populations within each variety/ploidal level*

Variety	Ploidy	N (pops)	P	A	H_S	H_O
intermedia	2x	6	44	1.5	0.11	0.10
intermedia	4x	16	20	1.2	0.04	0.07
elegans	4x	7	65	2.0	0.27	0.42

CONCLUSIONS

In the more than two decades since the publication of *The Genetics of Colonizing Species* (Baker and Stebbins, 1965) the enzyme electrophoresis revolution has provided a considerable amount of information on the population genetic structure of colonizing plants. Not surprisingly, given the variety of colonizing strategies that exists, a wide range of population genetic structures has been revealed, ranging from extensive areas of genetic uniformity in selfing, apomictic, and clonal species, to high levels of genetic diversity in outcrossing species. A recurrent theme in many colonizing species is the occurrence of marked founder effects, depauperate levels of genetic variation, pronounced interpopulation differentiation and well developed multilocus organization (Brown and Marshall, 1981). In addition, the associations between selfing, allopolyploidy, low genetic polymorphism, and high phenotypic plasticity in many agricultural weeds suggests that this combination of traits represents an adaptive strategy that has evolved independently numerous times among angiosperm families in response to selection pressures associated with habitat disturbance and repeated colonization events.

In a general review of this type it is inevitable that more questions are raised than answered. In particular, the association between polyploidy and colonizing success has stimulated vigorous discussion for many years, and, as we have seen, the problem is complex and fraught with difficulties. Nevertheless, before we conclude we cannot resist raising several additional questions and suggesting several lines of enquiry that could be profitably explored now that the mating systems and patterns of genetic variation in polyploids can be routinely assayed using electrophoretic techniques.

Isozyme studies have provided data indicating that polyploid individuals maintain high levels of biochemical diversity. Is this of direct adaptive value during colonizing episodes or is some other correlate of the polyploid condition of greater significance? Do isozymes simply provide markers of genomewide gene multiplication in polyploids, or are they of adaptive importance? Are the phenomena of phenotypic plasticity, developmental homeostasis, heterosis, and inbreeding depression in part determined by this biochemical diversity, or do they result from genetic phenomena unrelated to polyploidy? A positive relationship between heterozygosity at isozyme loci, developmental stability, growth rates, and fitness has been claimed for diploids (Mitton and Grant, 1984; see chapter by Mitton in this volume), but how do these arguments apply to allopolyploids with extensive monomorphism of individual loci but "fixed heterozygosity"? Studies of phenotypic plasticity, growth rates, and other correlates of fitness in natural and synthesized polyploids of contrasting ploidal level and heterozygosity would be valuable in assessing the relationships between biochemical diversity at the individual level and fitness. This type of work could provide more meaningful insights into the "general purpose genotype" concept in colonizing species (Baker, 1965) which, up to now, has been difficult to define or identify with any degree of precision.

The fact that allopolyploidy maintains individual heterozygosity in the face of high selfing rates in colonizing species suggests that it is the simultaneous contribution of high biochemical diversity and assured reproduction that may have led to the success of so many polyploid weeds. If allopolyploids depend on gene multiplication rather than allelic variation and outcrossing to attain heterozygosity, then the partial uncoupling of the mating system from individual heterozygosity can be viewed as the significant factor favoring colonizing ability in many polyploids.

Finally, what are the relationships between polyploidy and inbreeding depression, and how might such a relationship lead to increased colonizing success? Both theoretical and empirical studies of inbreeding depression and its converse, heterosis, have been undertaken in polyploid species (Lundquist, 1966; Busbice and Wilsie, 1966; Dewey, 1966, 1969; Bennett, 1976; Bingham, 1980; Gallais, 1984). Virtually all empirical studies have been conducted on synthetic polyploids of agronomic importance where yield has been the primary character of interest. The results have been variable, with polyploids exhibiting greater inbreeding depression than diploids in some instances and the reverse pattern occurring in others (Dewey, 1966, 1969). The relevance of these studies to natural polyploids is unclear because of the agricultural context and because levels of inbreeding depression exhibited by synthetic polyploids will be largely a function of the manner in which they are produced. Studies of inbreeding depression in natural polyploids and their diploid progenitors under field conditions are clearly required.

Lande and Schemske (1985) modelled the joint evolution of inbreeding depression and plant mating systems. Their model predicts that both autopolyploids and allopolyploids should exhibit lower equilibrium levels of inbreeding depression than diploids. If selfing is of selective importance in colonizing populations, it may arise more easily in polyploid populations, because inbreeding depression may not be of sufficient magnitude to prevent the spread of genes that confer high selfing rates. Barrett and Shore (1987) suggested that this process may explain the occurrence of self-compatible hexaploids at the margins of the geographical range of the *Turnera ulmifolia* complex. If inbreeding depression is indeed of reduced significance in polyploids (and see Hedrick, 1987), it is possible that polyploids are able to withstand marked reductions in population size and concomitant increased inbreeding, without suffering reductions in fertility due to the expression of deleterious genes. Data on mating systems, population size, and levels of inbreeding depression in related diploid and polyploid species would aid in the evaluation of these ideas.

ACKNOWLEDGMENTS

We thank Tony Brown, Brian Husband, Steve Novak, Suzanne Warwick, and Paul Wolf for allowing us to cite unpublished work and the Natural Sciences and Engineering Research Council of Canada for financial support.

LITERATURE CITED

Allard, R. W., G. R. Babbel, M. T. Clegg, and A. L. Kahler. 1972. Evidence for coadaptation in *Avena barbata*. *Proc. Natl. Acad. Sci. USA* 69: 3043–3048.

———, R. D. Miller and A. L. Kahler. 1978. The relationship between degree of environmental heterogeneity and genetic polymorphism. In A. H. J. Freysen and J. W. Woldendorp [eds.], *Structure and functioning of plant populations*, 49–70. North-Holland Publishing Co., New York.

Al Mouemar, A., and J. Gasquez. 1983. Environmental conditions and isozyme polymorphism in *Chenopodium album* L. *Weed Res.* 23: 141–149.

Antonovics, J. 1971. The effects of a heterogeneous environment on the genetics of natural populations. *Amer. Sci.* 59: 593–599.

Babbel, G. R., and R. P. Wain. 1977. Genetic structure of *Hordeum jubatum*. I. Outcrossing rates and heterozygosity levels. *Canad. J. Genet. Cytol.* 19: 143–152.

Baker, H. G. 1965. Characteristics and modes of origins of weeds. In H. G. Baker and G. L. Stebbins [eds.] *The genetics of colonizing species*, 147–172. Academic Press, New York.

———. 1967. The evolution of weedy taxa in the *Eupatorium microstemon* species aggregate. *Taxon* 16: 293–300.

———. 1974. The evolution of weeds. *Ann. Rev. Ecol. Syst.* 5: 1–24.

———, and G. L. Stebbins. 1965. *The genetics of colonizing species*. Academic Press, New York.

Barrett, S. C. H. 1978. Heterostyly in a tropical weed: the reproductive biology of the *Turnera ulmifolia* complex. *Canad. J. Bot.* 56: 1713–1725.

———. 1982. Genetic variation in weeds. In R. Charudattan and H. L. Walker [eds.], *Biological control of weeds with plant pathogens*, 73–98. John Wiley, New York.

———. 1983. Crop mimicry in weeds. *Econ. Bot.* 37: 255–282.

———. 1985. Floral trimorphism and monomorphism in continental and island populations of *Eichhornia paniculata*. *Biol. J. Linn. Soc.* 25: 41–60.

———. 1988a. Genetics and evolution of agricultural weeds. In M. Altieri and M. Liebman [eds.], *Weed management in agroecosystems: ecological approaches*, 57–75. CRC Press Inc., Boca Raton, FL.

———. 1988b. Evolution of breeding systems in *Eichhornia* (Pontederiaceae): a review. *Ann. Missouri Bot. Gard.* 75: 741–760.

———, and B. C. Husband. 1989. The genetics of plant migration and colonization. In A. H. D. Brown, M. T. Clegg, A. L. Kahler, and B. S. Weir [eds.], *Plant population genetics, breeding and genetic resources*. Sinauer Associates, Sunderland, MA.

———, and B. J. Richardson. 1986. Genetic attributes of invading species. In R. H. Groves and J. J. Burdon [eds.], *Ecology of biological invasions*, 21–33. Australian Academy of Science, Canberra.

———, and D. E. Seaman. 1980. The weed flora of Californian ricefields. *Aquat. Bot.* 9: 351–376.

———, and J. S. Shore. 1987. Variation and evolution of breeding systems in the *Turnera ulmifolia* complex (Turneraceae). *Evolution* 41: 340–354.

Bennett, J. H. 1968. Mixed self- and cross-fertilization in a tetrasomic species. *Biometrics* 24: 485–500.

———. 1976. Expectations for inbreeding depression on self-fertilization of tetraploids. *Biometrics* 32: 449–452.

Bingham, E. T. 1980. Maximizing heterozygosity in autoploids. In W. H. Lewis [ed.], *Polyploidy: biological relevance*, 471–490. Plenum Press, New York.

Bosbach, K., and H. Hurka. 1981. Biosystematic studies on *Capsella bursa-pastoris* (Brassicaceae):

enzyme polymorphisms in natural populations. *Pl. Syst. Evol.* 137: 73–94.

Brown, A. H. D. 1975. Efficient experimental designs for the estimation of genetic parameters in plant populations. *Biometrics* 31: 145–160.

―――. 1984. Multilocus organization of plant populations. In K. Wohrmann and V. Loeschcke [eds.], *Population biology and evolution*, 159–169. Springer-Verlag, Berlin.

―――, and J. J. Burdon. 1983. Multilocus diversity in an outbreeding weed, *Echium plantagineum* L. *Aust. J. Biol. Sci.* 36: 503–509.

―――, and ―――. 1987. Mating systems and colonizing success in plants. In A. J. Gray, M. J. Crawley, and P. J. Edwards [eds.], *Colonization, succession and stability*, 115–131. Blackwell Scientific Publications, Oxford.

―――, and D. R. Marshall. 1981. Evolutionary changes accompanying colonization in plants. In G. C. E. Scudder and J. L. Reveal [eds.], *Evolution today*, 351–363. Hunt Institute for Botanical Documentation, Carnegie-Mellon University, Pittsburgh.

―――, and B. S. Weir. 1983. Measuring genetic variability in plant populations. In S. D. Tanksley and T. J. Orton [eds.], *Isozymes in plant genetics and breeding*, Part A, 219–239. Elsevier, Amsterdam.

―――, M. W. Feldman, and E. Nevo. 1980. Multilocus structure of natural populations of *Hordeum spontaneum*. *Genetics* 96: 523–526.

Burdon, J. J. 1985. Pathogens and the genetic structure of plant populations. In J. White [ed.], *Studies on plant demography, a Festschrift for John L. Harper*, 313–324. Academic Press, London.

Busbice, T. H., and C. P. Wilsie. 1966. Inbreeding depression and heterosis in autotetraploids with application to *Medicago sativa* L. *Euphytica* 15: 52–67.

Clarke, B. C. 1976. The ecological genetics of host-parasite relationships. In A. E. R. Taylor and R. Muller [eds.], *Genetic aspects of host-parasite relationships*, 87–103. Blackwell Scientific Publications, London.

Clegg, M. T., and A. H. D. Brown. 1983. The founding of plant populations. In C. M. Schonewald-Cox, S. M. Chambers, B. MacBryde and W. L. Thomas [eds.], *Conservation Genetics*, 216–228. Benjamin/Cummins, CA.

Crawford, D. J, and E. B. Smith. 1984. Allozyme divergence and intraspecific variation in *Coreopsis grandiflora* (Compositae). *Syst. Bot.* 9: 219–225.

―――, and H. D. Wilson. 1979. Allozyme variation in several closely related diploid species of *Chenopodium* of the western United States. *Amer. J. Bot.* 66: 237–244.

Darlington, C. D. 1958. *The evolution of genetic systems.* Second Edition, Basic Books, New York.

Dewey, D. R. 1966. Inbreeding depression in diploid, tetraploid and hexaploid crested wheatgrass. *Crop Sci.* 6: 144–147.

―――. 1969. Inbreeding depression in diploid and induced autotetraploid crested wheatgrass. *Crop Sci.* 9: 592–595.

Ehrendorfer, F. 1965. Dispersal mechanisms, genetic systems, and colonizing abilities in some flowering plant families. In H. G. Baker and G. L. Stebbins [eds.], *The genetics of colonizing species*, 331–352. Academic Press, New York.

Ellstrand, N. C., and M. L. Roose. 1987. Patterns of genotypic diversity in clonal plant species. *Amer. J. Bot.* 74: 123–131.

Epes, D. A., and D. E. Soltis. 1984. An electrophoretic investigation of *Galax urceolata* (Diapensiaceae). (Abstract) *Amer. J. Bot.* 71: 165.

Epperson, B. K., and M. T. Clegg. 1986. Spatial-autocorrelation analysis of flower color polymorphisms within substructured populations of morning glory (*Ipomoea purpurea*). *Amer. Naturalist* 128: 840–858.

Gallais, A. 1984. An analysis of heterosis vs. inbreeding effects with an autotetraploid cross-fertilized plant: *Medicago sativa* L. *Genetics* 106: 123–127.

Giles, B. E. 1983. A comparison between quantitative and biochemical variation in the wild barley *Hordeum murinum*. *Evolution* 38: 34–41.

Glover, D. E., and S. C. H. Barrett. 1986. Variation in the mating system of *Eichhornia paniculata* (Spreng.) Solms (Pontederiaceae). *Evolution* 40: 1122–1131.

———, and ———. 1987. Genetic variation in continental and island populations of *Eichhornia paniculata* (Pontederiaceae). *Heredity* 59: 7–17.
Grant, V. 1971. *Plant speciation*. Columbia University Press, New York.
Grant, W. F. 1967. Cytogenetic factors associated with the evolution of weeds. *Taxon* 16: 283–293.
Grime, J. P. 1979. *Plant strategies and vegetation processes*. John Wiley and Sons, Ltd., New York.
Gustafsson, A. 1948. Polyploidy, life form and vegetative reproduction. *Hereditas* 34: 1–22.
Haldane, J. B. S. 1930. Theoretical genetics of autopolyploids. *J. Genet.* 22: 359–373.
Hamrick, J. L., Y. B. Linhart, and J. B. Mitton. 1979. Relationships between life history characteristics and electrophoretically-detectable genetic variation in plants. *Ann. Rev. Ecol. Syst.* 10: 173–200.
Harper, J. L. 1977. *Population biology of plants*. Academic Press, London.
Hauber, D. P. 1986. Autotetraploidy in *Haplopappus spinulosus* hybrids: evidence from natural and synthetic tetraploids. *Amer. J. Bot.* 73: 1595–1606.
Hedrick, P. W. 1987. Genetic load and the mating system in homosporous ferns. *Evolution* 41: 1282–1289.
———, M. E. Ginevan, and E. P. Ewing. 1976. Genetic polymorphism in heterogeneous environments. *Ann. Rev. Ecol. Syst.* 7: 1–32.
Helenurm, K., and F. R. Ganders. 1985. Adaptive radiation and genetic differentiation in Hawaiian *Bidens*. *Evolution* 39: 753–765.
Heiser, C. B. 1950. A comparison of the flora as a whole and the weed flora of Indiana as to polyploidy and growth habits. *Proc. Indiana Acad. Sci.* 59: 64–70.
Holm, L. G., D. L. Plucknett, J. V. Pancho, and J. P. Herberger. 1977. *The world's worst weeds: distribution and biology*. University Press of Hawaii, Honolulu.
Horak, M., and J. S. Holt. 1986. Isozyme variability and breeding system in populations of yellow nutsedge (*Cyperus esculentus*). *Weed Sci.* 34: 538–543.
Jackson, R. C. 1984. Chromosome pairing, hybrid sterility and polyploidy: comments on G. L. Stebbins's reply. *Syst. Bot.* 9: 121–123.
Jain, S. K. 1979. Adaptive strategies: polymorphism, plasticity and homeostasis. In O. T. Solbrig, S. K. Jain, G. B. Johnson, and P. H. Raven [eds.], *Topics in plant population biology*, 160–187. Columbia University Press.
———, and R. S. Singh. 1979. Population biology of *Avena*. VII. Allozyme variation in relation to genome analysis. *Bot. Gaz.* 140: 356–362.
Jefferies, R. L., and L. D. Gottlieb. 1982. Genetic differentiation of the microspecies *Salicornia europaea* L. (sensu stricto) and *S. ramosissima*, J. Woods. *New Phytol.* 92: 123–129.
———, and ———. 1983. Genetic variation within and between populations of the asexual plant *Puccinellia* × *phryganodes*. *Canad. J. Bot.* 61: 774–779.
Kahler, A. L., R. W. Allard, M. Krzakowa, C. F. Wehrhahn, and E. Nevo. 1980. Associations between isozyme phenotypes and environment in the slender wild oat (*Avena barbata*) in Israel. *Theor. Appl. Genet.* 56: 31–47.
Keeler, K. H. 1978. Intra-population differentiation in annual plants. II. Electrophoretic variation in *Veronica peregrina*. *Evolution* 32: 638–645.
Kihara, H., and T. Ono. 1926. Chromosomenzahlen und systematische Grupierung der *Rumex*-Arten Z Zellforsch u. mikroskop. *Anat.* 4: 475–481.
Koniuszek, J. W. J., and J. A. C. Verkeij. 1982. Genetic variation in two related annual *Senecio* species occurring in the same habitat. *Genetica* 59: 133–137.
Lande, R., and D. W. Schemske. 1985. The evolution of self-fertilization and inbreeding depression in plants. I. genetic models. *Evolution* 39: 24–40.
Les, D. H. 1988. Breeding systems, population structure and evolution in hydrophilous angiosperms. *Ann. Missouri Bot. Gard.* 75: 819–835.
Levin, D. A. 1975. Pest pressure and recombination systems in plants. *Amer. Naturalist* 109: 437–451.
———. 1983. Polyploidy and novelty in flowering plants. *Amer. Naturalist* 122: 1–25.
Levins, R. 1968. *Evolution in changing environments*. Monographs in population biology No. 2. Princeton University Press, Princeton.

Lewontin, R. C. 1984. Detecting population differences in quantitative characters as opposed to gene frequencies. *Amer. Naturalist* 123: 115–124.

Loveless, M. D., and J. L. Hamrick. 1984. Ecological determinants of genetic structure in plant populations. *Ann. Rev. Ecol. Syst.* 15: 65–95.

Lundquist, A. 1966. Heterosis and inbreeding depression in autotetraploid rye. *Hereditas* 56: 317–366.

Marshall, D. R., and P. W. Weiss. 1982. Isozyme variation within and among Australian populations of *Emex spinosa* (L.) Campd. *Austral. J. Biol. Sci.* 35: 327–332.

Martinez-Zapater, J. M., and J. L. Oliver. 1984. Genetic analysis of isozyme loci in tetraploid potatoes (*Solanum tuberosum* L.). *Genetics* 108: 669–679.

Mashburn, S. J., R. R. Sharitz, and M. H. Smith. 1978. Genetic variation among *Typha* populations of the southeastern United States. *Evolution* 32: 681–685.

Maun, M. A., and S. C. H. Barrett. 1986. The biology of Canadian weeds. 77. *Echinochloa crus-galli* (L.) Beauv. *Canad. J. Pl. Sci.* 66: 739–759.

Mayo, O. 1971. Rates of change in gene frequency in tetrasomic organisms. *Genetica* 42: 329–337.

McConnell, G., and J. L. Fyfe. 1975. Mixed selfing and random mating with polysomic inheritance. *Heredity* 34: 271–272.

McIntyre, S., and S. C. H. Barrett. 1985. A comparison of weed communities of rice in Australia and California. In J. R. Dodson and M. Westoby [eds.], *Are Australian ecosystems different?* 237–250. Proc. Ecol. Soc. of Australia, Vol. 14.

Mitton, J. B., and M. C. Grant. 1984. Associations among protein heterozygosity, growth rate and developmental homeostasis. *Ann. Rev. Ecol. Syst.* 15: 479–499.

Moran, G. F., and D. R. Marshall. 1978. Allozyme uniformity within and variation between races of the colonizing species *Xanthium strumarium* L. (Noogoora Burr). *Austral. J. Biol. Sci.* 31: 282–291.

———, ———, and W. J. Müller. 1981. Phenotypic variation and plasticity in the colonizing species *Xanthium strumarium* L. (Noogoora Burr). *Austral. J. Biol. Sci.* 34: 639–648.

Mulligan, G. A. 1965. Recent colonization by herbaceous plants in Canada. In H. G. Baker and G. L. Stebbins [eds.], *The genetics of colonizing species,* 127–146. Academic Press, New York.

Müntzing, A. 1936. The evolutionary significance of autopolyploidy. *Hereditas* 21: 263–378.

Nei, M. 1973. Analysis of gene diversity in subdivided populations. *Proc. Natl. Acad. Sci. USA* 70: 3321–3323.

Nei, M., T. Maruyama, and R. Chakraborty. 1975. The bottleneck effect and genetic variability in populations. *Evolution* 29: 1–10.

Ness, B. D., D. E. Soltis, and P. S. Soltis. 1986. Autopolyploidy in *Heuchera micrantha* Dougl. (Saxifragaceae). *Amer. J. Bot.* (Abstract) 73: 777.

Ornduff, R. 1987. Reproductive systems and chromosome races of *Oxalis pes-caprae* L. and their bearing on the genesis of a noxious weed. *Ann. Missouri Bot. Gard.* 74: 79–84.

Price, S. C., K. M. Shumaker, A. L. Kahler, R. W. Allard, and J. E. Hill. 1984. Estimates of population differentiation obtained from enzyme polymorphisms and quantitative characters. *J. Heredity* 75: 141–142.

Quiros, C. F. 1982. Tetrasomic segregation for multiple alleles in alfalfa. *Genetics* 101: 117–127.

Rice, K., and S. K. Jain. 1985. Plant population genetics and evolution in disturbed environments. In S. T. A. Pickett and P. A. White [eds.], *The ecology of natural disturbance and patch dynamics,* 287–303. Academic Press, New York.

Richards, A. J. 1986. *Plant breeding systems.* Allen and Unwin, London.

Rieseberg, L. H., and D. A. Warner. 1987. Electrophoretic evidence for hybridization between *Tragopogon mirus* and *T. miscellus* (Compositae). *Syst. Bot.* 12: 281–285.

Roose, M. L., and L. D. Gottlieb. 1976. Genetic and biochemical consequences of polyploidy in *Tragopogon*. *Evolution* 30: 818–830.

Rothera, S. L., and A. J. Davy. 1986. Polyploidy and habitat differentiation in *Deschampsia cespitosa*. *New Phytol.* 102: 449–467.

Sadul, H. 1987. *The effects of lesser snow goose grazing on subarctic coastal plant populations.* M. Sc. thesis, University of Toronto, Toronto, Canada.

Scribailo, R. W., K. Carey, and V. Posluszny. 1984. Isozyme variation and the reproductive biology of *Hydrocharis morsus-ranae* L. (Hydrocharitaceae). *Bot. J. Linn. Soc.* 89: 305–312.

Second, G. 1982. Origin of the genetic diversity of cultivated rice (*Oryza* spp.): study of the polymorphism scored at 40 isozyme loci. *Jap. J. Genet.* 57: 25–57.

Sharitz, R. R., S. A. Wineriter, M. H. Smith, and E. H. Lui. 1980. Comparison of isozymes among *Typha* species in the eastern United States. *Amer. J. Bot.* 67: 1297–1303.

Shore, J. S. 1986. *The genetics and evolution of breeding systems in the* Turnera ulmifolia *L. complex. (Turneraceae)*. Ph.D. thesis, University of Toronto, Toronto, Canada.

Sokal, R. R., and D. E. Wartenberg. 1983. A test of spatial autocorrelation analysis using an isolation-by-distance model. *Genetics* 105: 219–237.

Soltis, D. E., and L. H. Rieseberg. 1986. Autopolyploidy in *Tolmiea menziesii* (Saxifragaceae): genetic insights from enzyme electrophoresis. *Amer. J. Bot.* 73: 310–318.

Stebbins, G. L. 1950. *Variation and evolution in plants*. Columbia University Press, New York.

———. 1957. Self-fertilization and population variability in the higher plants. *Amer. Naturalist* 91: 337–354.

———. 1965. Colonizing species in the native California flora. In H. G. Baker and G. L. Stebbins [eds.], *The genetics of colonizing species*, 173–195. Academic Press, New York.

———. 1971. *Chromosomal evolution in higher plants*. Edward Arnold, London.

———. 1976. Chromosomes, DNA and plant evolution. In M. K. Hecht, W. C. Steere and B. Wallace [eds.], *Evolutionary Biology*, Vol. 9, 1–34. Plenum Press, New York.

———. 1984. Chromosome pairing, hybrid sterility and polyploidy: a reply to R. C. Jackson. *Syst. Bot.* 9: 119–121.

———. 1985. Polyploidy, hybridization and the invasion of new habitats. *Ann. Missouri Bot. Gard.* 72: 824–832.

Van Oostrum, H., A. A. Sterk, and H. J. W. Wijsman. 1985. Genetic variation in agamospermous microspecies of *Taraxacum* sect. *erythrosperma* and sect. *obliqua*. *Heredity* 55: 223–228.

Wain, R. P., W. T. Haller, and D. R. Martin. 1985. Isozyme studies in aquatic plants. *J. Aquat. Pl. Manage.* 23: 42–45.

Warwick, S. I. 1989. Allozyme variation and its correlation with other attributes in northwardly colonizing eastern North American weed species. *Pl. Syst. Evol.* (in press).

———, and L. D. Black. 1986. Genecological variation in recently established populations of *Abutilon theophrasti* (velvetleaf). *Canad. J. Bot.* 64: 1632–1643.

———, B. K. Thompson, and L. D. Black. 1987a. Genetic variation in Canadian and European populations of the colonizing weed species *Apera spica-venti*. *New Phytol.* 106: 301–317.

———, ———, and ———. 1987b. Life history and allozyme variation in populations of the weed species *Setaria faberi*. *Canad. J. Bot.* 65: 1396–1402.

Weir, B. S., and C. C. Cockerham. 1984. Estimating F-statistics for the analysis of population structure. *Evolution* 38: 1358–1370.

Wilson, H. D. 1981. Genetic variation among South American populations of tetraploid *Chenopodium* sect. *Chenopodium* subsect. *Cellulata*. *Syst. Bot.* 6: 380–398.

———, and C. B. Heiser, Jr. 1979. The origin and evolutionary relationship of "huauzontle" (*Chenopodium nuttalliae* Safford), domesticated chenopod of Mexico. *Amer. J. Bot.* 66: 198–206.

Wolf, P. G., C. H. Haufler, and E. Sheffield. 1988. Electrophoretic variation and mating system of the clonal weed *Pteridium aquilinum* L. Kuhn (Bracken). *Evolution* 42: 1350–1355.

———, P. S. Soltis, and D. E. Soltis. 1987. Autopolyploid evolution in *Heuchera grossulariifolia* (Saxifragaceae). *Amer. J. Bot.* (Abstract). 74: 767.

Wolff, S. L., and R. L. Jefferies. 1987a. Morphological and isozyme variation in *Salicornia europaea* (s.l.) (Chenopodiaceae) in northeastern North America. *Canad. J. Bot.* 65: 1410–1419.

———, and ———. 1987b. Taxonomic status of diploid *Salicornia europaea* (s.l.) (Chenopodiaceae) in northeastern North America. *Canad. J. Bot.* 65: 1420–1426.

Wright, S. 1951. The genetical structure of populations. *Ann. Eugen.* 15: 323–354.

———. 1969. *Evolution and the genetics of populations*, Vol. 2, *The theory of gene frequencies*. University of Chicago Press, Chicago.

CHAPTER 6

Physiological and Demographic Variation Associated With Allozyme Variation

Jeffry B. Mitton

Department of Environmental, Population and Organismic Biology
University of Colorado
Boulder, Colorado 80309

A corpus of empirical data, extending back more than a half century, documents the related phenomena of heterosis and inbreeding depression (Wright, 1977; Frankel, 1983). Because the majority of this experimental work has been with domesticated species of plants and animals, the observations and the controversy concerning the underlying mechanism(s) have had little impact upon the study of natural populations. Yet recent studies of enzyme polymorphisms suggest that the same continuum of inbreeding depression and heterosis occurs in natural populations. Further analyses of genetic variation and components of fitness in natural populations will not only reveal more about the dynamics of selection within populations, but they will also produce new insights into the mechanisms causing inbreeding depression and heterosis.

This chapter examines physiological and demographic correlates of allozyme variation in plant populations. It starts by reviewing inbreeding depression and heterosis, and contrasting the genetic mechanisms underlying them. Then, using new insights on the relationship between fitness and heterozygosity and a model of fitness determination called truncation selection, it introduces a new perspective on the fitness consequences of genetic variation. This is followed by a presentation of genetic data from annual ryegrass, *Lolium multiflorum,* and interpretation of those data to determine whether the natural selection measured at enzyme loci is consistent with selection acting upon those loci, or simply recorded by those loci. Results of this analysis are consistent with selection upon enzyme loci, and these results lead to a study of the physiological consequences of enzyme polymorphisms in perennial ryegrass, *Lolium perenne.*

INBREEDING DEPRESSION AND HETEROSIS

For many decades plant and animal breeders have recognized the disadvantages associated with inbreeding and utilized the advantages in yield and performance in the progeny of crosses between strains or varieties. When biologists impose inbreeding upon outcrossed species they typically observe a syndrome of deleterious effects called inbreeding depression. In plants, these effects include decreased proportions of filled seed, decreased germinabilty, decreased viability, decreased growth rate, and decreased yield or seed set. In animals, the effects include developmental problems, higher juvenile mortality, suppression of growth rate, and reductions in maximum size and fecundity (Lerner, 1954; Wright, 1977; numerous examples in Ralls and Ballou, 1983). The converse of inbreeding depression is heterosis. Viability, growth, and productivity in the progeny of crosses between strains often surpass those attainable in inbred strains or outcrossing populations (Frankel, 1983).

Despite decades of experimental studies, biologists continue to debate whether the dominance hypothesis or the overdominance hypothesis better explains inbreeding depression and heterosis. The dominance hypothesis ascribes inbreeding depression to the exposure of recessive, deleterious alleles.

Held at low frequencies by selection, deleterious alleles are carried in heterozygous condition in outbred populations, but are expressed increasingly in homozygous condition as inbreeding accumulates. From this perspective, heterosis is the superlative performance of genotypes whose recessive, deleterious alleles are masked by dominant alleles.

The overdominance hypothesis relies upon superior performance of heterozygous genotypes at a large number of loci. Inbreeding reduces heterozygosity below the range typically seen in outcrossing populations, dramatically reducing performance. From this perspective, heterosis is the consequence of high heterozygosity that cannot be maintained in randomly mating populations.

After many years of experimentation, neither of these hypotheses has been rejected, and it seems likely that both could contribute to the phenomena of inbreeding depression and heterosis. However, greater insights into these phenomena may be gained by considering new theoretical developments.

RECENT ADVANCES IN THEORY

Fitness Increases with Heterozygosity at Selected Loci

Computer simulations of additive, multiplicative, and epistatic models of fitness determination lead to a robust generalization concerning the relationship between fitness and heterozygosity (Ginzburg, 1979, 1983; Turelli and Ginzburg, 1983). For loci whose variation is balanced by natural selection, fitness generally increases with the number of heterozygous loci. Therefore, if we could genotype individuals at 20 polymorphisms balanced by selection, we would see that individuals heterozygous at 16 loci perform better than individuals with eight heterozygous loci, and both of these groups perform better than individuals with two heterozygous loci. Natural selection generally favors highly heterozygous individuals.

Truncation Selection

Diploid, eukaryotic genomes are composed of 5,000–60,000 gene loci, and a substantial proportion of these loci, perhaps 25–50%, are genetically variable (Hamrick et al., 1979). It is a virtual certainty that some proportion of this genetic variation is adaptively neutral and therefore not directly influenced by natural selection. Unfortunately, the relative proportions that are adaptively neutral and balanced by selection are unknown. The way in which loci interact to determine relative fitnesses of individuals, is, again, unknown, but most models employ additive, multiplicative, or epistatic models of fitness determinaton. These models share some properties that make them contrived and biologically unrealistic. Among these common properties is the generation of large amounts of genetic load. These high levels of genetic load give the impression that only a few loci could have a substantial impact upon fitness, and that the number of loci whose variation is maintained by selection must, therefore, be limited to a few dozen loci. Consideration of genetic load and the number of loci that might be maintained by selection led some biologists to propose that the majority of genetic variation was adaptively neutral (Crow and Kimura, 1970; Kimura and Ohta, 1971; Nei, 1975; Kimura, 1983).

Truncation selection (Milkman, 1978, 1982) is a model of fitness determination that is more biologically realistic than the models mentioned above, and it essentially eliminates the problems of high genetic load (Kimura and Crow, 1978; Wills, 1978, 1981; Crow and Kimura, 1979). This model of fitness determination first ranks individuals by fitness potential, then imposes a threshold above which individuals enjoy reproductive success, below which fitness is constrained to zero. Fitness potential may be constructed with additive, multiplicative, or epistatic interactions among loci. The biological reality of this model is in the placement of the threshold. A high threshold imposes strong selection and may limit population size, whereas a low threshold imposes mild selection. The variable threshold fits the experiences of population biologists, who witness fluctuations between salubrious and stringent environmental conditions. Models of truncation selection reveal the possibility that the majority of polymorphisms could be balanced by selection (Wills, 1978, 1981).

Outbreeding Depression

Although population biologists have long been aware of the deleterious consequences of inbreeding, evidence for more subtle, negative consequences of crossing between well-differentiated individuals is also seen in both plant and animal populations. For example, heterosis in *Drosophila* is often observed in crosses among local populations, but the progeny of crosses between more distant and presumably more differentiated populations do not fare so well (Dobzhansky and Wallace, 1953). Similarly, crosses among *Delphinium nelsoni* exhibit different seed sets as a function of the distance between individuals (Price and Waser, 1979; Waser and Price, 1983). Seed set of closely adjacent individuals (0–3 m) is presumably depressed by inbreeding depression. Crosses between distant individuals (greater than 50 m) may move genes between differentially adapted subpopulations (Waser and Price, 1985; Campbell and Waser, 1987), producing an outcrossing depression. Similarly, heterosis in corn hybrids generally increases with the differentiation of inbred parental lines (East, 1936; Moll et al., 1962; Heidrich-Sobrinho and Cordeiro, 1975; but see Price et al., 1986), but the relationship between heterozygosity and heterosis may not occur when races differ dramatically with respect to geographic origin (Moll et al., 1965) or pedigree (Frei et al., 1986). Although components of fitness are expected to increase with heterozygosity at balanced polymorphisms, this pattern does not continue indefinitely. Fitness may plateau at some level of heterozygosity, and may then decline at higher levels of heterozygosity (Vrijenhoek and Lerman, 1982). Thus, geneticists expect components of fitness to increase with heterozygosity within populations but outbreeding depression confounds predictions of the performance of progeny from crosses between populations.

RELEVANCE TO NATURAL POPULATIONS

The phenomena of inbreeding depression and heterosis are most commonly associated with cultivated species. Inbred strains are typically much more homozygous than wild progenitors, and therefore inbreeding depression is conspicuous. Crosses between inbred strains produce progeny heterozygous at an unknown but substantial proportion of the genome, and thus heterosis is also conspicuous.

Biologists working in natural populations have not expected the literature on inbreeding depression and heterosis to be relevant to their own work. The range in heterozygosity among inbred strains and the crosses among inbred strains is far greater than within a randomly outcrossing population. For example, consider a population with 2,000 independently segregating polymorphic loci, all with two alleles, and with all allelic frequencies equal to 0.5 (Fig. 6.1). Let us start with a population practicing random outcrossing. If selfing is imposed upon this population, heterozygosity decreases by half in a single generation, dropping far outside the range of heterozygosity seen in the panmictic population. In this example, 95% of the panmictic population will lie between 945 and 1055 heterozygous loci, but a single generation of selfing drops the mean heterozygosity to 500. Ten generations of selfing would eliminate virtually all of the heterozygosity, but crosses among homozygous lines might easily produce F_1 progeny with heterozygosity above the highest level seen in the panmictic population. Inbred strains and F_1 progeny are strikingly different in heterozygosity, and each is relatively homogeneous. In contrast, the individuals in a randomly outcrossing population are distributed along a narrow continuum of heterozygosity. Empiricists have not been sanguine about the prospects of identifying the most heterozygous and the most homozygous among the individuals in outcrossing populations. The narrow range in heterozygosity (Fig. 6.1) and the difficulty in identifying highly heterozygous and predominantly homozygous individuals has led population biologists to suspect that the literature on inbreeding depression and heterosis was either not relevant or could not be applied to the study of natural populations.

For the reasons listed above, population biologists were surprised to find that a phenomenon similar to the continuum between inbreeding depression and heterosis is apparent in natural populations. Whether the contrast between heterozygous and homozygous groups is based upon chromosomes, several protein polymorphisms, or a single protein polymorphism, heterozygous individuals often exhibit higher viability, greater developmental stability, and higher growth rates than do

130 Jeffry B. Mitton

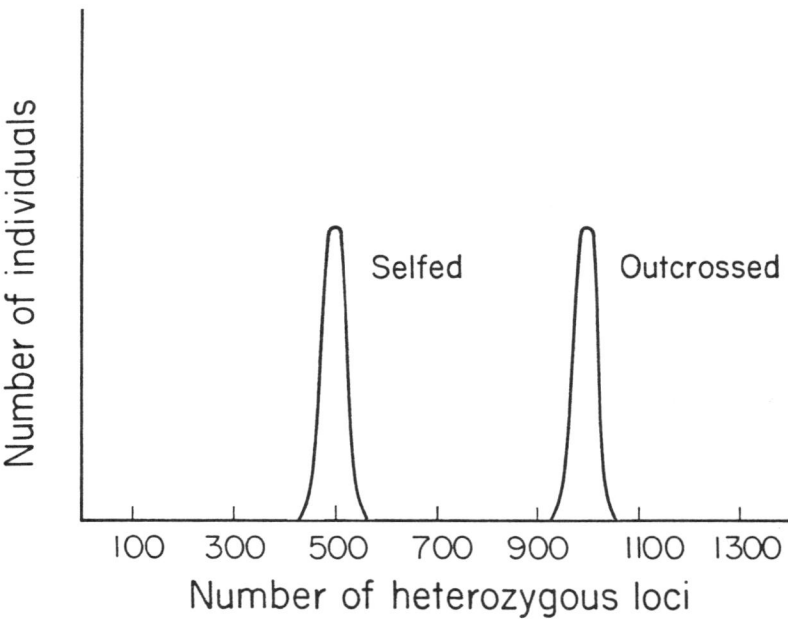

Fig. 6.1. The distribution of individual heterozygosity as a consequence of random outcrossing (O) and selfing (S). The simulation included 2000 polymorphic loci, each with 2 alleles, and with all allelic frequencies equal to 0.5.

homozygous individuals. For example, Robertson and Reeve (1952, 1955) found that *Drosophila melanogaster* heterozygous for their second chromosomes had lower phenotypic variation and greater fecundity than homozygous individuals from the same population. Comparisons of *Drosophila* heterozygous or homozygous for a particular chromosome generally demonstrate dramatically higher fitnesses in heterozygous individuals (Dobzhansky and Wallace, 1953; Tracey and Ayala, 1974; Seager and Ayala, 1982). For example, if the fitness of *D. melanogaster* heterozygous for either chromosome 2 or 3 is standardized to 1.0, homozygotes for either chromosome have a fitness of about 0.08 (Seager et al., 1982), approximately an order of magnitude lower. In addition to differences in viability and fecundity, heterozygotes may enjoy greater mating success as well (Anderson et al., 1979; Carson, 1987).

Single Locus Associations with Growth and Development

It is now apparent that differences among heterozygous and homozygous groups can be found when the groups are defined by a single locus. For example, viability differentials favoring heterozygotes have been reported at the *Lap* locus in the blue mussel, *Mytilus edulis* (Koehn et al., 1976) and at the *To* locus in the ribbed mussel, *Modiolus demissus* (Koehn et al., 1973). Lower morphological variation is associated with enzyme heterozygosity in the killifish, *Fundulus heteroclitus* (Mitton, 1978), in trout (Leary et al., 1983, 1984), and the blue mussel (Mitton and Koehn, 1985). Within populations, growth rate and developmental stability increase with enzyme heterozygosity in both invertebrates and vertebrates (Danzmann et al., 1987, 1988; reviewed in Mitton and Grant, 1984; Zouros and Foltz, 1987). Although these results are not universal, they appear to be general; highly heterozygous individuals often have growth rates and developmental stability superior to highly homozygous individuals from the same populations.

Single Locus Associations with Physiology

Evolutionary biologists are familiar with the story of the sickle-cell variant of hemoglobin. A hemoglobin modified by the replacement of one amino acid is favored in environments characterized by malarial infestations, and selected against in other environments (summarized in Wills, 1981; Templeton, 1982). All too often, evolutionary biologists naively proclaim that this is the only convincing case of an adaptive protein polymorphism. In fact, we now have a variety of well-documented cases of protein polymorphisms that produce adaptive physiological variation. For example, a pair of tightly linked genes code for hemoglobin subunits of the deer mouse, *Peromyscus maniculatus*. Deer mice span a remarkable range of elevations, from sea level to approximately 3,600 meters. Populations at opposite extremes of the elevational range are essentially monomorphic, with polymorphic populations at intermediate elevations. This clinal variation in allelic frequencies is adaptive; at each elevational extreme, the mice carry the hemoglobin that maximizes the P_{50} in that environment, maximizing metabolic rate and hence the level of activity (Snyder, 1981; Chappell and Snyder, 1984). Lactate dehydrogenase (LDH) variation in the killifish modifies the ratio of ATP per hemoglobin in red blood cells (Powers et al., 1979), and this variation modifies the amount of oxygen that can be delivered to tissues. Consequently, LDH variation is directly related to the time of development of eggs (DiMichele and Powers, 1982a) and to adult swimming endurance (DiMichele and Powers, 1982b). Leucine aminopeptidase (LAP) in the blue mussel plays a primary role in the regulation of cell volume by cleaving peptides into smaller but more numerous peptides when mussels are challenged by an increase in salinity (Moore et al., 1980). Selection acting upon LAP physiological phenotypes annually creates sharp clinal variation in allelic frequencies in areas of abrupt transition in salinity (Koehn et al., 1980; Koehn and Hilbish, 1987). Phosphoglucose isomerase (PGI) regulates flux through glycolysis and the pentose shunt in *Colias* butterflies, and the enzymes produced by alternate genotypes have different temperature optima (Watt, 1979). Consequently, butterflies with different PGI genotypes tend to fly at different temperatures (Watt, 1983). Several of the heterozygous genotypes have been characterized as "kinetically favored", and females preferentially mate with these genotypes (Watt et al., 1985, 1986). Additional examples of the physiological and demographic consequences of single-gene protein variation are summarized in Koehn et al. (1983). These case studies demonstrate that enzyme variation, just as in the familiar case of sickle-cell hemoglobin, can have a major impact upon physiological variation and components of fitness.

ASSOCIATIONS BETWEEN ENZYME GENOTYPES AND COMPONENTS OF FITNESS IN PLANT POPULATIONS

Protein Heterozygosity and Viability

Most examples of viability selection in plant populations reveal advantages accruing to heterozygotes. The first of these studies examined viability and fecundity selection with three protein polymorphisms in the slender wild oat, *Avena barbata* (Clegg and Allard, 1973). Large viability differentials favored heterozygotes, whereas fecundity selection favored homozygous genotypes. The average total fitness of heterozygotes was 1.59 (range 1.16–2.39) whereas the average total fitness of homozygotes was 0.92 (range 0.77–1.0).

An elaborate study of components of fitness and protein polymorphisms revealed a very complex pattern of selection in barley, *Hordeum vulgare* (Clegg et al., 1978). Strong selection was measured for each component of fitness, with the fitnesses of the best genotypes often three to five times those of the least favored genotypes. Selection seemed to favor alternate genotypes during successive bouts of selection, but, once again, heterozygotes generally exhibited the highest viabilities.

Viability selection also appears to favor heterozygotes in the perennial herb, *Liatris cylindracea* (Schaal and Levin, 1976). Highly heterozygous plants had greater viability and fecundity and also reproduced earlier.

A striking case of selection favoring heterozygous genotypes has been reported for yellow-poplar, *Liriodendron tulipifera* (Brotschol et al., 1986). Yellow-poplar reproduces with a mixed mating system, with 55% and 86% of the seed produced by outcrossing in two localities. Yet the genotypes of mature trees fit Hardy-Weinberg expectations. By removing homozygous genotypes, selection increases the proportion of heterozygous individuals as the cohort ages. Numerous other observations with forest trees report similar patterns of change within a life cycle (e.g., Phillps and Brown, 1977; Tigerstedt et al., 1982; Farris and Mitton, 1984). This pattern of genotypic change is most parsimoniously interpreted as selection against selfed and thus predominantly homozygous genotypes.

In addition to the pattern of F decreasing to zero within the life cycle, significant excesses of heterozygotes (summarized in Mitton and Jeffers, 1988) have been reported in mature stands of ponderosa pine, *Pinus ponderosa* (Linhart et al., 1981), jack pine, *Pinus banksiana* (Cheliak et al., 1985), black spruce, *Picea mariana* (Boyle and Morgenstern, 1986; Yeh et al., 1986), Monterey pine, *Pinus radiata* (Plessas and Strauss, 1986), Douglas-fir, *Pseudotsuga menziesii* (Shaw and Allard, 1982), and balsam fir, *Abies balsamea* (Neale and Adams, 1985). Whereas the decline in values of F from initial positive values to zero is consistent with selection against inbred individuals, the production of excesses of heterozygotes must be a different process, in which heterozygotes are favored (Shaw and Allard, 1982; Mitton and Jeffers, 1988). If selection acts only to eliminate selfed genotypes, then all outcrossed genotypes would have equal fitnesses. This view of selection is most consistent with the dominance model of inbreeding depression. But when selection produces excesses of heterozygotes, then fitness must increase with heterozygosity, even within a pool of genotypes produced by perfect outcrossing. Excesses of heterozygotes implicate the overdominance model of fitness determination.

Protein Heterozygosity and Growth Rate in Trees

Because their rate of growth is recorded in annual rings, temperate trees are convenient subjects for studies of the association between genotype and growth rate. The first relationship between protein heterozygosity and growth was reported for quaking aspen, *Populus tremuloides* (Mitton and Grant, 1980). Multiple regression was used to separate the effects of age, sex, elevation, and heterozygosity upon radial growth, which was estimated with cores from five ramets per clone. Growth rate increased with heterozygosity of three enzyme polymorphisms. Both ponderosa pine and lodgepole pine have complex relationships between heterozygosity and growth. Heterozygosity is not related to the mean growth rate of mature ponderosa pine, but the variance of growth rate increases, and the variance of female cone production decreases with heterozygosity (Knowles and Grant, 1981; Linhart and Mitton, 1985). The mean growth rate is not related to heterozygosity of lodgepole pine, but the variance of growth rate decreases with heterozygosity (Knowles and Mitton, 1980). There is a relationship between growth and heterozygosity in pitch pine, *Pinus rigida*, but it is highly dependent upon the age of the stand (Ledig et al., 1983). There is no clear relationship between growth and genotype in young stands, but in mature stands growth rate increases with heterozygosity. The differential capabilities of the various genotypes may only be expressed under the competitive conditions existing in mature stands of pitch pine (Ledig et al., 1983). The relationship in seedlings between radicle length and allozyme heterozygosity was tested under nine experimental conditions with jack pine (Govindaraju and Dancik, 1987). Radicle length increased significantly with heterozygosity in three of the nine experimental conditions and most clearly under conditions that produced water stress. Growth rate increases with heterozygosity in knobcone pine, *Pinus attenuata* (Strauss, 1986). The relationship is most pronounced in progeny produced by selfing, but a mild relationship is apparent in outcrosses as well. There is also a relationship in knobcone pine between growth variability and heterozygosity, but the relationship is complex (Strauss, 1987). Bongarten et al. (1985) examined the relationship between growth and heterozygosity in a plantation of 15-year-old Douglas-fir, but growth was not related to heterozygosity. These young trees were watered and not crowded, so they may be analogous to the young, relatively unstressed stands of pitch pine (Ledig et al., 1983), which exhibited no relationship between genotype and growth. Thus, although negative results have been reported, growth rates often increase with heterozygosity in trees. This relationship should be most evident in young trees (Mitton, 1983), when all energy is invested in growth. The patterns may be obscured in mature trees, which may vary their apportionment of energy to sex and growth (Linhart and Mitton, 1985).

Protein Heterozygosity and Yield in Corn

Heterosis of hybrid corn, *Zea mays*, is clearly associated with mitochondrial activity (Hanson et al., 1960) and complementation of mitochondria (McDaniel and Sarkissian, 1966, 1968; Sarkissian and Srivastava, 1969; McDaniel, 1972). There is also a substantial literature that examines the response of yield to allozyme heterozygosity.

There is some evidence that viability is related to heterozygosity in corn (Kahler et al., 1984). Approximately 10% of the seed of corn is produced by selfing, so seeds and seedlings exhibit a positive value of F. Yet when adult plants are sampled, the value of F has dropped to zero, revealing selection against homozygous genotypes. This raises the question of whether selection is acting against selfed genotypes, perhaps genotypes with deleterious recessive alleles carried in homozygous condition, or whether selection is simply eliminating homozygous genotypes, regardless of whether those homozygotes were produced by selfing or outcrossing.

Studies in F_2 populations: The genetic resources of corn are preserved in numerous inbred lines, and corn is brought to the market in the form of F_1 crosses between inbred lines. Although corn is predominantly outcrossed, the F_2 would undoubtedly carry a substantial level of linkage disequilibrium, for they would have experienced a single bout of recombination between the inbred lines. The performance of F_2 is much more heterogeneous than the performance of F_1.

Two studies that measured yield, morphological characters, and protein polymorphisms revealed that yield is highly correlated with protein heterozygosity in F_2 populations. Kahler and Wehrhan (1986) measured 11 morphological and yield characters and summarized the relationships between these variables with principal components analysis. The first axis of variation, which captured 42% of the variation, was primarily an axis describing variation in yield. Protein heterozygosity was highly correlated with this axis, and heterozygotes generally performed better than homozygotes. The results suggested that most of the enzyme loci were linked to or associated with genes that contribute to high seed yield. A similar study of yield and protein heterozygosity was performed with the F_2 from a different set of inbred strains (Edwards et al., 1987). Yield increased dramatically with heterozygosity at 15 polymorphic loci (Fig. 6.2), so that the most heterozygous individuals produced about 50% more seed than the most homozygous individuals.

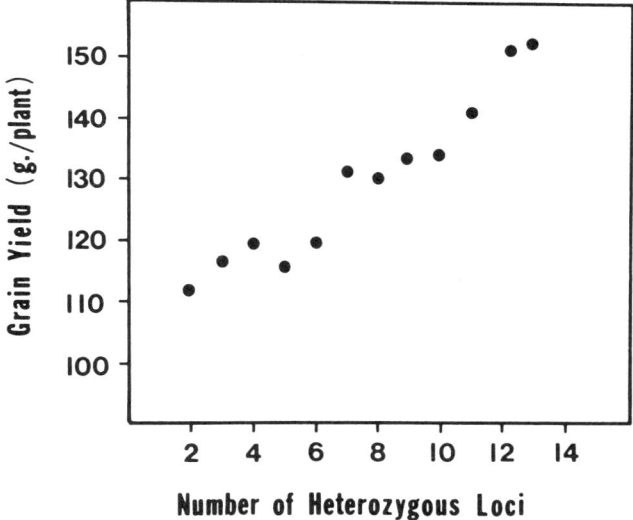

Fig. 6.2. The relationship of mean grain yield to the number of heterozygous enzyme loci in an F_2 population of corn (from Edwards et al., 1987).

Studies of hybrid corn: Whether protein polymorphisms directly influence yield in corn or whether heterozygosity at these loci inadvertently predicts heterozygosity at loci directly controlling yield, protein heterozygosity should be associated with yield in F_1 hybrids. Several studies of heterozygosity and yield of F_1 have been conducted, and the results are mixed. In general, early studies inferred the degree of genetic differentiation from either morphological characters or the geographic distance between sites of origin of the strains. These early studies indicated that yield increased with the degree of differentiation between parental strains (Moll et al., 1962). This general result broke down when crosses between highly differentiated strains were included in the analyses (Moll et al., 1965). Studies using protein polymorphisms to estimate the degree of differentiation of parental strains have also returned mixed results. Some studies have reported positive correlations between yield and heterozygosity (Heidrich-Sobrinho and Cordeiro, 1975; Gonella and Peterson, 1978; Tsaftaris and Efthimiadis, 1987), others have reported weak correlations or no association at all (Hunter and Kannenberg, 1971; Price et al., 1986; Lamkey et al., 1987). These contrasting results may be resolved when the pedigree or genealogical lineage of the inbred strain is considered (Frei et al., 1986). When F_1 yield was examined as a function of both allozyme heterozygosity at 21 loci and pedigree or germplasm similarity, both pedigree and allozymes had significant results. Yield was 37% higher in the F_1 of dissimilar pedigrees than similar pedigrees, and the relationship between heterozygosity and yield depended upon pedigree. Yield increased with heterozygosity when parental strains were of similar pedigree, but not when parental strains differed in pedigree. These results suggest that there may be some subtle outbreeding depression in corn, and they may also help us to understand the conflicting results on the relationship between yield and protein heterozygosity in F_1s. Although yield continues to increase with the differentiation of dissimilar germplasm sources, the increase of yield with protein heterozygosity disappears. Ultimately, crosses between the most distinct lines exhibit no heterosis, and occasionally perform more poorly than the inbred strains (Moll et al., 1965).

These results, like many others, bring us back to a persistent dilemma! Does yield increase with protein heterozygosity as a direct consequence of enzyme phenotypes, or do protein genotypes fortuitously reveal genotypes at other, perhaps linked, loci that directly influence yield? Several studies of directional selection suggest that the enzymes themselves play an important role. Allelic frequencies at protein loci changed with selection for yield in two separate studies of corn (Stuber and Moll, 1972; Kahler, 1983). A close examination of the allelic changes in the first study revealed that they were greater than what would be expected due to genetic drift (Stuber et al., 1980), suggesting that selection directly influenced these loci. Finally, when a population of otherwise unselected corn was constructed to have allozyme frequencies similar to a strain selected for high yield, there was a significant increase in yield (Stuber et al., 1982).

The *Adh* Polymorphism in Corn

Perhaps the most comprehensive biochemical analysis of an enzyme polymorphism in plants is the series of studies focusing on the *Adh* polymorphism in corn. Two common alleles segregate at the Adh_1 locus, F and Cm. The subunits produced by these alleles are inactive, but they assemble *in vivo* and *in vitro* to form active dimers. Thus, homozygotes exhibit single bands, and heterozygotes exhibit three bands on starch gels (Schwartz, 1960). The subunits produced by the alternate alleles differ biochemically. The F allele has high activity but is labile, losing activity quickly at either high temperature or high pH. Alternatively, the Cm allele has lower activity but is more resistant to denaturation by either temperature or pH. The activities of the three bands in heterodimers illustrate the biochemical differences of the subunits and document the advantage of the heterodimer over the homodimers. The heterodimer captures the advantages of each of the subunits, exhibiting both high activity and stability (Schwartz and Endo, 1966; Schwartz and Laughner, 1969). This enhanced stability of the F subunit appears to spring not from the strength of interdimeric bonds but from some configurational change during the linking of F and Cm subunits (Schwartz, 1973). These studies clearly demonstrate biochemical differences among genotypes, and they invite speculation on the adaptive significance of the variation. The enhanced activity and stability of the enzymes of heterozygotes may permit lower protein turnover rates, providing an energetic advantage to heterozygotes (Hawkins et al., 1986). The fixation of an Adh^{FCm} duplication in a strain of Colombian maize (Schwartz and Endo, 1966) suggests some fitness advantage of heterozygotes. However, we do not have the empirical data demonstrating a physiological consequence of this variation, nor has anyone shown an effect of the *Adh* polymorphism upon either viability or fecundity.

Although these studies are a necessary first step in determining the significance of the protein polymorphisms and provide the incentive to press further, they provide no definitive answers. Decisive evidence for the selective maintenance of protein polymorphisms comes from kinetic, physiological, and demographic studies of polymorphic loci (Koehn, 1978), but these programs are so labor intensive and time consuming that few are willing to amass the evidence needed to demonstrate unambiguously the adaptive role of enzyme polymorphisms. We need a simple, direct test to indicate whether specific polymorphisms warrant a major investment in time, either for kinetic and physiological studies or for incorporation into breeding programs.

The Adaptive Distance Model

A recent theoretical development (Smouse, 1986) provides a test to make inferences concerning the presence or absence of selection at polymorphic loci. This test can be used as an initial assessment of whether allelic frequencies are consistent with estimates of fitness at allegedly overdominant loci. The test is particularly useful when correlations between heterozygosity and components of fitness lead an investigator to wonder whether selection is acting directly at the locus in question or whether there is simply a correlation between genotypes at the marker locus and the direct targets of selection.

If an investigator is willing to make three important assumptions concerning the locus of interest, there is a clear relationship between allelic frequencies and relative fitnesses. These assumptions are:

(1) selection is acting directly upon the polymorphism;
(2) there is overdominance or heterozygote superiority at the locus;
(3) allelic frequencies are at equilibrium.

If natural selection is acting upon genotypes to produce marginal overdominance, then the fitnesses of those genotypes are predicted by the allelic frequencies. Consider a locus with two alleles, A and a, whose frequencies are p and q.

	AA	Aa	aa
frequency	p^2	$2pq$	q^2
fitness	$1-s$	1	$1-t$
adaptive distance	$1/p$	0	$1/q$

If we consider the classical case of overdominance, in which the fitness of the AA genotype is depressed by s and the fitness of the aa genotype is depressed by t relative to the heterozygote, then the equilibrium allelic frequencies are $p=t/s+t$ and $q=s/s+t$, and the segregational load is $l=st/s+t$. Clearly, the fitness of the more common homozygote exceeds the fitness of the rarer homozygote. Smouse (1986) pointed out that allelic frequencies are related to relative fitnesses in the following way:

$\ln$ (fitness AA) $= \ln$ (adaptive distance AA) $= -s = -1/p$
$\ln$ (fitness Aa) $= \ln$ (adaptive distance Aa) $= 0$
$\ln$ (fitness aa) $= \ln$ (adaptive distance aa) $= -t = -1/q$.

These relations permit us to test the fit of estimated fitnesses to fitnesses predicted from allelic frequencies.

If the locus is not influenced by natural selection, there should be no significant differences among the observed fitnesses, and there should be no relationship between observed fitnesses and fitnesses predicted from allelic frequencies. Although there will always be the possibility that the marker locus is neutral but in linkage disequilibrium with a locus directly influenced by selection, linkage disequilibrium appears to be a rare phenomenon in outbreeding species (Mukai et al., 1971; Charlesworth and Charlesworth, 1976; Hedrick et al., 1978; Clegg et al., 1980), and therefore this test may be used to make a weak inference concerning the action of selection upon the locus.

The adaptive distance model has been applied to empirical data on pitch pine to study protein polymorphisms (Bush et al., 1987). Radial growth rate of pitch pine increases with heterozygosity of proteins, although this relationship was only observed in mature stands. The initial report (Ledig et al., 1983) presented growth rate as a function of heterozygosity class with regressions run on class means rather than on individuals. When the regression was run with individuals rather than class means, the positive relationship between growth and heterozygosity was no longer significant (Bush et al., 1987). The adaptive distance model was then applied to these data to determine whether a higher proportion of the variation in growth rate might be explained. Using multiple regression, they tested the relationship of age-standardized growth rate to adaptive distance at eight polymorphic loci. This multiple regression explained higher proportions of the variances in growth rates than did linear regressions on the number of heterozygous loci. For example, the proportion of variance of growth explained in two of the populations was $R^2=0.49$ and 0.28. The authors concluded that specific genotypes did play a role in the growth of pitch pine.

Viability Differentials in Annual Ryegrass

The method applied to pitch pine will now be illustrated with estimates of seedling mortality in annual ryegrass, *Lolium multiflorum*. Population samples were taken at two sites, A and B, on the eastern shore of Lake Berryessa, in Napa County, California. These sites are separated by approximately 10 km, are not accessible by road, and receive minimal disturbance from cattle. Panicles were taken from randomly chosen plants sampled in an area of approximately 50 × 100 m at each site. Seeds were sown as families in a greenhouse, and genotypes at five enzyme polymorphisms were obtained with starch gel electrophoresis of homogenates of seedling tissue. The enzyme polymorphisms employed were peroxidase (PER), phosphoglucose isomerase (PGI), acid phosphatase (AP), phosphoglucomutase (PGM), and glutamate oxaloacetate transaminase (GOT). Nine seedlings were run from each wind-pollinated family, and these progeny arrays were analyzed with the maximum-likelihood estimation procedure of Brown and Allard (1970) to estimate the proportion of seed set by outcrossing and to assign the maternal genotype to the family. Thus, for each family, nine seedling genotypes and one maternal genotype were available at each of five polymorphic loci (Table 6.1).

Table 6.1. *Genotypic frequencies, F values, and relative viabilities in* Lolium multiflorum

	Enzyme[b]	Sample	Frequencies					Relative viabilities		
			Genotype					Genotype		
			11	12	22	F[c]	χ^2[a]	11	12	22
Site A	PER	seedlings	257	1355	2954	0.088	***	0.59	1	0.87
		adults	19	171	323	−0.027				
	AP	seedlings	4247	411	46	0.137	***	0.64	1	0
		adults	455	68	0	−0.069				
	PGM	seedlings	4425	251	35	0.19	***	0.66	1	0.19
		adults	481	41	1	0.006				
	PGI	seedlings	583	1997	1958	0.031	*	0.94	0.98	1
		adults	62	222	221	0.024				
	GOT	seedlings	47	737	3872	0.026	NS	0.78	0.86	1
		adults	4	72	441	0.024				
Site B	PER	seedlings	278	626	758	0.178	***	0.57	1	0.72
		adults	22	87	76	−0.028				
	AP	seedlings	1529	123	4	0.147	***	0.85	1	0.54
		adults	168	16	1	0.066				
	PGI	seedlings	132	644	890	0.025	NS	0.83	0.86	1
		adults	13	66	106	0.045				

[a] χ^2 tests homogeneity of genotypic frequencies in seedling and adults with a rows by column test of independence. * = $P < 0.05$; *** = $P > 0.001$; NS = Nonsignificant.
[b] Insufficient genetic variation at GOT and PGM at Site B to infer maternal genotypes.
[c] F is the inbreeding coefficient.

If we assume that the populations of ryegrass were in genetic equilibrium, maternal and seedling genotypes can be used to estimate relative fitnesses. The assumption of equilibrium implies that the distribution of maternal genotypes is representative of the genotypes that would have survived to set seed later that same year. This is equivalent to assuming that the genotypes of plants setting seed change little from year to year. The transition from seedling to maternal genotypes can be used to estimate relative viabilities of the genotypes.

Homogeneity of genotypic frequencies is tested with a row by columns test of independence and is reported as a value of chi-square (Table 6.1). Statistical significance indicates that the genotypes of seedling and adults are different. In these cases, relative viabilities have been estimated (Hedrick, 1983), and the viability of the most fit genotype has been assigned a value of 1.0.

Several points can be made from these data. First, the selection coefficients are large. Whereas evolutionary geneticists often model selective processes with selection differentials of 0.001 or 0.01, many of the selection coefficients estimated in annual ryegrass exceed 0.20. The magnitude of these selection coefficients will surprise theoretical population geneticists, but they are within the range of selection coefficients measured by empiricists studying natural populations (Endler, 1986).

A second point is that the ranking of genotypes is relatively consistent in sites A and B. For example, at both sites, the selection intensities at the Per locus are very large, and the ranking of relative fitnesses is $12 > 22 > 11$.

Third, in every case in which the differences between seedling and adult genotypic frequencies are highly significant, heterozygotes have the highest fitnesses. The point most relevant to the theme of this paper is that the patterns of fitnesses conform to the predictions of the adaptive distance model (Smouse, 1986).

The fit of the empirical data to the predictions of the adaptive distance model suggests that selection in ryegrass is not simply removal of selfed individuals, but that viability is dependent upon genotype. That heterozygotes exhibit the highest viability comes as no surprise, for this is expected with both the adaptive distance model and a model of selection against selfed individuals. Approximately 10–20% of the seed set by annual ryegrass are produced by selfing (Allard et al., 1977), and among the most highly homozygous genotypes are the selfed genotypes. In five of the eight cases, and in each case where a striking deficiency of heterozygotes was observed in seedlings, the heterozygotes exhibited the highest fitnesses. Thus, selective mortality preferentially culled homozygous genotypes, increasing the frequency of heterozygotes at each of the polymorphic loci. However, the pattern of fitnesses at these loci is not explained simply by lower viability of homozygous genotypes; the fitnesses of the homozygous genotypes are predicted by the frequencies of the alleles that they carry (Fig. 6.3). Common homozygotes have high fitnesses, rare homozygotes have low fitnesses.

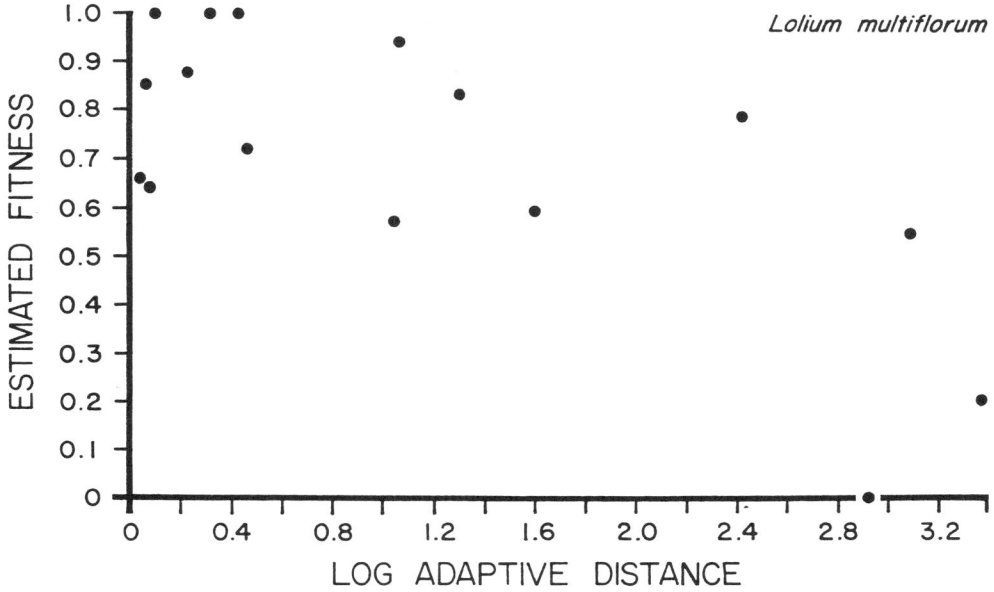

Fig. 6.3. The relationship of viability to adaptive distance in annual ryegrass.

Although the data clearly indicate that mortality is genotypically dependent, the adaptive distance model does not definitively identify the precise targets of selection. Selection may be acting upon these enzyme phenotypes, or the enzyme genotypes may be recording stronger selection directed against loci in linkage disequilibrium with the enzyme loci. Selection would actually have to be stronger at the associated loci, for imperfect correlation between adaptive loci and enzyme loci would underestimate the intensity of selection. It is difficult to conceive of selection being stronger than that indicated in Table 6.1, but neither this inference nor the adaptive distance model can reject the hypothesis that selection is actually acting upon linked loci rather than the marker loci.

It is not a simple matter to determine whether a protein polymorphism is adaptive, that is, that genetic variation is maintained by selection discriminating among the genotypes at that particular locus. The data needed to demonstrate the adaptive significance of a protein polymorphism were outlined by Koehn (1978):

(1) functional differences among genotypes;
(2) physiological differences among genotypes;
(3) physiological variation must be important in the ecological context of the population and;
(4) variation in fitness as a consequence of the variation in physiology.

It appears that enzyme genotypes are associated with viability in ryegrass (Table 6.1, Fig. 6.3). If that variation in fitness is directly attributable to the enzyme phenotypes, then there should be physiological variaton reliably associated with the enzyme genotypes. This reasoning led to studies of physiological variation among enzyme genotypes in *Lolium*.

Physiological Studies of Enzyme Genotypes in *Lolium*

Physiological studies of enzyme genotypes were conducted with perennial ryegrass, *L. perenne*, rather than annual ryegrass, to take advantage of the superb physiological and demographic studies of perennial ryegrass (Jones et al., 1978; Wilson, 1982; Wilson and Jones, 1982; Day et al., 1984). Rates of dark respiration vary among individuals of perennial ryegrass, and this variation is heritable. Heritability is high; one estimate of heritability is 50%, another is 100% (Wilson, 1982; Wilson and Jones, 1982). Furthermore, the rate of dark respiration is negatively correlated with growth rate (Wilson, 1982; Wilson and Jones, 1982), so that the productivity of a pasture can be enhanced by planting it with individuals selected for their low rates of dark respiration. Biochemical studies of the control of dark respiration show that the variation does not reside in mitochondria, suggesting that it is in glycolysis (Day et al., 1984).

Our studies of dark respiration (Rainey et al., 1987) revealed substantial variation in both dark respiration and the Q_{10} of dark respiration, measured between 20–35°C. Our measurements, like previous measurements, estimated the mean Q_{10} to be 2.0. But far more interesting than the mean is the range in values among individuals—from 1.4 to 4.1. This variation in Q_{10} was related to genotypic variation at one locus for each of the enzymes phosphoglucomutase and 6-phosphogluconate dehydrogenase. At both loci, the Q_{10} was highest in the most common homozygote, intermediate in the heterozygote, and lowest in the rare homozygote (Fig. 6.4). A comparison of individuals heterozygous at both loci with individuals homozygous for the common alleles at both loci revealed that the Q_{10} was 20% lower in double heterozygotes (Fig. 6.5).

The biological relevance of variation in Q_{10} values was tested with a heat stress experiment (Rainey et al., 1987). Plants doubly homozygous for the common alleles at *Pgm* and *6-Pgd* and double heterozygotes were exposed to high temperatures (43°C day, 40°C night) for 5 days. The condition of plants, estimated with the proportion of wilted leaves, was correlated with both the Q_{10} of dark respiration and enzyme genotypes. Doubly heterozygous plants, with lower values of Q_{10}, survived the heat stress in better condition—less leaf-yellowing, less wilting—than homozygous plants (Fig. 6.5).

Associations between enzyme genotypes and both demographic and physiological variation in ryegrass are striking. In annual ryegrass, large viability differentials are associated with enzyme

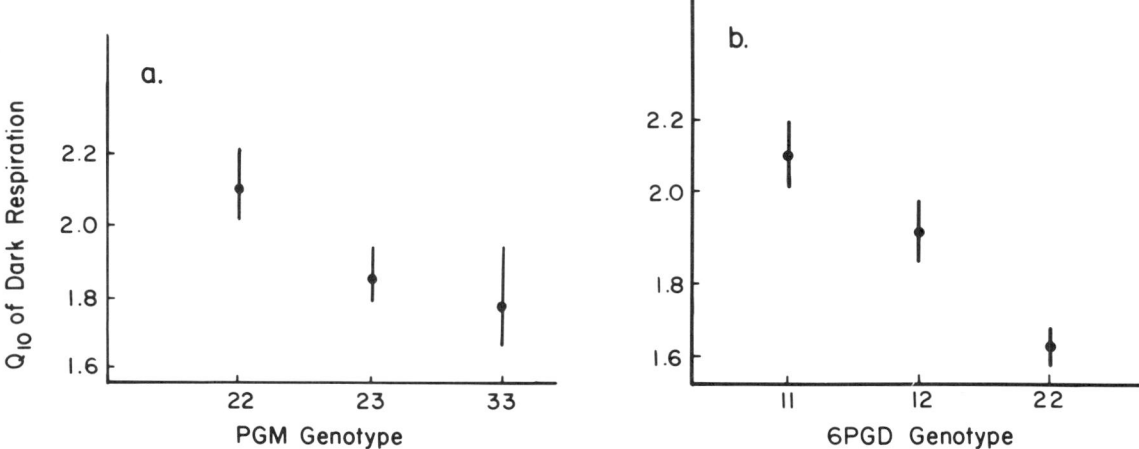

Fig. 6.4. Relationships between PGM (a) and 6PGD (b) genotypes and the Q_{10} of dark respiration in perennial ryegrass. (from Rainey et al., 1987).

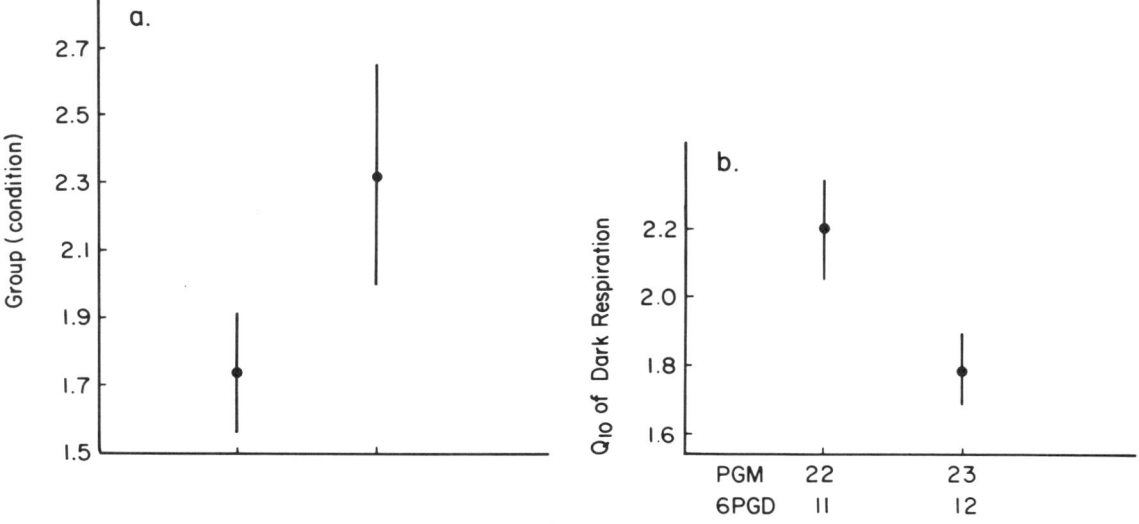

Fig. 6.5. Condition after heat stress (a) and Q_{10} of dark respiration (b) for perennial ryegrass doubly homozygous and doubly heterozygous at the PGM and 6PGD loci.

genotypes, and the patterns of fitness variation at a locus are consistent with predictions based on assumptions of overdominance at those loci. In perennial ryegrass, enzyme genotypes are associated with both variation in the Q_{10} of dark respiration and condition after heat stress. While none of these studies definitively demonstrates the adaptive significance of protein polymorphisms, they certainly demonstrate physiological variation, and they encourage further physiological studies and studies of enzyme kinetics of enzyme polymorphisms in plants.

CONCLUSIONS

We now have sufficient empirical evidence to state unequivocally that at least some enzymes have a major impact upon physiology (Koehn et al., 1983). The most familiar examples are sickle-cell hemoglobin in man (Templeton, 1982), leucine aminopeptidase in the blue mussel, (Koehn et al., 1980; Hilbish and Koehn, 1985), lactate dehydrogenase in the killifish, (DiMichele and Powers, 1982a, b) phosphoglucose isomerase in *Colias* butterflies (Watt, 1979, 1983; Watt et al., 1985), alcohol dehydrogenase in *Drosophila melanogaster* (McDonald, 1983), and the closely linked hemoglobin loci in the deer mouse, (Snyder 1981; Chappel and Snyder 1984). The most thoroughly documented case for biochemical differences among genotypes of an enzyme polymorphism in plants involves alcohol dehydrogenase in corn (Schwartz, 1960, 1973; Schwartz and Laughner, 1969). Another example demonstrates physiological differences among Adh genotypes in *Bromus mollis* (Brown et al., 1976). Despite this paucity of examples, there is no reason to suspect that protein polymorphisms will affect the physiology and demography of plants less than they do in animals. The physiological and demographic consequences of protein polymorphisms have simply received less attention in plants than in animals. Although there are numerous examples of selection in plants recorded at protein polymorphisms (Clegg and Allard, 1973; Clegg et al., 1978; Mitton and Jeffers, 1988), there is not yet a case study in plants that includes the physiological and demographic data needed (Koehn, 1978) to demonstrate selection acting directly upon a protein polymorphism. Physiological studies of enzyme genotypes in ryegrass (Table 6.1, Figs. 6.4, 6.5; Rainey et al., 1987) suggest that at least some enzyme polymorphisms have a substantial impact upon plant physiology, and they encourage further studies of the physiological and demographic consequences of enzyme polymorphisms in plants.

Although we are still uncertain about the mode and intensity of natural selection in natural populations, recent observations and theoretical developments provide new insight into the ways that selection acts in natural populations. Because numerous observations are consistent with selection against inbred individuals, many biologists have presumed that the dominance hypothesis was the most parsimonious choice for a model of selection in species with mixed mating systems and/or restricted powers of dispersal. But some recent observationss are more compatible with the overdominance model of fitness determination. For example, selection produces excesses of heterozygotes in the mature stands of several species of conifers. These heterozygote excesses cannot be produced by selection against inbred individuals (Mitton and Jeffers, 1988). The adaptive distance model (Smouse, 1986) makes predictions of genotypic fitnesses from allelic frequencies, and allows inferences to be made about the target of selection. When the adaptive distance model is applied to data from pitch pine (Bush et al., 1987) and annual ryegrass (Table 6.1, Fig. 6.3), the results suggest that specific genotypes, not just a dichotomy of inbred and outcrossed genotypes, play a role in determining fitness. In ryegrass, the highest fitness is generally captured by heterozygotes. This observation supports the overdominance hypothesis and is consistent with the general theoretical result (Ginzburg 1979; Turelli and Ginzburg, 1983) predicting that fitness should generally increase with the number of heterozygous loci.

ACKNOWLEDGMENTS

Comments on the manuscript were contributed by Yan Linhart, William Schuster, and Michael Grant. Partial support was provided by NSF grant BSR 8614937.

LITERATURE CITED

Allard, R. W., A. Kahler and M. C. Clegg. 1977. Estimation of mating cycle components of selection in plants. *In* F. B. Christiansen and T. M. Fenchel [eds.], *Measuring selection in natural populations,* 1–19. Springer-Verlag, New York.

Anderson W. W., L. Levine, O. Olivera, J. R. Powell, M. E. de la Rosa, V. M. Salceda, M. I. Gaso, and J. Guzman. 1979. Evidence for selection by male mating success in natural populations of *Drosophila pseudoobscura. Proc. Natl. Acad. Sci. USA.* 76: 1519–1523.

Bongarten, B. C., N. C. Wheeler, and K. S. Jech. 1985. *Isozyme heterozygosity as a selection criterion for yield improvement in Douglas-fir.* Proceedings, Canadian Tree Improvement Association, 121–128.

Brotschol, J. V., J. H. Roberds and G. Namkoong. 1986. Allozyme variation among North Carolina populations of *Liriodendron tulipifera. Silvae Genet.* 35: 131–138.

Brown, A. H. D., and R. W. Allard. 1970. Estimation of mating systems in open-pollinated maize populations using isozyme polymorphisms. *Genetics* 66: 133–145.

———, D. R. Marshall, and J. Munday. 1976. Adaptedness of variants at an alcohol dehydrogenase locus in *Bromus mollis* L. (soft bromegrass). *Austral. J. Biol. Sci.* 29: 389–396.

Boyle, T. J. B., and E. K. Morgenstern. 1986. Estimates of outcrossing rates in six populations of black spruce in central New Brunswick. *Silvae Genet.* 35: 102–106.

Bush, R. M., P. E. Smouse, and F. T. Ledig. 1987. The fitness consequences of multiple-locus heterozygosity: the relationship between heterozygosity and growth rate in pitch pine (*Pinus rigida* Mill.). *Evolution* 41: 787–798.

Campbell, D. R., and N. M. Waser. 1987. The evolution of plant mating systems: multilocus simulations of pollen dispersal. *Amer. Naturalist.* 129: 593–609.

Carson H. L. 1987. High fitness of heterokaryotypic individuals segregating naturally within a long-standing laboratory population of *Drosophila silvestris. Genetics* 116: 415–422.

Chappel, M. A., and L. R. G. Snyder. 1984. Biochemical and physiological correlates of deer mouse alpha-chain hemoglobin polymorphisms. *Proc. Natl. Acad. Sci. USA* 81: 5484–5488.

Charlesworth, B., and D. Charlesworth. 1976. An experiment on recombinational load in *Drosophila melanogaster. Genet. Res. Camb.* 25: 267–274.

Cheliak, W. M., B. P. Dancik, K. Morgan, F. C. H. Yeh, and C. Strobeck. 1985. Temporal variation of the mating system in a natural population of jack pine. *Genetics* 109: 565–584.

Clegg, M. T., and R. W. Allard. 1973. Viability versus fecundity selection in the slender wild oat, *Avena barbata* L. *Science* 181: 667–668.

———, A. L. Kahler, and R. W. Allard. 1978. Estimation of life cycle components of selection in an experimental plant garden. *Genetics* 89: 765–792.

———, J. F. Kidwell, and C. R. Horch. 1980. Dynamics of correlated genetic systems. V. Rates of decay of linkage disequilibria in experimental populations of *Drosophila melanogaster. Genetics* 217–234.

Crow, J. F., and M. Kimura. 1970. *An introduction to population genetics theory.* Harper and Row, New York.

———, and ———. 1979. Efficiency of truncation selection. *Proc. Natl. Acad. Sci. USA* 76: 396–399.

Danzmann, R. G., M. M. Ferguson, and F. W. Allendorf. 1987. Heterozygosity and oxygen consumption rate as predictors of growth and developmental rate in rainbow trout. *Physiol. Zool.* 60: 211–220.

———, ———, and ———. 1988. Heterozygosity and components of fitness in a strain of rainbow trout. *Biol. J. Linn. Soc.* 39: 285–304.

Day, D., O. C. De Vos, D. Wilson, and H. Lanbers. 1984. Regulation of respiration in the leaves and roots of two *Lolium perenne* populations with two contrasting mature leaf respiration rates and crop yields. *Pl. Physiol.* 78: 678–683.

DiMichele, L., and D. A. Powers. 1982a. LDH-B genotype-specific hatching times of *Fundulus heteroclitus* embryos. *Nature* 296: 563–565.

———, and ———. 1982b. Physiological basis for swimming endurance differences between LDH-B genotypes of *Fundulus heteroclitus. Science* 216: 1014–1016.

Dobzhansky, T., and B. Wallace. 1953. The genetics of homeostasis in *Drosophila. Proc. Natl. Acad. Sci. USA* 39: 162–171.

East, E. M. 1936. Heterosis. *Genetics.* 21: 375–397.

Edwards, M. D., C. W. Stuber, and J. F. Wendel. 1987. Molecular-marker-facilitated investigations of quantitative-trait loci in maize. I. Numbers, genomic distribution and types of gene action. *Genetics* 116: 113–125.

Endler, J. A. 1986. *Natural selection in the wild.* Princeton University Press, NJ.
Farris, M. A., and J. B. Mitton. 1984. Population density, outcrossing rate, and heterozygote superiority in ponderosa pine. *Evolution* 38: 1151–1154.
Frankel, R. 1983. *Heterosis: reappraisal of theory and practice.* Springer-Verlag, Berlin.
Frei, O. M., C. W. Stuber, and M. M. Goodman. 1986. Use of allozymes as genetic markers for predicting performance in maize single cross hybrids. *Crop Sci.* 26: 37–42.
Ginzburg, L. R. 1979. Why are heterozygotes often superior in fitness? *Theor. Pop. Biol.* 15: 264–267.
———. 1983. *Theory of natural selection and population growth.* Benjamin/Cummings Publishing Co., Menlo Park, CA.
Gonella, J. A., and P. A. Peterson. 1978. Isozyme relatedness of inbred lines of maize and performance of their hybrids. *Maydica* 23: 55–61.
Govindaraju, D. R., and B. P. Dancik. 1987. Allozyme heterozygosity and homeostasis in germinating seeds of jack pine. *Heredity* 59: 279–283.
Hamrick, J. L., Y. B. Linhart, and J. B. Mitton. 1979. Relationship between life history parameters and electrophoretically-detectable genetic variability in plants. *Ann. Rev. Ecol. Syst.* 10: 173–200.
Hanson, J. B., R. H. Hageman, and M. E. Fisher. 1960. The association of carbohydrates with the mitochondria of corn scutellum. *Agron. J.* 52: 49–52.
Hawkins, A. J. S., B. L. Bayne, and A. J. Day. 1986. Protein turnover, physiological energetics and heterozygosity in the blue mussel, *Mytilus edulis*: the basis of variable age-specific growth. *Proc. R. Soc. Lond.* B 229: 161–176.
Hedrick, P. W. 1983. *Genetics of populations.* Science Books International, Boston.
Hedrick, P., S. Jain, and L. Holden. 1978. Multilocus systems in evolution. *Evol. Biol.* 11: 101–184.
Heidrich-Sobrinho, E., and A. R. Cordeiro. 1975. Codominant isoenzymic alleles as markers of genetic diversity correlated with heterosis in maize (Zea mays). *Theor. App. Genet.* 46: 197–199.
Hilbish, T. J., and R. K. Koehn. 1985. Dominance in physiological phenotypes and fitness at an enzyme locus. *Science* 229: 52–54.
Hunter, R. B., and L. W. Kannenberg. 1971. Isozyme characteristics of corn (Zea mays) inbreds and its relationship to single cross hybrid performance. *Canad. J. Genet. Cytol.* 13: 649–655.
Jones, M. B., E. L. Leafe, W. Stiles, and B. Collett. 1978. Pattern of respiration of a perennial ryegrass crop in the field. *Ann. Bot.* 42: 693–703.
Kahler, A. L. 1983. Effect of half-sib and S_1 recurrent selection for increased grain yield on allozyme polymorphisms in maize. *Crop Sci.* 23: 572–576.
———, C. O. Gardner, and R. W. Allard. 1984. Nonrandom mating in experimental populations of maize. *Crop Sci.* 24: 350–354.
———, and C. F. Wehrhan. 1986. Associations between quantitative traits and enzyme loci in the F_2 population of a maize hybrid. *Theor. Appl. Genet.* 72: 15–26.
Kimura, M. 1983. *The neutral theory of molecular evolution.* Cambridge University Press, Cambridge.
———, and J. F. Crow. 1978. Effect of overall phenotypic selection on genetic change at individual loci. *Proc. Natl. Acad. Sci. USA* 75: 6168–6171.
———, and T. Ohta. 1971. *Theoretical aspects of population genetics.* Princeton University Press, Princeton, NJ.
Knowles, P., and M. C. Grant. 1981. Genetic patterns associated with growth variability in ponderosa pine. *Amer. J. Bot.* 68: 942–946.
———, and J. B. Mitton. 1980. Genetic heterozygosity and radial growth variability in *Pinus contorta. Silvae Genet.* 29: 114–117.
Koehn, R. K. 1978. Physiology and biochemistry of enzyme variation: the interface of ecology and population genetics. In P. Brussard [ed.], *Ecological genetics: the interface,* 51–72. Springer-Verlag, New York.
———, and T. J. Hilbish. 1987. The adaptive importance of genetic variation. *Amer. Sci.* 75: 134–141.
———, R. Milkman, and J. B. Mitton. 1976. Population genetics of marine pelecypods. IV. Selection, migration and genetic differentiation in the blue mussel *Mytilus edulis. Evolution* 30: 2–32.

_____, R. I. Newell, and F. Immerman. 1980. Maintenance of aminopeptidase allelic frequency cline by natural selection. *Proc. Natl. Acad. Sci. USA* 77: 5385–5389.

_____, F. J. Turano, and J. B. Mitton. 1973. Population genetics of marine pelecypods. II. Genetic differences in microhabitats of *Modiolus demissus. Evolution* 27: 100–105.

_____, A. J. Zera, and J. G. Hall. 1983. Enzyme polymorphism and natural selection. *In* M. Nei and R. K. Koehn [eds.], *Evolution of genes and proteins*, 115–136. Sinauer Associates, Inc., Sunderland, MA.

Lamkey, K. R., A. R. Hallauer, and A. L. Kahler. 1987. Allelic differences at enzyme loci and hybrid performance in maize. *J. Hered.* 78: 231–234.

Leary, R. F., F. W. Allendorf, and K. L. Knudsen. 1983. Developmental stability and enzyme heterozygosity in rainbow trout. *Nature* 301: 71–72.

_____, _____, and _____. 1984. Superior developmental stability of heterozygotes at enzyme loci in salmonid fishes. *Amer. Naturalist* 124: 540–551.

Ledig, F. T., R. P. Guries, and B. A. Bonefeld. 1983. The relation of growth to heterozygosity in pitch pine. *Evolution* 37: 1227–1238.

Lerner, I. M. 1954. *Genetic homeostasis.* Oliver and Boyd, Edinburgh.

Linhart, Y. B., J. B. Mitton, K. B. Sturgeon, and M. L. Davis. 1981. Genetic variation in space and time in a population of ponderosa pine. *Heredity* 46: 407–426.

Linhart, Y. B., and J. B. Mitton. 1985. Relationships among reproduction, growth rate, and protein heterozygosity in ponderosa pine. *Amer. J. Bot.* 72: 181–184.

McDaniel, R. G. 1972. Mitochondrial heterosis and complementation as biochemical measures of yield. *Nature, New Biol.* 236: 190–191.

_____, and I. V. Sarkissian. 1966. Heterosis: complementation by mitochondria. *Science* 152: 1640–1642.

_____, and _____. 1968. Mitochondrial heterosis in maize. *Genetics* 59: 465–475.

McDonald, J. F. 1983. The molecular basis of adaptation: a critical review of relevant ideas and observatons. *Ann. Rev. Ecol. Syst.* 14: 77–102.

Milkman, R. 1978. Selectiion differentials and selection coefficients. *Genetics* 88: 391–403.

_____. 1982. Toward a unified selection theory. *In* R. Milkman [ed.], *Perspectives on evolution,* 105–118. Sinauer Associates, Inc., Sunderland, MA.

Mitton, J. B. 1978. Relationship between heterozygosity for enzyme loci and variation of morphological characters in natural populations. *Nature* 273: 661–662.

_____. 1983. Conifers. *In* S. Tanksley and T. Orton [eds.], *Isozymes in plant genetics and breeding, Part B,* 443–472. Elsevier, Amsterdam.

_____, and M. C. Grant. 1980. Observations on the ecology and evolution of quaking aspen, *Populus tremuloides,* in the Colorado Front Range. *Amer. J. Bot.* 67: 1040–1045.

_____, and _____. 1984. Relationships among protein heterozygosity, growth rate, and developmental stability. *Ann. Rev. Ecol. Syst.* 15: 479–499.

_____, and R. M. Jeffers. 1988. The genetic consequences of mass selection for growth rate in Engelmann spruce. *Silvae Genet.:* (In press).

_____, and R. K. Koehn. 1985. Shell shape variation in the blue mussel, *Mytilus edulis,* and its association with enzyme heterozygosity. *J. Exp. Mar. Biol. Ecol.* 90: 73–80.

Moll, R. H., J. H. Lonnquist, J. Velez Fortuno, and E. C. Johnson. 1965. The relationship of heterosis and genetic divergence in maize. *Genetics* 52: 139–144.

_____, W. S. Salhuana, and H. F. Robinson. 1962. Heterosis and genetic diversity in variety crosses in maize. *Crop Sci.* 2: 197–198.

Moore, M. N., R. K. Koehn, and B. L. Bayne. 1980. Leucine aminopeptidase (aminopeptidase-I), N-acetyl-β-hexosaminidase and lysosomes in the mussel, *Mytilus edulis* L., in response to salinity changes. *J. Exp. Zool.* 214: 239–249.

Mukai, T., L. E. Mettler, and S. Chigusa. 1971. Linkage disequilibrium in a local population of *Drosophila melanogaster. Proc. Natl. Acad. Sci. USA* 68: 1065–1069.

Neale, D. B., and W. T. Adams. 1985. Allozyme and mating-system variation in balsam fir (*Abies balsamea*) across a continuous elevational transect. *Canad. J. Bot.* 63: 2448–2453.

Nei, M. 1975. *Molecular population genetics and evolution.* North-Holland, Amsterdam.

Phillips, M. A., and A. H. D. Brown. 1977. Mating systems and hybridity in *Eucalyptus peacittona. Austral. J. Biol. Sci.* 30: 337–344.

Plessas, M. E., and S. H. Strauss. 1986. Allozyme differentiation among populations, stands, and cohorts in Monterey pine. *Canad. J. For. Res.* 16: 1155–1164.

Powers, D. A., G. S. Greaney, and A. R. Place. 1979. Physiological correlation between lactate dehydrogenase genotype and haemoglobin function in killifish. *Nature* 277: 240–241.

Price, M. V., and N. M. Waser. 1979. Pollen dispersal and optimal outcrossing in *Delphinium nelsoni. Nature* 277: 294–296.

Price, S. C., A. L. Kahler, A. R. Hallauer, P. Charmley, and D. A. Giegel. 1986. Relationships between performance and multilocus heterozygosity at enzyme loci in single-cross hybrids of maize. *J. Hered.* 77: 341–344.

Rainey, D. Y., J. B. Mitton, and R. K. Monson. 1987. Associations between enzyme genotypes and dark respiration in perennial ryegrass, *Lolium perenne* L. *Oecologia* 74: 335–338.

Ralls, K., and J. Ballou. 1983. Extinction: lessons from zoos. In C. M. Schonewald-Cox, S. Chambers, B. MacBryde, and W. Thomas [eds.], *Genetics and conservation. A reference for managing wild animal and plant populations,* 164–184. Benjamin/Cummings Publishing Co., Menlo Park, CA.

Robertson, F. W., and E. C. R. Reeve. 1952. Heterozygosity, environmental variation and heterosis. *Nature* 170: 286–287.

———, and ———. 1955. Studies in quantitative inheritance. VIII. Further analysis of heterosis in crosses between inbred lines of *Drosophila melanogaster. Z. Indukt. Abstammungs Vererbunglsehre* 86: 439–458.

Sarkissian, I. V., and H. K. Srivastava. 1969. High efficiency, heterosis, and homeostasis in mitochondria of wheat. *Proc. Nat. Acad. Sci. USA* 63: 302–309.

Schaal, B. A., and D. A. Levin. 1976. The demographic genetics of *Liatris cylindracea* Michx. *Amer. Naturalist* 110: 191–206.

Schwartz, D. 1960. Genetic studies on mutant enzymes in maize: synthesis of hybrid enzymes by heterozygotes. *Proc. Natl. Acad. Sci. USA* 46: 1210–1215.

———. 1973. Single gene heterosis for alcohol dehydrogenase in maize: the nature of the subunit interaction. *Theor. Appl. Genet.* 43: 117–120.

———, and T. Endo. 1966. Alcohol dehydrogenase polymorphism in maize—simple and compound loci. *Genetics* 53: 709–715.

———, and W. J. Laughner. 1969. A molecular basis for heterosis. *Science* 166: 626–627.

Seager, R. D., and F. J. Ayala. 1982. Chromosome interactions in *Drosophila melanogaster.* I. Viability studies. *Genetics* 102: 467–483.

———, ———, and R. W. Marks. 1982. Chromosome interactions in *Drosophila melanogaster.* II. Total fitness. *Genetics* 102: 485–502.

Shaw, D., and R. W. Allard. 1982. Isozyme heterozygosity in adult and open-pollinated embryo samples of Douglas-fir. *Silvae Fenn.* 16: 115–121.

Smouse, P. E. 1986. The fitness consequences of multiple-locus heterozygosity under the multiplicative overdominance and inbreeding depression models. *Evolution* 40: 946–957.

Snyder, L. R. G. 1981. Deer mouse hemoglobins: is there genetic adaptation to high altitude? *BioScience* 31: 299–304.

Strauss, S. H. 1986. Heterosis at allozyme loci under inbreeding and crossbreeding in *Pinus attenuata. Genetics* 113: 115–134.

———. 1987. Heterozygosity and developmental stability under inbreeding and crossbreeding in *Pinus attenuata. Evolution* 41: 331–339.

Stuber, C. W., M. M. Goodman, and R. H. Moll. 1982. Improvement of yield and ear number resulting from selection at allozyme loci in a maize population. *Crop Sci.* 22: 737–740.

———, and R. H. Moll. 1972. Frequency changes of isozyme alleles in a selection experiment for grain yield in maize (*Zea mays* L.). *Crop Sci.* 12: 337–340.

———, ———, M. M. Goodman, H. E. Schaffer, and B. S. Weir. 1980. Allozyme frequency changes associated with selection for increased grain yield in maize (*Zea mays* L.). *Genetics* 95: 225–236.

Templeton, A. R. 1982. Adaptation and the integration of evolutionary forces. In R. Milkman [ed.], *Perspectives on evolution,* 15–31, Sinauer Associates, Inc. Sunderland, MA.

Tigerstedt, P. M. A., D. Rudin, T. Niemela, and J. Tammisola. 1982. Competition and neighbouring effect in a naturally regenerating population of Scots pine. *Silva Fennica* 16: 122–129.

Tracey, M. L., and F. J. Ayala. 1974. Genetic load in natural populations: is it compatible with the hypothesis that many polymorphisms are maintained by natural selection? *Genetics* 77: 569–589.

Tsaftaris, A. S., and P. Efthimiadis. 1987. F_1 heterosis and heterozygosity for isozymic structural loci in maize. In M. C. Rattazzi, J. G. Scandalios, and G. S. Whitt [eds.], *Isozymes: current topics in biological and medical research,* Vol. 16, Agriculture, physiology and medicine, 157–174. Alan R. Liss, Inc., New York.

Turelli, M., and L. Ginzburg. 1983. Should individual fitness increase with heterozygosity? *Genetics* 104: 191–209.

Vrijenhoek, R. C., and S. Lerman. 1982. Heterozygosity and developmental stability under sexual and asexual breeding systems. *Evolution* 36: 768–776.

Waser, N. M., and M. V. Price. 1983. Optimal and actual outcrossing, and the nature of plant-pollinator interaction. In C. E. Jones and R. J. Little [eds.], *Handbook of experimental pollination biology,* 341–359. Van Nostrand Reinhold, New York.

———, and ———. 1985. Reciprocal transplant experiments with *Delphinium nelsonii* (Ranunculaceae): evidence for local adaptation. *Amer. J. Bot.* 72: 1276–1732.

Watt, W. B. 1979. Adaptation at specific loci. I. Natural selection in phosphoglucose isomerase of *Colias* butterflies: biochemical and population aspects. *Genetics* 87: 177–194.

———. 1983. Adaptation at specific loci. II. Demographic and biochemical elements in the maintenance of the *Colias* PGI polymorphism. *Genetics* 103: 691–724.

———, P. A. Carter, and S. M. Blower. 1985. Adaptation at specific loci. IV. Differential mating success among glycolytic allozyme genotypes of colias butterflies. *Genetics* 109: 157–175.

———, ———, and K. Donohue. 1986. Females' choice of "good genotypes" as mates is promoted by an insect mating system. *Science* 233: 1187–1190.

Wills, C. 1978. Rank-order selection is capable of maintaining all genetic polymorphisms. *Genetics* 89: 403–414.

———. 1981. *Genetic variability.* Clarendon Press, Oxford.

Wilson D. 1982. Response to selection for dark respiration rate of mature leaves in *Lolium perenne* and its effects on growth of young plants and similar simulated swards. *Ann. Bot.* 49: 303–312.

———, and J. B. Jones. 1982. Effect of selection for dark respiration rate of mature leaves on crop yields of *Lolium perenne* cv s23. *Ann. Bot.* 49: 313–320.

Wright, S. 1977. *Evolution and the genetics of populations,* Vol. 3. Experimental results and evolutionary deductions. University of Chicago Press, Chicago.

Yeh, F. C., M. A. K. Khalil, Y. A. El-Kassaby, and D. C. Trust. 1986. Allozyme variation in *Picea mariana* from Newfoundland: genetic diversity population structure, and analysis of differentiation. *Canad. J. For. Res.* 16: 713–720.

Zouros, E., and D. W. Foltz. 1987. The use of allelic isozyme variation for the study of heterosis. In M. C. Rattazzi, J. G. Scandalios, and G. S. Whitt [eds.], *Isozymes: current topics in biological and medical research,* Vol. 13. Alan K. Liss, Inc., New York.

CHAPTER 7

Enzyme Electrophoresis and Plant Systematics

Daniel J. Crawford

Department of Botany
The Ohio State University
Columbus, OH 43210

This chapter discusses the use of enzyme electrophoresis for addressing questions in plant systematics. The present treatment updates previous reviews (Gottlieb, 1977a, 1981a, 1982, 1983, 1984a, 1986; Crawford, 1983, 1985; Giannasi and Crawford, 1986) and also discusses: (1) the factors involved in the initial applications of enzyme electrophoresis to plant systematics; (2) the kinds of studies employing electrophoresis to be emphasized in the future; and (3) the future of enzyme electrophoresis in plant systematics in light of the increasing emphasis on comparative studies of DNA.

The nature of electrophoretic data and comparisons with other information routinely employed by the plant systematist is discussed. Generalizations and patterns emerging from the use of allelic data for examining a variety of systematic questions are presented. Consideration will be given to the value of gene number (or isozyme number) for addressing phylogenetic questions in plants.

Exhaustive treatments of particular topics are not attempted. Rather, one to several examples suffice to illustrate a point. Additional studies will be cited but not discussed, and the reader is referred to the more extensive reviews cited earlier for more in-depth treatments. The present review is from a more general perspective and also includes several of my own perceptions and opinions (biases?) about enzyme electrophoresis and plant systematics.

NATURE OF ELECTROPHORETIC DATA AND RATIONALE FOR USE IN PLANT SYSTEMATICS

The data generated from enzyme electrophoresis differ fundamentally from other information routinely employed by plant systematists because the banding patterns in gels produced by staining for specific enzymes (see Shields et al., 1983; Vallejos, 1983; Soltis et al., 1983; Wendel and Stuber, 1984; Morden et al., 1987; and particularly the chapter by Wendel and Weeden in this volume, for detailed methods) may be interpreted in genetic terms. Different banding patterns may be equated to different alleles at a gene locus or to alleles at different loci. *Allozymes* are different forms of an enzyme specified by alternative alleles at one gene locus, whereas *isozymes* may be used to designate forms of an enzyme specified by structural genes at different loci. Allozymes are inherited as codominants in a simple Mendelian fashion, which allows one to ascertain allelic frequencies for a population of plants, for species, etc. From these data, one can quantify the similarity and differences between populations, groups of populations, species, etc. (Gottlieb, 1977a, 1981a). Electrophoresis also allows one to ascertain the number of isozymes (and therefore the number of gene loci) of particular enzymes included in a study.

Application of the electrophoretic technique for addressing a variety of problems in plant systematics was developed and discussed by Gottlieb in several reviews and original research papers (Gottlieb, 1971a, 1973a, b, 1977a, as examples). His salient points will be considered briefly. The first, mentioned above, is the ability to quantify genetic similarities and differences using enzyme electrophoresis. A second point emerging from Gottlieb's early work (Gottlieb, 1973a, 1974; Gottlieb and Pilz, 1976) was that divergence at genes specifying soluble enzymes is often uncorrelated with plant speciation. That is, speciation may occur but the divergence between the species is no greater than is normally found between populations of a single taxon. Conversely, allozyme divergence may occur between populations in the absence of speciation (Whalen, 1979; Crawford and Bayer, 1981; Roberts, 1983; Gottlieb, 1984b, as examples). Although these findings are perhaps not surprising, they are fortunate for the plant systematist because if allozyme divergence were linked inextricably to speciation, electrophoretic data would be useful solely as taxonomic markers. As will be discussed later in this chapter, electrophoretic data may be employed to investigate a variety of problems in plant systematics that are more interesting and significant than strict taxonomic delimitation.

For a variety of enzymes normally examined electrophoretically there is a highly conserved minimal number of isozymes in diploid plants (Gottlieb, 1982). This means that any increase in isozyme number is indicative of gene duplication at the diploid level or duplication due to polyploidy. Gottlieb (1983) discussed the utility of isozyme number for studying phylogenetic relationships in plants; this topic will be considered later in the chapter.

ELECTROPHORETIC DATA AND PLANT TAXONOMY

Most plant systematists or taxonomists generate some kind of data in addition to the usual observations and analyses (phenetic and/or cladistic) of morphology. Beginning in the late 1950s and early 1960s, secondary chemistry (particularly flavonoid compounds) was a part of many taxonomic investigations that were otherwise largely classical. In contrast to flavonoid compounds, data from enzyme electrophoresis have been employed infrequently as taxonomic markers or characters in more classical taxonomic studies, and the question arises as to why this is so.

One obvious reason why allozyme data are not routine taxonomic tools is the point alluded to earlier, namely speciation may occur with a lack of divergence at enzyme loci whereas divergence in allozymes may take place without speciation. Thus, in some instances allozymes may prove "unreliable" as "good" taxonomic characters.

There are other important factors, primarily historical in nature, that explain why secondary chemistry gained such immediate popularity among "rank and file" taxonomists whereas over a decade passed between Gottlieb's early studies and the present popularity of enzyme electrophoresis among plant systematists. The first widespread use of flavonoids involved employing them as additional characters for distinguishing species. In addition, these marker compounds were also employed for documenting interspecific hybridization. Comparative secondary chemistry was utilized initially by plant taxonomists to address primarily taxonomic problems. In contrast, enzyme electrophoresis was initially exploited by population biologists to quantify genetic variation within populations and species. The method was then adopted by workers who could best be classified as plant evolutionists and geneticists, and subsequently it was used by those plant systematists whose research objectives were concerned more with ascertaining relationships and elucidating evolution than with classifying plants.

Another factor delaying the incorporation of electrophoretic data into plant systematics was the prevalent yet mistaken opinion that there was too much variation within species for the data to be useful for assessing relationships among species. This view originated from the early demonstration of the high levels of genetic variation within populations. In general, plant systematists consider characters varying within populations to be of little use for systematic purposes, and this misconception carried over to allozymes. This may be due in part to the fact that plant systematists typically have not been well versed in population biology. Indeed, the idea of "population samples" and thinking in terms of populations may alienate some. It is now clear that this electrophoretic variation can be interpreted from a genetic perspective, and therein lies its systematic utility.

Comparative flavonoid chemistry requires the collection of bulked, dried material, a procedure similar to gathering material for preparing herbarium specimens. In contrast, electrophoretic studies require living material (or material frozen soon after collection); this often necessitates culturing plants from seed in much the same manner as is done in biosystematic and other experimental studies. Therefore, even with regard to the nature of the plant material required, electrophoretic studies are similar to experimental evolutionary investigations whereas flavonoid analyses more closely parallel taxonomic studies.

Despite the fact that enzyme electrophoresis has not been widely employed in taxonomic studies, the data have been useful in certain instances; several examples will suffice to demonstrate this point. Jefferies and Gottlieb (1982) studied two diploid species of *Salicornia* that are nearly indistinguishable morphologically. The taxa occupy slightly different habitats in salt marshes, but may be found intermixed. The question is whether these represent two distinct species (gene pools) or one variable population. Jefferies and Gottlieb (1982) detected no allozyme variation among individuals within each of the species, but found the two taxa to be monomorphic for alternative alleles at six of 30 genes. No evidence of hybridization was detected, and the allozyme data provide compelling evidence for recognizing two species.

Crawford and Wilson (1979) and Crawford (1979) examined allozyme variation within and among several closely related species of *Chenopodium* that are diploid, annual weeds. Taxonomic treatments of the complex had varied, and certain taxa seemingly intergrade morphologically when found growing together. The electrophoretic data indicated that particular species viewed as doubtfully distinct do indeed contain alternative alleles at several genes, with no evidence of hybridization based on isozyme loci. Allelic data were also useful for showing that plants previously recognized as *C. incognitum* in reality represent two biological entities each of which is conspecific with another species in the complex.

Although enzyme electrophoresis has not been employed routinely for investigating difficult taxonomic problems in plants, it can be a powerful method. As will be discussed in the next section, congeneric species are often (but not always!) divergent at genes specifying soluble enzymes; thus allozymic data are useful for ascertaining whether two different morphological types represent the same or different gene pools.

ALLELIC DATA

General Comments and Studies of Conspecific Plant Populations

Allelic frequency data are usually employed to calculate the similarity or differences among plant populations, infraspecific taxa, species, etc. In the present chapter, the genetic identity statistic of Nei (1972) is used as a measure of similarity. For any two populations, this value can range from 0.0 to 1.0; no alleles in common gives a value of 0.0 whereas the same alleles in the same frequencies produce a value of 1.0. Measures of genetic variation, including percentage of loci polymorphic per population, observed heterozygosity, mean number of alleles per locus within populations, etc., are also employed, but they are not considered in detail here (see Gottlieb, 1981a, for further discussion).

One generalization that has emerged from studies of plant populations belonging to the same taxon is that they are very similar allozymically. The mean identities for these populations are often above 0.90 (see discussions and tabulations in Gottlieb, 1977a, 1981a; Crawford, 1983; Giannasi and Crawford, 1986). For outcrossing plants, the numbers and frequencies of alleles detected in any one population are often very similar to another population of the same taxon. That is, one population may well be nearly representative of the species as a whole. This should not, however, be used as an argument for limited populational sampling.

The number of plants to be examined from each population will depend on how important it is to detect low-frequency alleles. If only genetic similarities are to be estimated among populations or species, then low-frequency alleles have little effect on the values. Small samples could lead to incorrect conclusions on whether a particular allele is present in one species and not in another. Marshall and Brown (1975) suggest that in most instances 30 diploid individuals should provide a 95% probability of detecting all alleles at a locus if 20 alleles were present in a frequency of 0.05 each.

The mean genetic identities among populations of self-pollinating plants are similar to those of outcrossers (Gottlieb, 1981a; Crawford, 1983), however, genetic diversity is typically partitioned among rather than within populations of the species. In selfing species, alternative alleles may be either in high frequencies or fixed in different populations. Another pattern sometimes found in self-pollinating species is the virtual lack of variation either within or among populations.

Several studies (see Crawford, 1983; Giannasi and Crawford, 1986, for more complete discussions) have assessed allozymic similarity among geographical races of a species. These are often recognized taxonomically as subspecies or varieties. In the majority of cases, populations of infraspecific taxa are as similar electrophoretically as populations of the same taxon. In certain instances, however, genetic identities for the infraspecific taxa are lowered to the level routinely seen for congeneric species.

The nonuniformity of allozyme similarities detected among geographic races in different studies is not surprising given that the biological situations are no doubt quite different in various cases. Different subspecies of conifers typically exhibit the same high identities found among populations of a single taxon (Wheelar and Guries, 1982; Wheelar et al., 1983). The lack of genetic divergence among subspecies may result from factors such as outcrossing, long distance dispersal of pollen and seeds, and recent separation of the subspecies (Hamrick et al., 1981; Wheelar et al., 1983).

In several instances where lowered genetic identities have been reported for infraspecific taxa (e.g., *Solanum,* Whalen, 1979; *Coreopsis,* Crawford and Bayer, 1981; *Limnanthes,* McNeill and Jain, 1983), the taxa are largely or completely allopatric. Genetic divergence in these instances could result from lack of the homogenizing effect of gene flow; also, the geographic divergence could be a reflection of greater time since the interruption of gene flow between the populations as compared to taxa that are not allopatric. Thus, one or a combination of these factors, lack of gene flow and time, likely accounts for lowered identities.

Differences in morphological features that lead to the recognition of infraspecific taxa often develop without divergence at isozyme gene loci. This is best seen in domesticated plants and their wild progenitors, often recognized as distinct at the infraspecific level, where considerable morphological divergence can occur with little or no differentiation in allozymes. Several examples include *Capsicum* (McLeod et al., 1983), *Zea* (Doebley et al., 1984), and *Zizania* (Warwick and Aiken, 1986) (see chapter by Doebley in this volume). It appears that strong selection by humans for certain features associated with domestication can produce phenotypic changes so rapidly that there is not time for mutations to occur at gene loci specifying soluble enzymes.

Diploid Species and Divergent (or Primary) Speciation

Gottlieb (1977a) compiled mean genetic identities for populations of congeneric species and pointed out that these values are often considerably lower than conspecific populations. The grand mean for available data a decade ago was 0.67. Tabulations of additional data during the past 10 years (see Gottlieb, 1981a; Crawford, 1983; Giannasi and Crawford, 1986) have changed this grand mean very little, if at all. These results strengthen the general conclusion that populations of congeners are *on the average* much less similar to each other allozymically than are populations of the same species (Gottlieb, 1977a).

What is perhaps more impressive than the maintenance of this average during the past decade, however, is the *range* of mean genetic identity values reported for different congeneric species. These range from a low of about 0.25 to complete identity of 1.0, with nearly every value in between having been reported. It follows, therefore, that if one were given only the information that two species are in the same genus it would be impossible to predict with reasonable certainty the mean genetic identity for populations of the two taxa. The question is whether genetic identity can be related to other data available for the taxa. The answer is clearly in the affirmative for the majority of cases examined in sufficient detail, and the best illustrations are provided by taxa exhibiting very high identities and those with low values.

High genetic identities have been reported for species pairs from a number of genera thought to be related as progenitor-derivative (Gottlieb, 1984a, provides an excellent discussion; Crawford, 1983, 1985, and Giannasi and Crawford, 1986, also consider the topic). The species (mostly annuals) are usually very similar morphologically yet are often isolated by chromosomal repatterning and/or other factors. The derivative species typically contain a limited extraction of the allozyme variation found in the progenitors. Results from electrophoretic studies of progenitor-derivative species pairs have suggested that speciation has been recent and that the genetic differences between the species may not be as profound as has often been assumed (Gottlieb, 1984c; Stebbins, 1986).

Consistently high genetic identities among congeners have also been found on oceanic islands (Helenurm and Ganders, 1985; Lowrey and Crawford, 1985; Crawford et al., 1987a; see Crawford et al., 1987b, for general discussion). The mean genetic identities for populations of different species are frequently as high (approximately 0.90 and above) as conspecific populations. The situation for these species is quite different from the progenitor-derivative pairs because the island species are perennial, very distinct morphologically, and have few if any post-mating isolating factors. The one thing insular species appear to have in common with progenitor-derivative pairs is recent divergence. The ability to determine the age of oceanic islands makes it possible to set upper limits on the ages of endemic species, and it seems highly likely that many island species are more recent than well-differentiated congeners in continental areas (Crawford et al., 1987b). The totality of data, therefore, suggests that speciation has been rapid and recent on oceanic islands, and the electrophoretic information is concordant with this hypothesis.

An electrophoretic study of the Hawaiian silversword alliance (Compositae-Madiinae), a fascinating, monophyletic assemblage of plants, revealed a different pattern of interspecific variation. The group has been examined cytogenetically and in other ways in several excellent studies by Carr and Kyhos (1981, 1986) and Carr (1985a, b). Despite spectacular differences in morphology (three genera are recognized) and ecology, many species are highly interfertile, although some species have reduced fertility due to chromosomal differences (Carr and Kyhos, 1981, 1986). The hybridization data argue for the monophyletic nature of the whole alliance.

Certain species in the silversword alliance do exhibit much lower genetic identities as compared to previous values obtained for *Bidens, Tetramolopium,* and *Dendroseris* (Witter, 1986, 1987). In particular, those species of *Dubautia* on the oldest island (Kauai) show lowest identities among themselves. Carr and Kyhos (1986) postulated these taxa to be the oldest in the genus based on cytogenetic data. Species on younger islands exhibit much higher genetic similarities. Witter's results are particularly significant because they implicate time as the critical factor for allozyme divergence, and demonstrate that with sufficient time divergence among congeners on islands is comparable to continental taxa.

Rapid radiation of continental taxa may also result in high interspecific genetic identities, as demonstrated in an electrophoretic study of genetic variation within and among four species belonging to the genus *Heuchera* (Soltis, 1985). The same species had been accorded a detailed biosystematic study by Wells (1979), and her data provided the perspective from which the allozymic information could be viewed. Based on information from morphology, flowering time, and fertility of synthetic hybrids, Wells (1979) suggested that *Heuchera americana* and *H. pubescens* form one closely related species pair and *H. parviflora* and *H. villosa* compose a similar pair. *Heuchera pubescens* is a montane species preferring cooler climates than *H. americana,* but when they occur together swarms of highly fertile hybrids are produced. *Heuchera parviflora* and *H. villosa* are isolated by ecological factors such as substrate preference; synthetic hybrids between the two are highly fertile but hybridization under natural conditions apparently does not occur. Wells (1979) demonstrated that hybrids between species belonging to different pairs have very reduced fertility relative to hybrids between species of the same pair.

Mean genetic identities of 0.99 and 0.98 were reported for populations of the two species in each pair (Soltis, 1985). Pair-wise comparisons for populations of species from different pairs were in the range of 0.83 to 0.85. The high allozymic similarities for these two pairs of diploid, herbaceous, outcrossing plant species are concordant with the hypothesis that they are in the process of diverging as they adapt to differing ecological conditions. That is, from an evolutionary perspective, it appears that the pairs of species are in early stages of divergence and with time additional barriers to gene flow could develop, just as they have between species of the different pairs.

It seems fair to conclude that available data indicate that congeneric species exhibiting genetic identities similar to conspecific populations are the result of recent speciation. That is, these are relatively "new" species as compared to species exhibiting much lower identities. Defining "new" or "recent" in absolute terms is difficult or impossible to do in most instances, although data for certain taxa suggest it could be as short as 5,000 to 20,000 years (Lowrey and Crawford, 1985; Crawford et al., 1985; Stutz, 1978) to perhaps as long as one million years (Witter, 1986, 1987). Speciation may have occurred more recently in other plants that have been studied electrophoretically, but there is no way to document the time.

Congeneric species of homosporous ferns generally have very low genetic identities compared to flowering plants (Haufler, 1985; Werth et al., 1985a; see the chapter by Soltis and Soltis in this volume), and the antiquity of many of these taxa may again be the critical factor. Low genetic identities approaching those in homosporous ferns have sometimes been found for congeners of flowering plants. For example, genetic similarities among populations of five species in the genus *Mabrya* of the Scrophulariaceae are relatively low (Elisens and Crawford, 1988). The genus consists of herbaceous, perennial plants occurring in small, isolated populations on rock walls in canyons and barrancas of the southwestern U.S. and Mexico. A variety of data suggested that species of *Mabrya* represent relictual endemics, and that they have been isolated since before the Pleistocene (see discussion in Elisens and Crawford, 1988). Electrophoretic data are concordant with this hypothesis because mean identities for pair-wise comparisons of populations of different taxa range from 0.459 to 0.715, with half the comparisons falling below 0.60.

Electrophoretic data have been interpreted from the perspective of mode of plant speciation (Crawford, 1985). It is clear from the previous discussion that speciation may occur without divergence in allozymes, and in these instances all available data suggest a rapid process (quantum speciation in the general concept of Grant, 1981). The progenitor-derivative species pairs and the insular plants discussed earlier ostensibly represent examples of quantum speciation.

The geographical model of speciation assumes a more gradual divergence of populations in a variety of features through time. Electrophoretic studies of plants thought to be examples of geographic speciation reveal a contrasting pattern of divergence compared to examples of quantum speciation. Electrophoretic results for species of diploid ($n = 7$) *Layia* (Warwick and Gottlieb, 1985) are concordant with the geographical model of speciation proposed for these taxa by Clausen et al. (1941). In short, the six species may be placed in three groups on the basis of a variety of features including the fertility of synthetic hybrids. Mean genetic identities for populations of species within each of the groups range from 0.824 to 0.901 whereas values for the different groups are 0.576, 0.576, and 0.642 (Warwick and Gottlieb, 1985). If the species of *Layia* have diverged gradually through time, then one might expect concordance between allozyme divergence and differences in other features, which indeed is the case.

Studies of Hybrid Speciation at the Diploid Level and the Origin of Polyploids

The earlier discussion concerned use of allozymes to examine divergence at the diploid level; the present section will briefly consider allelic data for studies of hybrid speciation at the diploid level and for elucidating the origin of polyploids. Raven (1976, 1980), Grant (1981), and others have argued that hybridization between congeneric diploid taxa (often species) followed by selection for one or more recombinant derivatives can result in the formation of new species. Despite the general view that hybridization at the diploid level is a potentially important evolutionary factor in plants (at least in certain groups), detailed studies of species thought to represent hybrid derivatives are lacking. Notable exceptions include species of *Delphinium* (Lewis and Epling, 1959), *Lasthenia* (Ornduff, 1969), and *Stephanomeria* (Gottlieb, 1971b).

Gottlieb (1972) provided a critical discussion of the difficulties of employing various kinds of characters for documenting hybridization. The advantages of allozymic data for investigating hybridization were presented initially by Gallez and Gottlieb (1982) and also were discussed by Crawford (1983, 1985); thus, a brief review of the salient points will suffice. Because alleles of a gene locus produce different allozymes and these are inherited as codominants, it is possible to ascertain whether a presumed hybrid plant does or does not contain allozymes unique to or characteristic of its putative parental taxa. In addition to additivity at one locus, it is possible to document additivity *among loci*. That is, a plant could contain an allele unique to one parent at one gene and have an allele diagnostic of the other parent at another locus. It is ideal for the presumed parents to be monomorphic for alternative alleles at a number of different loci or for them to have alternative high-frequency alleles. It should also be noted that since species of hybrid origin probably represent different recombinant types, the ability to detect additivity across (or among) loci is important because the likelihood of detecting additivity at each gene locus is low.

Gottlieb (1971b) hypothesized that *Stephanomeria diegensis* represents a stabilized hybrid derivative of *Stephanomeria exigua* and *S. virgata*; Gallez and Gottlieb (1982) tested this electrophoretically. Despite the two putative parental taxa being nondivergent at 16 of the 20 gene loci examined, the remaining four loci provided supporting data for the hybrid origin of *S. diegensis*. The presumed hybrid possessed the alleles of one parent at two genes and the other parent at the remaining two loci. In addition, *S. diegensis* contained only one unique allele in one individual. These results demonstrate that allozymic data may be useful for studying hybridization even when the putative parental taxa are not highly divergent.

The hypothesis of hybrid origin of a species could be rejected by electrophoretic data if the presumed hybrid did not combine species-specific alleles from the parents and if it contained alleles not found in either parental species. Results from a recent study of allozymes in *Lasthenia conjugens*, *L. fremontii*, and *L. burkei* by D. J. Crawford and R. Ornduff (unpublished data) do not support the hypothesis of Ornduff (1969) that the latter species is a stabilized hybrid derivative of the former two species. Ornduff (1969) employed data from morphology and chromosome structure, and compared naturally occurring *L. burkei* with synthetic hybrids between *L. conjugens* and *L. fremontii* to suggest that *L. burkei* is a hybrid derivative. Although the data seemed convincing, Ornduff (1969) cautioned that *L. burkei* could represent an intermediate between the other two species in a linear sequence of evolution. The three species are similar electrophoretically, which is not unexpected since Ornduff (1969) viewed them as, "a closely knit phylad in which the interspecific differences are few and probably have a relatively simple genetic basis." Despite their similarity, the three species each contain several unique alleles. In addition, *L. burkei* does not exhibit any combination of the alleles unique to its putative parental species. The electrophoretic data, therefore, reject the hybridization hypothesis.

Polyploidy is very common in plants as contrasted to animals; papers in Lewis (1980) provide a series of comprehensive reviews. Classically, polyploidy has been divided into two types, auto- and allopolyploidy, but it is well-known that the distinctions are imperfect in many instances (see critical discussion by Stebbins, 1980). Soltis and Rieseberg (1986) provide a brief and lucid consideration of this complex situation. Autopolyploids consist of genomes that are identical (or seemingly so) chromosomally and quite similar genetically. Allopolyploids, in contrast, contain genomes that are well differentiated chromosomally and genetically. This brief discussion explores the ways in which enzyme electrophoresis can contribute to understanding the origin and nature of polyploids.

Allopolyploids appear unquestionably to be more common than autopolyploids in nature and will be treated first. The earlier discussion of allozymes for studying hybrid speciation at the diploid level also applies to elucidating the origin of allopolyploids in the sense that the latter also involves hybridization between genetically differentiated plants (usually species); thus the polyploid combines species-specific alleles from the parents. Polyploids differ from diploid hybrid recombinants because they show fixed heterozygosity for the diagnostic allozymes. The heterozygosity results from the pairing of homologous chromosomes, which assures that each gamete receives the alternative alleles inherited from each parent. Two factors combine to make allozymes more useful for elucidating the origin of allopolyploids as compared to diploid species of hybrid origin. First, allopolyploids probably result from crosses between plants that are more divergent genetically than are the parents of diploid hybrid species, presumably because polyploidy can circumvent hybrid sterility. This means that the parents of allopolyploids are usually highly divergent allozymically, making detection of hybridization easier. Also, the fixed heterozygosity enforced by polyploidy keeps these additive allelic combinations at numerous genes intact. In diploid hybrids, additivity at individual loci is broken up by recombination.

Two of the classic studies of allopolyploidy (in noncultivated plants) involve *Tragopogon* (Ownbey, 1950) and *Asplenium* (Wagner, 1954). These authors used various kinds of data to elucidate the origins of several polyploid species. Roose and Gottlieb (1976) electrophoretically examined the

Tragopogon polyploids and their parental diploid taxa, and Werth et al. (1985a, b) did the same for the Aspleniums. Electrophoretic data confirmed completely the parentage of the allopolyploids in both genera; they exhibited fixed heterozygosity for the alternative alleles characteristic of or unique to the presumed parents.

Gastony (1986) used electrophoretic data to test two competing hypotheses for the origin of the tetraploid species *Asplenium plenum*. Briefly, one hypothesis is that this species originated by hybridization between the two diploid species *A. cristatum* and *A. abscissum* with concomitant chromosome doubling. The alternative hypothesis is that hybridization between the tetraploid *A. verecundum* and the diploid *A. abscissum* produced the sterile triploid *A. curtissii*. Subsequently, an unreduced spore from the triploid produced a gametophyte which in turn provided viable sperm for backcrossing to eggs of *A. abscissum*. Clearly, the first hypothesis is more parsimonious and would be favored on this basis alone; morphology does not allow one to choose between the two (Gastony, 1986).

The electrophoretic data provide a clear choice between the competing hypotheses; the less parsimonious explanation involving unreduced spores is supported against the simpler model. Allozymes unique to *A. verecundum* are found in *A. plenum* whereas none of the alleles of *A. cristatum* were detected in *A. plenum*. Also, the triploid *A. curtissii* has the same allozymes as the tetraploid *A. plenum* with only dosage differences seen, which is exactly what would be expected from the hypothesis. Gastony's (1986) results are an elegant example of the power of enzyme electrophoresis for documenting the origin of an allopolyploid species.

Allozyme data provided some confirming data on the origin of two allopolyploid complexes in *Antennaria* (Bayer and Crawford, 1986). Allozymes did not furnish strong, independent support for the diploid progenitors of the polyploids because the diploids are not highly divergent at isozyme loci.

Consider next the use of electrophoretic data for examining cases of autopolyploidy. If a polyploid is composed of genomes that are very similar or identical genetically and chromosomally, then one would expect it to contain the same alleles that occur in the presumed diploid progenitor(s). Also, if the polyploid is of recent origin, it should exhibit no unique alleles relative to the diploid plants from which it originated, however, the polyploid may not incorporate all of the alleles present in the diploid(s) from which it originated (see Soltis and Rieseberg, 1986). An autopolyploid will not have the fixed heterozygosity and disomic inheritance of allozymes found in allopolyploids. Rather, because autotetraploids, for example, contain four sets of highly homologous chromosomes, there are several equally probable ways in which pairing can occur among each of the four chromosomes (see lucid discussion and diagrams in Hauber, 1986; more detailed discussions are provided by Jackson, 1982, 1984; Jackson and Casey, 1982; Jackson and Hauber, 1982), and this results in tetrasomic inheritance (see illustration in Soltis and Rieseberg, 1986, p. 311). Tetrasomic inheritance has been considered a good criterion for documenting that a plant is an autotetraploid; the problem has been to find characters that could be used to document tetrasomic inheritance. Enzyme electrophoresis does this very well because of the simple codominant inheritance of the allozymes.

Several electrophoretic studies have been carried out on taxa thought to represent autopolyploids, including *Galax urceolata* (Epes and Soltis, 1984), *Coreopsis grandiflora* var. *longipes* (Crawford and Smith, 1984), *Zea perennis* (Doebley et al., 1984), *Tolmiea menziesii* (Soltis and Rieseberg, 1986), *Heuchera grossulariifolia* (Wolf et al., 1987), *Heuchera micrantha* (Ness et al., 1986), and *Dactylis glomerata* (Lumaret, 1986; Ardouin et al., 1987). The results of all have certain features in common. First, the polyploids contain very few, if any, alleles not found in their progenitor diploids. Secondly, enzyme multiplicity occurs in the polyploids, which is a reflection of duplicated genes. Thirdly, fixed heterozygosity is not found in the polyploids. D. Soltis and P. Soltis (1988a) documented tetrasomic inheritance for several enzymes in *Tolmiea,* and Hauber (1986) demonstrated tetrasomic inheritance of allozymes in a putative autotetraploid in *Haplopappus*.

Enzyme electrophoresis has demonstrated that both allo- and autotetraploids maintain higher levels of heterozygosity than their diploid progenitors, with allotetraploids achieving it by fixed heterozygosity whereas autotetraploids employ tetrasomic inheritance. Increased heterozygosity at various loci may contribute to the success of polyploids. With regard to isozymes, the biochemical diversity created by polyploidy, including novel enzymes, may be important in the establishment and spread of polyploids (see discussion in Soltis and Rieseberg, 1986).

In summary, enzyme electrophoresis provides a powerful method for exploring the genetic situation in polyploid plants and for elucidating their origin. The data are useful in testing whether the plants behave genetically as allo- or autopolyploids. It would be desirable to obtain data on the adaptive value of enzyme multiplicity produced by polyploidy.

ISOZYME NUMBER AND STUDIES OF PLANT EVOLUTION

As mentioned earlier, diploid plants have a highly conserved minimal number of isozymes for many of the enzymes routinely included in electrophoretic studies (Gottlieb, 1982). Therefore, plants with additional genes specifying these enzymes are of phylogenetic interest because they may be either diploids with one or more gene duplications or they could represent polyploids. The two most recent discussions of the utility of isozyme number for studying plant phylogeny were provided by Gottlieb (1983) and Giannasi and Crawford (1986). This chapter considers points raised in those reviews as well as in more recent studies. The probable chromosomal mechanisms for generating gene duplications in diploids will not be considered; Gottlieb (1984a) should be consulted.

A duplication for the cytosolic form of phosphoglucose isomerase (PGI) in diploid members of the genus *Clarkia* has been studied extensively from a genetic and phylogenetic perspective by Gottlieb and collaborators (Gottlieb, 1977b, 1983, 1986; Gottlieb and Weeden, 1979). The distribution of the duplication was superimposed on the phylogeny of *Clarkia* hypothesized by Lewis and Lewis (1955). When this was done, an interesting pattern emerged: those sections with the PGI duplication represented ones Lewis and Lewis (1955) viewed as highly specialized. However, the Lewis and Lewis phylogeny for *Clarkia* would dictate that the duplication arose several times because the sections having it are at the ends of different branches. Lewis (1980) recently revised the phylogeny of *Clarkia* so that the four sections containing the extra isozyme form a monophyletic group (see Gottlieb, 1984a). The isozyme number data stimulated a reevaluation of the phylogeny because the presumed mechanism by which the duplication was generated involves a series of low probability events, thus arguing against multiple origins of the duplications.

Clarkia rostrata represents an exception to the previous statements about gene number and phylogeny because it lacks the duplication yet is placed in the advanced section *Peripetasma* in which all other members exhibit the extra isozyme. Two possible explanations for this are that the species never possessed the duplication and has been misplaced taxonomically in section *Peripetasma* or that the duplication in *C. rostrata* has been silenced. It is not possible to test these hypotheses electrophoretically because one can only assay for the presence or absence of the gene product and cannot ascertain the cause of its absence. Given the vagaries sometimes associated with using morphology for inferring phylogenetic relationships, the problem was unresolved until restriction site analyses of chloroplast DNA (Sytsma and Gottlieb, 1986a) demonstrated that *C. rostrata* clearly belongs in section *Peripetasma*; thus, expression of the duplicate gene has been silenced in this species.

The PGI duplication is also of phylogenetic significance with regard to whether two species sometimes segregated as the separate genus *Eucharidium* should be included in *Clarkia*, as suggested by Lewis and Lewis (1955). These species contain the extra PGI isozyme, which argues for their inclusion in *Clarkia* (Gottlieb and Weeden, 1979).

An additional problematical taxon thought to be closely related to *Clarkia* yet very distinct in certain morphological features is *Heterogaura heterandra*, a monotypic genus. This species has the duplicate gene for cytosolic PGI (L. D. Gottlieb, personal communication), which argues for a very close relationship to *Clarkia*, if not inclusion within the genus. Data from chloroplast DNA show that *Heterogaura* has evolved within *Clarkia* section *Peripetasma* (Sytsma and Gottlieb, 1986b).

Two isozymes for 6-phosphogluconate dehydrogenase (6PGD), one in the cytosol and another in the plastids, typically are found in diploid plants (Gottlieb, 1982), yet Odrzykoski and Gottlieb (1984) demonstrated the presence of four isozymes in most diploid members of *Clarkia*. Two isozymes each were detected in the plastids and the cytosol, indicating that a duplication has occurred for the gene specifying each of the isozymes (Odrzykoski and Gottlieb, 1984). Four species lack the cytosolic duplication, with all four (*C. epilobioides, C. cylindrica, C. lewisii,* and *C. rostrata*) assigned to the advanced section *Peripetasma*. Two of the species, *C. epilobioides* and *C. rostrata,* also have only one gene specifying plastid 6PGD. The widespread distribution of the two duplications in the genus suggests that this is the ancestral condition *within Clarkia* whereas the loss or silencing of the duplications is the derived situation; the fact that all species with reduced isozyme number belong to a section comprising specialized species adds strength to this argument.

Because these four species have been viewed as closely related, it is reasonable to assume that loss of duplicate gene expression for the cytosolic form of 6PGD probably occurred in a plant ancestral to the four species. With regard to loss of the duplication for plastid 6PGD, the situation was not so apparent initially (see Odrzykoski and Gottlieb, 1984) because the two species lacking it, *C. epilobioides* and *C. rostrata,* had been assigned to different subsections. The former was included as the only member of a subsection whereas the latter was placed in the same subsection as *C. cylindrica* and *C. lewisii,* both of which have the duplication for the plastid isozyme. The question, then, was whether loss of the duplication for the plastid form occurred in a common ancestor to *C. cylindrica* and *C. lewisii,* which would not be in agreement with the prevailing taxonomic treatment, or there had been two independent losses, which would be concordant with the accepted taxonomy of the section. A subsequent study employing chloroplast DNA restriction site analyses (Sytsma and Gottlieb, 1986a) left little doubt that loss of the duplication for plastid 6PGD was a single event, and thus *C. epilobioides* and *C. rostrata* have close phylogenetic affinities. Clearly the former species should be placed in the same subsection as the other three species (Sytsma and Gottlieb, 1986a; Gottlieb, 1986).

More recently, Soltis et al. (1987) have reported gene duplications for both plastid and cytosolic isozymes of phosphoglucomutase (PGM) in *Clarkia*. All diploid species in the genus, save two in distantly related sections, have a duplication for plastid PGM. This suggests that the duplication has been lost independently in the two taxa (Soltis et al., 1987). All species in two sections of *Clarkia* (*Godetia* and *Myxocarpa*) exhibit an extra cytosolic isozyme for PGM, and one species (*C. arcuata*) in section *Rhodanthos* likewise has two isozymes. These data suggest that sections *Godetia* and *Myxocarpa* may have originated from plants ancestral to *C. arcuata* of section *Rhodanthos*. One cannot, on the basis of available data, dismiss the possibility of independent duplications in the three groups (Soltis et al., 1987). It is important to note that the distribution of duplications for PGM is concordant with phylogenetic interpretations inferred from gene duplications for other enzymes.

Contrasting the results for the PGI and 6PGD duplications in *Clarkia* illustrates the necessity of viewing duplications in the context of their distribution within the taxa under investigation. In the case of PGI, presence of the duplication is clearly the derived condition (except for *C. rostrata*) because it is restricted to specialized species comprising several sections of the genus. The ancestral condition for PGI in *Clarkia* is also the common condition for plants as a whole, that is, one gene for the cytosolic form. In contrast, the vast majority of species in *Clarkia* has the duplications for both forms of 6PGD whereas only four highly specialized species lack one or both of them. This means that the uncommon condition for vascular plants as a whole is the common or ancestral state for *Clarkia* as a whole. One must ascertain the specialized or advanced character state within the context of the taxonomic group under study. Gene duplications for isozymes can be both generated and silenced (or lost) during evolution, and this is why it is necessary to view each situation carefully. This is also the reason why isozyme number cannot be used as a totally independent source of data for inferring phylogenies.

Additional gene duplications have been found, both in *Clarkia* and a number of other genera; Gottlieb (1986) and Soltis et al. (1987) should be consulted for a partial list. Lumaret (1986) has reported the expression of two loci for cytosolic PGI in diploid plants of *Dactylis glomerata*. In contrast to the PGI duplications in *Clarkia* discussed above, the two genes are tightly linked, suggesting a tandem duplication produced by unequal crossing over (Lumaret, 1986). *Dactylis* is a monospecific genus, and the duplication is found in all diploid subspecies of the genus; it has not been detected in other genera of the same tribe, suggesting that it originated early in the evolution of *Dactylis*. Duplications may occur as polymorphisms in populations, be restricted to a species, characterize groups of species within a genus, be present in a genus, or be distributed among genera. This means (assuming they result from single events) that gene duplications have occurred at different times in various evolutionary lineages. Because of this, they are useful for circumscribing monophyletic groups of different sizes and ages.

Gottlieb (1981b) used isozyme number to investigate the question of whether certain species in the tribe Astereae of the Compositae are diploids or polyploids. Chromosome numbers in the tribe range from $n = 2$ to $n = 9$, with numbers of $n = 4$ and 5 being quite common. Two hypotheses have been advanced to explain this interesting distribution of chromosome numbers. Raven et al. (1960) viewed the plants with $n = 9$ as diploids, with aneuploid reduction accounting for the range of numbers now present. Turner et al. (1961) and Turner and Horn (1964) argued that the plants with $n = 9$ are allopolyploids resulting from hybridization between the $n = 4$ and $n = 5$ plants.

In most instances in flowering plants, it is possible to infer from chromosome number alone whether species are diploid or polyploid. This was not possible with the Astereae, so Gottlieb (1981b) used the number of genomes present rather than the chromosome number for inferring ploidal level. This fresh perspective on the problem then allowed the question of ploidy to be investigated more explicitly with enzyme electrophoresis. In particular, if the plants with $n = 9$ are allotetraploids they should show increased isozyme numbers relative to the normal minimal conserved numbers found in diploids because of extensive gene duplications (assuming the plants are not ancient polyploids that have experienced extensive gene silencing). In contrast, if plants with all chromosome numbers in the Astereae are diploids and the lower numbers result from aneuploid reduction, then all species should exhibit the typical minimal conserved number of isozymes. Species with all chromosome numbers had the same isozyme number, and the electrophoretic data, therefore, lend support to the hypothesis of aneuploid reduction in the Astereae (Gottlieb, 1981b).

Several electrophoretic studies of homosporous ferns have asked the same question addressed by Gottlieb and have demonstrated that species with the lowest chromosome numbers in a genus have the typical diploid number of isozymes (Gastony and Darrow, 1983; Gastony and Gottlieb, 1985; Haufler, 1985; Haufler and Soltis, 1986; Werth et al., 1985a). This is so despite the actual chromosome numbers being so high that they would be viewed as polyploids in flowering plants. A similar situation has been found in *Equisetum* (Soltis, 1986), lycopods (D. Soltis and P. Soltis, 1988b), and *Psilotum* (P. Soltis and D. Soltis, 1988). This interesting situation will be discussed by Soltis and Soltis in another chapter of this volume.

MAJOR CONTRIBUTIONS OF ENZYME ELECTROPHORESIS TO PLANT SYSTEMATICS

There seems little doubt that the most important contribution of electrophoretic data to plant systematics is that it allows quantification of genetic similarity and differences among populations and taxa. This has facilitated comparative studies of genetic similarity with other biological information for species including such important factors as probable time of origin, mode of origin, etc. No other method has been so readily available to the plant systematist for accomplishing this. It seems fair to say that electrophoretic data have been employed most often for studying primary or divergent speciation in diploid plants.

Because polyploidy is such a common phenomenon in plant evolution, it is rather surprising that electrophoresis has not been used more frequently to examine the origin of polyploids, particularly allopolyploids. The few studies that have been done document the utility of the method, particularly for providing unambiguous genetic documentation of the genomes present in the polyploids. The method has also been useful for demonstrating the multiple origins of the same polyploid species.

The very few studies of autotetraploids have been enlightening in showing a genetic situation that contrasts sharply with allotetraploids. These differences include tetrasomic inheritance, lack of fixed heterozygosity, and presence of alleles typical of only one diploid species.

Although hybridization at the diploid level is ostensibly a significant evolutionary mechanism in plants, few electrophoretic studies have been used to address the situation. The basic rationale for using enzyme electrophoresis has been spelled out, and in most instances it should be possible either to support or falsify the hypothesis of the hybrid origin of plants. Electrophoretic studies of polyploidy and diploid hybrid speciation no doubt give the most unambiguous results if the events (hybridization and increase in chromosome number) are recent relative to the ages of the diploid species involved. This reduces the probability of divergence at genes specifying the enzymes subsequent to the origin of the new entities. This problem, however, is not unique to electrophoretic studies; it is true for any method of investigation.

The study of isozyme (or gene) number has been a useful approach for answering other kinds of questions than those just discussed for allelic data. In particular, the increase in isozyme number in diploid plants resulting from a gene duplication may be of utility for circumscribing monophyletic assemblages. That is, the data are more useful than allelic information for addressing phylogenetic questions within and among genera. Some care must be taken in using isozyme number, however, because genes can be silenced or lost as well as duplicated. In certain instances, gain of duplicate gene expression may be derived whereas with other plants loss of the expression may be the advanced condition.

The number of isozymes expressed may be employed to ascertain whether a plant is diploid or polyploid genetically. This may be very clear in many instances because the polyploid will exhibit duplicate gene expression for many enzymes whereas a diploid will have the minimal conserved number for most enzymes. An "extra" gene for one or two enzymes would be viewed as the result of gene duplication in a diploid. Each situation must be studied carefully, particularly in plants exhibiting several duplications because the extra isozymes could result from duplications at the diploid level or "residual" expression in an old polyploid.

FUTURE ENZYME ELECTROPHORETIC STUDIES IN PLANT SYSTEMATICS

One question concerns whether it is desirable to continue routine systematic electrophoretic studies of congeneric species. Such investigations are most interesting if done within the framework of some biological or evolutionary question about the plants. For example, Stutz (1978) suggested that the perennial plants representing species of *Atriplex* may be undergoing rapid evolution in the Intermountain West of the United States. Likewise, Stebbins (1982) discussed evidence in support of recent speciation in woody perennials of the genus *Ceanothus*. It would be interesting to examine allozyme divergence among species of woody plants in these and other genera where the data suggest recent speciation. Data in the literature have documented high genetic identities between species of annuals (i.e., the progenitor-derivative species pairs) that appear to have diverged recently, and the same is true for congeners on islands. Future studies could be directed at the question of whether perennial plants thought to have speciated recently in continental areas exhibit the same high genetic identities.

Electrophoretic studies of taxonomically difficult groups of species may be helpful in defining species. This approach may be most useful with plants that are phenotypically plastic and difficult to classify on the basis of morphology.

More studies of the type carried out by Witter (1986, 1987) for the silverswords on the Hawaiian Islands are needed. Endemic species from older as well as younger islands in an archipelago should be examined electrophoretically. The Hawaiian Islands are a useful system for such studies because of the documented age differences among the islands. If a similar pattern of greater divergence among species on older islands could be documented for other genera it would indicate that time is a critical, if not sole, factor in allozyme divergence.

The electrophoretic method is useful for distinguishing auto- and allopolyploids, and such studies should be done prior to more detailed comparative examinations of the biology of diploids and their derivative polyploids. In some instances, allelic data may provide novel insights into the genomes present in an allopolyploid. There is little doubt that electrophoretic studies will provide valuable information on the origin (or perhaps multiple origins in some cases) of allopolyploids.

As indicated earlier, isozyme number is increased in polyploids because of the resulting gene duplications. Gene number was the basis for Gottlieb's (1981b) inferences about the diploid or polyploid nature of plants in the Astereae. A question arising from such studies is whether polyploids will continue to exhibit duplicate gene expression with age or rather extensive silencing of the "extra" genes will take place. Discussion of the evolutionary implications of gene number in ferns is presented by Soltis and Soltis in another chapter of this volume. Wilson et al. (1983) have documented loss of duplicate gene expression for leucine aminopeptidase in tetraploid plants of *Chenopodium*. It would be of interest to ascertain whether flowering plants with higher chromosome numbers indicative of polyploidy do in fact contain duplicate genes. For example, many of the so-called archaic or primitive woody plants have chromosome numbers suggesting polyploidy, and it would be useful to know whether they have additional isozymes for many of their enzymes. Certain members of these groups have additional isozymes (D. E. Soltis and P. S. Soltis, unpublished data). Genera thought to consist of plants of polyploid origin because of chromosome number or other factors should be examined for isozyme number. Elisens and Crawford (1988) found almost no duplicate gene expression in the genus *Mabrya* (Scrophulariaceae) despite chromosome numbers indicative of tetraploid plants. Of interest in this regard is the fact that the genus is viewed as quite old based on geological and other data.

Although measures such as the genetic identity statistic are useful for visualizing overall similarity at isozyme genes, they may obscure useful information. Future studies should place additional emphasis on considering the factors generating particular identity values rather than just reporting values. For example, do two populations have a low identity because they contain different (unique) alleles or because they have the same alleles in very different frequencies? Witter (1986) and others have emphasized that divergence at genes for isozymes can develop in two ways. If a species consists of large populations containing high levels of genetic variation, then new or founder populations can diverge first by changes in the frequencies of alleles, partly or wholly by drift. The second process, the accumulation of new mutations in two populations, is presumably a much slower process. It is important that genetic identities be evaluated or considered from this perspective. Witter (1986) has correctly pointed out that the same value for different taxa may be achieved by different processes requiring different amounts of time. She argued, for example, that genetic divergence between island taxa accumulates as a result of mutations because the ancestor (or ancestors) of these plants contained little or no variation at genes specifying soluble enzymes. Thus, future studies should place more emphasis on understanding the causes of particular identity values rather than just computing values.

The value of future systematic electrophoretic studies in plants depends in large measure on knowing as much as possible about the plants themselves. That is, unless basic data on chromosome number, crossing relationships, ecology (including the age of the habitats), and other important biological parameters are known, the value of the electrophoretic information is likewise diminished. As indicated earlier, allozymic and isozymic data are of greatest utility when viewed within an evolutionary-phylogenetic framework rather than a taxonomic context. For example, the results of Gottlieb and collaborators from their electrophoretic studies of *Clarkia* are particularly important because they may be interpreted within the context of the wealth of biological data gathered for the genus earlier by Harlan Lewis and associates.

A question that arises with regard to the future of enzyme electrophoresis in plant systematics is whether it will be replaced largely or completely by studies of DNA. Although predictions are risky at best, it appears that isozymes will continue to be used for addressing a variety of problems in plant systematics. This is so because DNA and enzyme electrophoresis can provide complementary information in certain cases whereas in other instances DNA may not be as useful as allozymes.

Allelic frequency data for allozymes will continue to be used for studying variation within and among populations of the same species and congeneric species, genetic divergence among populations and congeners, and natural hybridization. The genes for ribosomal RNA (rDNA) are useful for studying some of the same problems in plant systematics for which allozymes are employed (e.g., Saghai-Maroof et al., 1984; Doyle and Beachy, 1985; Doyle et al., 1985; Flavell et al., 1986; Learn and Schaal, 1987; Schaal et al., 1987). Comparative studies of variation in rDNA (in restriction sites and/or repeat lengths) and gene loci encoding soluble enzymes will no doubt appear in the future. There are no apparent reasons for rDNA *totally replacing* isozymes for studies at the generic level and lower.

Chloroplast DNA (cpDNA) and isozymes are useful for studying different problems in plant systematics. Restriction site analyses of cpDNA have been valuable for constructing phylogenies among congeneric species, genera, etc. (see Palmer, 1987, for review and discussion). The number of alleles found at isozyme loci, both within and among populations, make it difficult to use the data for constructing phylogenies. Richardson et al. (1986, chapter 12) discuss the use of allelic data for phylogenetic reconstruction and possible pitfalls with the method. Thus, restriction site analyses of the chloroplast genome are useful for phylogenetic reconstruction whereas allelic data from enzyme electrophoresis are less than optimal for this purpose.

The chloroplast genome has proven to be too conservative to assess genetic variation within and among plant populations and to study speciation (Palmer, 1987). Allelic data, in contrast, are perhaps of greatest utility for looking at genetic variation at the populational level. The plant systematist, therefore, can use the two methods for studying evolution and phylogeny at different levels.

It appears that enzyme electrophoresis and studies of DNA (both rDNA and cpDNA) provide valuable complementary data for plant systematics. The questions being asked will dictate the methods to be employed. This means that although comparative studies of DNA in plants will increase in number (as well they should), it does not mean that there will (or should) be a corresponding reduction in enzyme electrophoretic investigations.

CONCLUSIONS

Enzyme electrophoresis has provided useful data for addressing a wide variety of questions in plant systematics and evolution. The primary advantage of the electrophoretic method is that genetic similarities (and differences) may be quantified and the number of gene loci included in the study may be ascertained.

Allelic data have been most useful for studying congeneric species. Problems addressed in primary or divergent speciation include the mode of speciation, whether speciation was relatively recent and which species represent the progenitor and the derivative taxa. Allelic data are most valuable when combined with information from other areas such as morphology, cytogenetics, and geographical distribution.

Allelic data can provide a robust test of hybrid speciation at the diploid level provided that the putative parental taxa are divergent at isozyme loci. The simple, codominant inheritance of allozymes allows one to determine whether alleles characteristic of the suspected parents are or are not combined in the putative hybrid plants.

At the polyploid level, allelic data may contribute very useful evidence for determining whether plants are auto- or allopolyploids. In tetraploids, for example, disomic inheritance and fixed heterozygosity characterize alloploids whereas tetrasomic inheritance would be expected in autoploids.

For many enzymes included in electrophoretic studies, a minimal highly conserved number of isozymes is found in diploid plants. "Extra" isozymes, presumably resulting from gene duplications, may be useful for phylogenetic purposes. The relative rarity of gene duplications means that taxa sharing the extra isozymes may represent monophyletic assemblages. Caution must be exercised in assessing the phylogenetic implications of duplications because extra isozymes may be lost as well as gained during the evolution of given lineages. Isozyme number may also be employed for testing whether plants with particular chromosome numbers are diploid or polyploid. Extensive duplications would favor a polyploid hypothesis.

Enzyme electrophoresis assumes its greatest value when biosystematic and other information are available for the plants under investigation. This allows one to use the enzyme data to address particular hypotheses or question about the plants.

Many interesting problems in plant systematics await input from enzyme electrophoresis. The increasing popularity of using DNA in systematic studies will not diminish the value of enzyme electrophoresis; the two approaches are complementary in many instances.

ACKNOWLEDGMENTS

Financial support from National Science Foundation grants BSR-8306436, DEB 80-11055 and DEB 82-04073 is appreciated. Linda Crawford typed (word processed) the manuscript.

LITERATURE CITED

Ardouin, P., M. Jay, and R. Lumaret. 1987. Etude d'une situation de sympatrie diploides et tetraploides de *Dactylis glomerata* (Gramineae), sur la base du polymorphisme enzymatique et phenolique. *Canad J. Bot.* 65: 526–531.

Bayer, R. J., and D. J. Crawford. 1986. Allozyme divergence among five diploid species of *Antennaria* (Asteraceae: Inuleae) and their allopolyploid derivatives. *Amer. J. Bot.* 73: 287–296.

Carr, G. D. 1985a. Habitat variation in the Hawaiian Madiinae (Heliantheae) and its relevance to generic concepts in the Compositae. *Taxon* 34: 22–25.

———. 1985b. Monograph of the Hawaiian Madiinae (Asteraceae): *Argyroxiphium, Dubautia* and *Wilkesia. Allertonia* 4: 1–123.

———, and D. W. Kyhos. 1981. Adaptive radiation in the Hawaiian silversword alliance (Compositae: Madiinae). I. Cytogenetics of spontaneous hybrids. *Evolution* 35: 343–356.

———, and ———. 1986. Adaptive radiation in the Hawaiian silversword alliance (Compositae: Madiinae). II. Cytogenetics of artificial and natural hybrids. *Evolution* 40: 959–976.

Clausen, J., D. D. Keck, and W. M. Hiesey. 1941. Experimental taxonomy. *Carnegie Inst. Wash. Yearbook* 40: 160–170.

Crawford, D. J. 1979. Allozyme studies in *Chenopodium incanum*: intraspecific variation and comparison with *C. fremontii. Bull. Torrey Bot. Club.* 106: 256–266.

———. 1983. Phylogenetic and systematic inferences from electrophoretic studies. In S. O. Tanksley and T. J. Orton [eds.], *Isozymes in plant genetics and breeding*, Part A, 257–287. Elsevier, Amsterdam.

———. 1985. Electrophoretic data and plant speciation. *Syst. Bot.* 10: 405–416.

———, and R. J. Bayer. 1981. Allozyme divergence in *Coreopsis cyclocarpa* (Compositae). *Syst. Bot.* 6: 373–379.

———, R. Ornduff, and M. C. Vasey. 1985. Allozyme variation within and between *Lasthenia minor* and its derivative species, *L. maritima* (Asteraceae). *Amer. J. Bot.* 72: 1177–1184.

———, and E. B. Smith. 1984. Allozyme divergence and intraspecific variation in *Coreopsis grandiflora* (Compositae). *Syst. Bot.* 9: 219–225.

———, T. F. Stuessy, and M. Silva. 1987a. Allozyme divergence and evolution of *Dendroseris* (Compositae: Lactuceae) on the Juan Fernandez Islands. *Syst. Bot.* 12: 435–443.

———, R. Whitkus, and T. F. Stuessy. 1987b. Plant evolution and speciation on oceanic islands. In K. Urbanska [ed.], *Differentiation patterns in higher plants*, 183–199. Academic Press, London.

———, and H. D. Wilson. 1979. Allozyme variation in several closely related diploid species of *Chenopodium* of the western United States. *Amer. J. Bot.* 66: 237–244.

Doebley, J. F., M. M. Goodman, and C. W. Stuber. 1984. Isoenzymatic variation in *Zea* (Gramineae). *Syst. Bot.* 9: 203–218.

Doyle, J. J., and R. N. Beachy. 1985. Ribosomal gene variation in soybean (*Glycine*) and its relatives. *Theor. Appl. Genet.* 70: 369–376.

———, D. E. Soltis, and P. S. Soltis. 1985. Ribosomal RNA gene sequence variation: *Tolmiea, Tellima*, and their intergeneric hybrid. *Amer. J. Bot.* 72: 1388–1391.

Elisens, W. J., and D. J. Crawford. 1988. Genetic variation and differentiation in the genus *Mabrya* (Scrophulariaceae-Antirrhineae): systematic and evolutionary inferences. *Amer. J. Bot.* 75: 85–96.

Epes, D. A., and D. E. Soltis. 1984. An electrophoretic investigation of *Galax urceolata* (Diapensiaceae). *Amer. J. Bot.* (Abstract) 71: 165.

Flavell, R. B., M. O'Dell, P. Sharp, E. Nevo, and A. Beiles. 1986. Variation in the intergenic spacer at ribosomal DNA of wild wheat, *Triticum dicoccoides*, in Israel, *Mol. Biol. Evol.* 3: 547–558.

Gallez, G. P., and L. D. Gottlieb. 1982. Genetic evidence for the hybrid origin of the diploid plant *Stephanomeria diegensis*. *Evolution* 36: 1158–1167.

Gastony, G. J. 1986. Electrophoretic evidence for the origin of fern species by unreduced spores. *Amer. J. Bot.* 73: 1563–1569.

———, and D. C. Darrow. 1983. Chloroplastic and cytosolic isozymes of the homosporous fern *Athyrium filix-femina*. *Amer. J. Bot.* 70: 1409–1415.

———, and L. D. Gottlieb. 1985. Genetic variation in the homosporous fern *Pellaea andromedifolia*. *Amer. J. Bot.* 72: 257–267.

Giannasi, D. E., and D. J. Crawford. 1986 Biochemical systematics II. A reprise. *In* M. K. Hecht, B. Wallace, and G. T. Prance [eds.], *Evolutionary Biology* Vol. 20, 25–248. Plenum Press, New York.

Gottlieb, L. D. 1971a. Gel electrophoresis: new approach to the study of evolution. *BioScience* 21: 939–944.

———. 1971b. Evolutionary relationships in the outcrossing diploid annual species of *Stephanomeria* (Compositae). *Evolution* 25: 312–329.

———. 1972. Levels of confidence in the analysis of hybridization in plants. *Ann. Missouri Bot. Gard.* 59: 435–446.

———. 1973a. Genetic differentiation, sympatric speciation, and the origin of a diploid species of *Stephanomeria*. *Amer. J. Bot.* 60: 545–553.

———. 1973b. Enzyme differentiation and phylogeny in *Clarkia franciscana*, *C. rubicunda* and *C. amoena*. *Evolution* 27: 205–214.

———. 1974. Genetic confirmation of the origin of *Clarkia lingulata*. *Evolution* 28: 244–250.

———. 1977a. Electrophoretic evidence and plant systematics. *Ann. Missouri Bot. Gard.* 64: 161–180.

———. 1977b. Evidence for duplication and divergence of the structural gene for phosphoglucose isomerase in diploid species of *Clarkia*. *Genetics* 86: 289–307.

———. 1981a. Electrophoretic evidence and plant populations. *Prog. Phytochem.* 7: 1–46.

———. 1981b. Gene number in species of Astereae that have different chromosome numbers. *Proc. Natl. Acad. Sci. USA* 78: 3726–3729.

———. 1982. Conservation and duplication of isozymes in plants. *Science* 216: 373–380.

———. 1983. Isozyme number and plant phylogeny. *In* U. Jensen and D. E. Fairbrothers [eds.], *Proteins and nucleic acids in plant systematics*, 210–221. Springer-Verlag, Berlin.

———. 1984a. Isozyme evidence and problem solving in plant systematics. *In* W. F. Grant [ed.], *Plant biosystematics*, 343–357. Academic Press. Orlando. FL.

———. 1984b. Electrophoretic analysis of the phylogeny of the self-pollinating populations of *Clarkia xantiana*. *Pl. Syst. Evol.* 147: 91–102.

———. 1984c. Genetics and morphological evolution in plants. *Amer. Naturalist* 123: 681–709.

———. 1986. Genetic differentiation, speciation and phylogeny in *Clarkia* (Onagraceae). *In* K. Iwatsuki, P. H. Raven and W. J. Bock [eds.], *Modern aspects of species*, 145–160. University of Tokyo Press. Tokyo.

———, and G. Pilz. 1976. Genetic similartiy between *Gaura longifolia* and its obligately outcrossing derivative *G. demareei*. *Syst. Bot.* 1: 181–187.

———, and N. F. Weeden. 1979. Gene duplication and phylogeny in *Clarkia*. *Evolution* 33: 1024–1039.

Grant, V. 1981. *Plant speciation*. Second Edition, Columbia Univ. Press, New York.

Hamrick, J. L., J. B. Mitton, and Y. B. Linhart. 1981. Levels of genetic variation in trees: influence of life history characteristics. *In* M. T. Conkle [ed.], *Proceedings, symposium on isozymes of North American forest trees and forest insects*. 35–41. USDA Forest Service Gen. Tech. Rep. PSW-48.

Hauber, D. P. 1986. Autotetraploidy in *Haplopappus spinulosus* hybrids: evidence from natural and synthetic tetraploids. *Amer. J. Bot.* 73: 1595–1606.

Haufler, C. H. 1985. Enzyme variability and modes of evolution in the fern genus *Bommeria*. *Syst. Bot.* 10: 92–104.

―――, and D. E. Soltis. 1986. Genetic evidence suggests that homosporous ferns with high chromosome numbers are diploid. *Proc. Natl. Acad. Sci. USA.* 83: 4389–4393.

Helenurm, K., and F. R. Ganders. 1985. Adaptive radiation and genetic differentiation in Hawaiian *Bidens. Evolution* 39: 753–765.

Jackson, R. C. 1982. Polyploidy and diploidy: new perspectives on chromosome pairing and its evolutionary implications. *Amer. J. Bot.* 69: 1512–1523.

―――. 1984. Chromosome pairing in species and hybrids. *In* W. F. Grant [ed.], *Plant biosystematics,* 67–86. Academic Press, Orlando, FL.

―――, and J. Casey. 1982. Cytogenetic analysis of autopolyploids: models and methods for triploids to octoploids. *Amer. J. Bot.* 69: 487–501.

―――, and D. P. Hauber. 1982. Autotriploid and autotetraploid cytogenetic analyses: correction coefficients for proposed binomial models. *Amer. J. Bot.* 69: 644–646.

Jefferies, R. L., and L. D. Gottlieb. 1982. Genetic differentiation of the microspecies *Salicornia europea* L. (sensu stricto) and *S. ramosissima* J. Woods. *New Phytol.* 92: 123–129.

Learn, G. H., and B. A. Schaal. 1987. Population subdivision for ribosomal DNA repeat variants in *Clematis fremontii. Evolution* 41: 433–438.

Lewis, H. 1980. The mode of evolution in *Clarkia* (Onagraceae). *Second. Int. Congress. Syst. and Evol. Biol.* (Abstract) Vancouver, B. C.

―――, and C. Epling. 1959. *Delphinium gypsophilum,* a diploid species of hybrid origin. *Evolution* 13: 511–525.

―――, and M. E. Lewis. 1955. The genus *Clarkia. Univ. Calif. Publ. Bot.* 20: 241–392.

Lewis, W. H., [ed.]. 1980. *Polyploidy: biological relevance.* Plenum Press, New York.

Lowrey, T. K. and D. J. Crawford. 1985. Allozyme divergence and evolution in *Tetramolopium* (Compositae: Astereae) on the Hawaiian Islands. *Syst. Bot.* 10: 64–72.

Lumaret, R. 1986. Doubled duplication of the structural gene for cytosolic phosphoglucose isomerase in the *Dactylis glomerata* L. polyploid complex. *Mol. Biol. Evol.* 3: 499–521.

Marshall, D. R., and A. H. D. Brown. 1975. Optimum sampling strategies in general conservation. *In* O. H. Frankel and J. G. Hawkes [eds.], *Genetic resources for today and tomorrow.* 53–80. Cambridge University Press, Cambridge.

McLeod, M. J., S. I. Guttman, W. H. Eshbaugh, and D. E. Rayle. 1983. An electrophoretic study of evolution in *Capsicum* (Solanaceae). *Evolution* 37: 562–574.

McNeill, C. I., and S. K. Jain. 1983. Genetic differentiation studies and phylogenetic inference in the plant genus *Limnanthes* (section *Inflexae*). *Theor. Appl. Genet.* 66: 257–269.

Morden, C. W., J. Doebley, and K. F. Schertz. 1987. *A Manual of Techniques for Starch Gel Electrophoresis of* Sorghum *Isozymes.* Texas Agricult. Exper. Station. MP-1635. 10 pp.

Nei, M. 1972. Genetic distance between populations. *Amer. Naturalist* 106: 283–293.

Ness, B. D., D. E. Soltis, and P. S. Soltis. 1986. Autopolyploidy in *Heuchera micrantha* Dougl. (Saxifragaceae). *Amer. J. Bot.* (Abstract) 73: 777–778.

Odrzykoski, I. J., and L. D. Gottlieb. 1984. Duplications of genes coding 6-phosphogluconate dehydrogenase in *Clarkia* (Onagraceae) and their phylogenetic implications. *Syst. Bot.* 9: 479–489.

Ornduff, R. 1969. The origin and relationships of *Lasthenia burkei* (Compositae). *Amer. J. Bot.* 56: 1042–1047.

Ownbey, M. 1950. Natural Hybridization and amphiploidy in the genus *Tragopogon. Amer. J. Bot.* 37: 487–499.

Palmer, J. D. 1987. Chloroplast DNA evolution and biosystematic uses of chloroplast DNA variation. *Amer. Naturalist* 130: S6–S29.

Raven, P. R. 1976. Systematics and plant population biology. *Syst. Bot.* 1: 284–316.

―――. 1980. Hybridization and the nature of species in higher plants. *Canad. Bot. Assoc. Bull. Suppl.* 13: 3–10.

―――, O. T. Solbrig, D. W. Kyhos, and R. Snow. 1960. Chromosome numbers in Compositae. I. Astereae. *Amer. J. Bot.* 47: 124–132.

Richardson, B. J., P. R. Baverstock, and M. Adams. 1986. Allozyme electrophoresis—a handbook for animal systematics and population studies. Academic Press, Orlando, FL

Roberts, M. L. 1983. Allozyme variation in *Bidens discoidea* (Compositae). *Brittonia* 35: 239–247.

Roose, M. L., and L. D. Gottlieb. 1976. Genetic and biochemical consequences of polyploidy in *Tragopogon*. *Evolution* 30: 818–830.

Saghai-Maroof, M. A., K. M. Soliman, R. A. Jorgensen, and R. A. Allard. 1984. Ribosomal spacer-length polymorphisms in barley: Mendelian inheritance, chromosomal location, and population dynamics. *Proc. Natl. Acad. Sci. USA* 81: 8014–8018.

Schaal, B. A., W. J. Leverich, and J. Nieto-Sotelo. 1987. Ribosomal DNA variation in the native plant *Phlox divaricata*. *Mol. Biol. Evol.* 4: 611–621.

Shields, C. R., T. J. Orton, and C. W. Stuber. 1983. An outline of general resource needs and procedures for the electrophoretic separation of active enzymes from plant tissue. In S. O. Tanksley and T. J. Orton [eds.], *Isozymes in plant genetics and breeding*, Part A, 443–468. Elsevier, Amsterdam.

Soltis, D. E. 1985. Allozymic differentiation among *Heuchera americana*, *H. parviflora*, *H. pubescens*, and *H. villosa* (Saxifragaceae). *Syst. Bot.* 10: 193–198.

_____. 1986. Genetic evidence for diploidy in *Equisetum*. *Amer. J. Bot.* 73: 908–913.

_____, C. H. Haufler, D. C. Darrow and G. J. Gastony. 1983. Starch gel electrophoresis of ferns: a compilation of grinding buffers, gel and electrode buffers and staining schedules. *Amer. Fern J.* 73: 9–27.

_____, and L. H. Rieseberg. 1986. Autopolyploidy in *Tolmiea menziesii* (Saxifragaceae): genetic insights from enzyme electrophoresis. *Amer. J. Bot.* 73: 310–318.

_____, and P. S. Soltis. 1988a. Electrophoretic evidence for tetrasomic segregation in *Tolmiea menziesii* (Saxifragaceae). *Heredity* 60: 375–382.

_____, and _____. 1988b. Are lycopods with high chromosome numbers ancient polyploids? *Amer. J. Bot.* 75: 238–247.

Soltis, P. S., and D. E. Soltis. 1988. Electrophoretic evidence for genetic diploidy in *Psilotum nudum*. *Amer. J. Bot.* 75: 1667–1671.

_____, _____, and L. D. Gottlieb. 1987. Phosphoglucomutase gene duplications in *Clarkia* (Onagraceae) and their phylogenetic implications. *Evolution* 41: 667–671.

Stebbins, G. L. 1980. Polyploidy in plants: unsolved problems and prospects. In W. H. Lewis [ed.], *Polyploidy: biological relevance*. Plenum Press, New York.

_____. 1982. Plant speciation. In C. Barigozzi [ed.], *Mechanisms of speciation*, 21–39. Alan R. Liss, New York.

_____. 1986. Gene action and morphogenesis in plants. In J. P. Gustafson, G. L. Stebbins, and F. J. Ayala [eds.], *Genetics, development, and evolution*, 29–49, Plenum, New York.

Stutz, H. C. 1978. Explosive evolution of perennial *Atriplex* in Western North America. In K. T. Harper and J. L. Reveal [eds.], *Intermountain biogeography: a symposium*. Great Basin Naturalist Memoirs No. 2., 161–168.

Sytsma, K. J., and L. D. Gottlieb. 1986a. Chloroplast DNA evolution and phylogenetic relationships in *Clarkia* sect. *Peripetasma* (Onagraceae). *Evolution* 40: 1248–1261.

_____, and _____. 1986b. Chloroplast DNA evidence for the origin of the genus *Heterogaura* from a species of *Clarkia* (Onagraceae): *Proc. Natl. Acad. Sci. USA* 83: 5554–5557.

Turner, B. L., W. L. Ellison, and R. M. King. 1961. Chromosome numbers in the Compositae IV. North American species with phyletic interpretations. *Amer. J. Bot.* 47: 216–223.

_____, and D. Horn. 1964. Taxonomy of *Machaeranthera* sect. *Psilactis* (Compositae-Astereae). *Brittonia* 16: 316–331.

Vallejos, D. E. 1983. Enzyme activity staining. In S. O. Tanksley and T. J. Orton [eds.]. *Isozymes in plant genetics and breeding*, Part A, 469–516. Elsevier, Amsterdam.

Wagner, W. H., Jr. 1954. Reticulate evolution in the Appalachian Aspleniums. *Evolution* 8: 103–118.

Warwick, S. I., and S. G. Aiken. 1986. Electrophoretic evidence for the recognition of two species in annual wild rice (*Zizania*, Poaceae). *Syst. Bot.* 11: 464–473.

_____, and L. D. Gottlieb. 1985. Genetic divergence and geographic speciation in *Layia* (Compositae). *Evolution* 39: 1236–1241.

Wells, E. F. 1979. Interspecific hybridization in eastern North American *Heuchera* (Saxifragaceae). *Syst. Bot.* 4: 319–338.

Wendel, J., and C. W. Stuber. 1984. Plant isozymes: enzymes studied and buffer systems for their electrophoretic resolution in starch gels. *Isozyme Bull.* 17: 4–11.

Werth, C. R., S. I. Guttman, and W. H. Eshbaugh. 1985a. Electrophoretic evidence of reticulate evolution in the Appalachian *Asplenium* complex. *Syst. Bot.* 10: 184–192.

_____, _____, and _____. 1985b. Recurring origins of allopolyploid species in *Asplenium*. *Science* 228: 731–733.

Whalen, M. D. 1979. Allozyme variation and evolution in *Solanum* section *Androceras*. *Syst. Bot.* 4: 203–222.

Wheelar, N. C., and R. P. Guries. 1982. Population structure, genic diversity, and morphological variation in *Pinus contorta* Dougl. *Canad. J. For. Res.* 12: 595–606.

_____, _____, and D. M. O'Malley. 1983. Biosystematics of the genus *Pinus*, subsection *Contortae*. *Biochem. Syst. Evol.* 11: 333–340.

Wilson, H. D., S. C. Barber, and T. Walters. 1983. Loss of duplicate gene expression in tetraploid *Chenopodium*. *Biochem. Syst. Ecol.* 11: 7–13.

Witter, M. S. 1986. *Adaptive radiation and genetic differentiation in the Hawaiian silversword alliance* (Compositae-Madiinae). Chapter 3, Ph.D. thesis, University of Hawaii, Honolulu.

_____, 1987. Adaptive radiation and genetic differentiation in the Hawaiian silversword alliance (Compositae: Madiinae). *Amer. J. Bot* (Abstract) 74: 766.

Wolf, P. G., P. S. Soltis, and D. E. Soltis. 1987. Autopolyploid evolution in *Heuchera grossulariifolia* (Saxifragaceae). *Amer. J. Bot.* (Abstract) 74: 767.

CHAPTER 8

Isozymic Evidence and the Evolution of Crop Plants

John Doebley

Department of Botany
University of Minnesota
St. Paul, Minnesota 55108

There is nothing in nature that faintly resembles such human creations as an ear of corn, a head of cauliflower, or the curiously sculptured fruit of an ornamental gourd. While these are extreme examples, they typify a problem that has unnerved crop evolutionists for over a century, namely that crop species often show such wide morphological departures from their nearest wild relatives that the link between them cannot be established with certainty. Another problem that has muddied the waters surrounding the origin of many domesticated plants is the existence of weedy intermediates between the wild and cultivated forms. Do these represent escapes from cultivation, hybrid derivatives of the cultigen and wild type, or true evolutionary intermediates? The question is apt to have different answers for different crops.

Over the past century, crop evolutionists have employed a diverse arsenal of techniques to unravel these and other mysteries surrounding the origin and evolution of domesticated plants. Classical taxonomy, biogeography, cytology, archaeology, and classical genetics have all made important contributions. Although much has been learned, the origins of many crops, even some of the most important ones, remain obscure.

In this chapter, I review the contributions made by isozyme analysis to the study of the origin and evolution of domesticated plants. The many advantages of this approach have been frequently acclaimed: (1) laboratory methods are relatively simple; (2) many plants can be assayed at reasonable cost in time and materials; (3) numerous loci scattered over the genome act as markers for variation in the entire genome; (4) levels of variation in isozymes are appropriate for studies within genera, species, or populations; and (5) if genetic analyses are performed, the frequencies or presence/absence of electrophoretically identifiable alleles can be determined. Perhaps the most significant value of isozyme analysis in sorting out the evolution of cultivated plants is that isozymes were never under direct human selection, and therefore they provide a measure of relatedness among crops and their relatives that is not apt to have been severely altered by the domestication process.

The literature on isozymes of crop plants is so extensive that no single chapter could summarize it adequately. Thus, this chapter is restricted to papers bearing on the evolution of crop plants. Reports on genetics, cultivar identification, physiology, and developmental expression of the isozymes of crop plants are excluded. I have focused on the contributions made by isozyme analysis to answering the following questions: (1) What wild taxon is ancestral to the cultigen?; (2) Is there evidence for introgression between crops and their wild relatives?; (3) What are the origins of weedy crop relatives?; (4) Has there been a loss of genetic variation as a result of domestication?; (5) Has there been a reapportionment of genetic variation in crops compared to their wild relatives?; (6) Can isozyme analysis be employed to identify centers of genetic diversity for crops?; and (7) What is the relationship between morphological and isozymic variation in genera containing crop species?

THE ORIGIN OF CROP SPECIES

Two principal concerns of crop evolutionists have been to determine taxonomic relationships among crops and their wild relatives and to identify, as nearly as possible, which wild species gave rise to the cultigen. Despite a century of effort, the accuracy with which progenitors of our most important crop plants can be identified is uneven. For a few of the 200 or so domesticated plants, wild ancestors have been identified through multiple lines of investigation including morphology, archaeology, biochemical methods (flavonoids, seed proteins, isozymes, etc.), cytology, and biogeography. However, many crop species, including some of considerable economic importance, have not been thoroughly investigated, and their progenitors are totally unknown or have been questionably identified only through conflicting or uncorroborated lines of evidence.

In this section, I discuss the contribution isozymic studies have made to identifying the progenitors of crop species. Ideally, such studies should include population samples of both the crop and its wild relatives from throughout their ranges. These population samples should be well documented, i.e., full locality data should be available, number of individuals contributing to the sample should be known, and voucher specimens should be deposited in herbaria. The isozyme analysis should involve a large number of loci (greater than 20), and genetic control of the isozymes should have been fully investigated.

In combing the literature, I uncovered isozyme studies of the origins of only 15 crop species that I felt were sufficiently extensive or definitive to merit prolonged discussion. The origins of another 15 or so crops have received more limited investigation by isozymic methods, but often without definitive results. These studies are listed but are not discussed.

One must consider how isozyme analyses can aid in the identification of the progenitors of crop species. The answer differs depending on whether the crop is of polyploid origin or of the same ploidy as its wild ancestors. In the former case, authors have attempted to identify those wild diploids whose combined isozymes yield an isozyme profile similar or identical to the polyploid crop. This approach has proven of utility in wheat, tobacco, and banana. In the case of crops at the same ploidal level as their progenitors, detailed statistical analyses have been employed. These analyses attempt to determine if the crop falls within the range of variation found within the presumed wild ancestor. This criterion stems from the fact that crops are relatively young taxonomic entities (less than 10,000 years old), whereas their wild progenitors are apt to be much older. Thus, isozyme variation found within the crop should not exceed that found in the progenitor, although drift, selection, and introgression from other wild species during or after the domestication process can upset this expectation.

One approach used to compare the relative amounts of variation in crops and their presumed progenitors is multivariate statistical analysis of isozyme data such as principal component, principal coordinate, or cluster analyses. If a wild taxon represents the progenitor of a crop, then one expects that populations of the crop and progenitor will not be fully separated by these multivariate analyses. This approach has been employed successfully with peppers (Jensen et al., 1979), rice (Second, 1982, 1985a), maize (Doebley et al., 1984), and squash (Decker and Wilson, 1987).

Another approach is to examine the allelic constitutions and general levels of polymorphism in the crop and its presumed progenitor. The crop and its progenitor should possess the same most common alleles at the vast majority of their loci, and, with the exception of a few rare alleles, the crop should contain only a subset of the alleles present in the progenitor. As can be seen from Table 8.1, most crops show a reduction in levels of polymorphism as compared to their presumed progenitors. In the one case where a crop clearly does not (*Cucurbita pepo*—*C. texana*), there is serious doubt from other lines of evidence that this wild form is the sole, true progenitor.

Finally, a determination of whether a particular wild form is genetically similar enough to a crop to be considered a potential progenitor can be made using Nei's (1972) measure of genetic identity (I). I between populations of a crop and those of its presumed progenitor should be within the range of variation among populations of the progenitor. Because populations within a species generally show a mean I of 0.90 or greater (Gottlieb, 1981), one can expect that the mean I between crops and their progenitors will usually be 0.90 or greater. As can be seen from Table 8.2, this expectation has generally been realized, and the exceptions to this rule each have explanations. For example, the identity between *Capsicum pubescens* and *C. eximum* and that between *Cucurbita pepo* and *C. texana* are lower than expected. There is serious doubt from other lines of evidence whether these wild forms are the true progenitors of these crops (see below). The value of *I* between *Cucurbita mixta* and *C. sororia* is only 0.89, but this study examined only seven loci. *I* between *Lens culinaris* subsp. *culinaris* and subsp. *orientalis* is only 0.87, but *I* among populations within these two taxa is also quite low. The same is true of Asiatic rice (*Oryza sativa*) and its progenitor (*O. rufipogon*) (Table 8.2). Thus, it appears that the degree of genetic identity between crops and their wild relatives may be used as a guide in determining whether a wild taxon is genetically similar enough to a crop to be considered its progenitor.

In the following summaries, one can see how various authors have applied the above criteria to identify the progenitors of some 15 crop plants. In each case the isozyme studies have enhanced our knowledge of the evolution of these crops.

Banana

The genus *Musa* contains over 30 species of wild, diploid, perennial plants native to southeastern Asia and the Pacific. The genus also includes the cultivated bananas, which are believed to have originated both as sterile selections within the diploid, *M. acuminata* (A genome) and as hybrids between this species and another wild diploid, *M. balbisiana* (B genome). Today, most cultivated bananas are sterile triploids with one of the following genome constitutions: AAA, AAB, or ABB.

Jarret and Litz (1986) examined isozymic variation for nine enzyme systems among a limited selection of *M. balbisiana*, *M. acuminata*, and some cultivated forms. They have not performed genetic analyses, but the comparative banding patterns of wild and cultivated forms proved informative. *Musa balbisiana* and *M. acuminata* gave distinctive patterns for several enzymes. The triploid cultivar 'Valery' (AAA) gave patterns nearly identical to *M. acuminata* as expected. A second cultivar, 'Chato' (ABB), gave isozyme patterns that were a composite of *M. acuminata* and *M. balbisiana* bands, confirming its bispecific origin. Together, these results tend to confirm the polyphyletic origin of cultivated bananas by polyploid development both within *M. acuminata* and between this species and *M. balbisiana*.

Barley

The genus *Hordeum* includes over 30 species of diploid and polyploid grasses native to four continents. The cultigen, *H. vulgare* subsp. *vulgare* (barley), is among these. Grains of this cereal recovered from archaeological sites in the Fertile Crescent of the Middle East have been dated to about the year 8,000 B.C., making barley the oldest of the domesticated cereals. Barley has one very near wild relative, *H. vulgare* subsp. *spontaneum* (wild barley), which is also native to the Middle East. As the conspecific status of these two taxa suggests, they are fully interfertile, and subsp. *spontaneum* has generally been regarded as the progenitor of cultivated barley (Zohary, 1969).

Barley is among the most intensively studied plants for isozyme variation (see Brown, 1983). Most isozyme studies on barley have involved a large number of loci (greater than 20) and large population samples. Genetic control and linkage relationships for barley isozyme genes have been analyzed in detail (Brown 1983). Extensive studies of the amount and apportionment of allozymic variation in both wild and cultivated barley have been conducted (see below).

Jorgensen (1986) examined isozyme variation among 32 species of *Hordeum* for six enzyme systems representing 12 loci. Jorgensen deliberately chose loci that showed little variation within species, but considerable variation among species, because the goal was to examine relationships among a diverse group of species. These data were subjected to both phenetic and cladistic analyses. Jorgensen's (1986) analysis divides the genus into three groups, one of which (the "vulgare" group) contains *H. vulgare* (both subspecies), *H. bulbosum*, and *H. murinum*. The remaining 29 species showed only a distant relationship to cultivated barley. *Hordeum vulgare* subsp. *vulgare* and subsp. *spontaneum* were not distinguished by Jorgensen's analysis, which is consistent with their conspecific status and the hypothesis that subsp. *spontaneum* is ancestral to cultivated barley.

Table 8.1. Genetic diversity statistics for some crop plants and their nearest wild relatives. H_t is total heterozygosity for the taxon. H_e is expected heterozygosity per population averaged over all populations. PLP is the proportion of polymorphic loci per population averaged over all populations. PLT is the proportion of polymorphic loci in the species or taxon over all. A/L/T is the number of alleles per locus in the taxon

Species	H_t	H_e	PLP	PLT	A/L/T	No. Pop.	No. Plants/Pop.	No. Loci	Source
Capsicum annuum var. annuum	0.074	—	—	0.31	1.39	—	—	26	McLeod et al., 1983b
Capsicum frutescens	0.085	—	—	0.39	1.54	—	—	26	
Capsicum chinense	0.083	—	—	0.42	1.39	—	—	25	
Capsicum annuum var. aviculare	0.105	—	—	0.42	1.54	—	—	26	
Capsicum baccatum var. pendulum	0.009	—	—	0.15	1.19	—	—	26	McLeod et al., 1983b
Capsicum baccatum var. baccatum	0.023	—	—	0.23	1.31	—	—	25	
Capsicum pubescens	0.047	—	—	0.35	1.54	—	—	26	McLeod et al., 1983b
Capsicum eximium	0.111	—	—	0.42	1.50	—	—	23	
Chenopodium quinoa subsp. quinoa	—	0.013	0.05	0.57	1.67	98	—	21	Wilson, 1988
Chenopodium quinoa subsp. milleanum	—	0.018	0.06	0.62	1.57	69	—	21	
Cucurbita mixta	—	0.067	0.22	0.57	1.57	11	12–36	7	Decker, 1986
Cucurbita sororia	—	0.079	0.27	0.43	1.43	11	5–16	7	
Cucurbita pepo var. ovifera	—	0.089	0.32	0.58	1.67	17	12	12	Decker and Wilson, 1987
Cucurbita texana	—	0.031	0.14	0.50	1.67	17	5–24	12	
Glycine max	0.244	—	—	0.36	2.08[a]	109	—	23–55	Kiang and Gorman, 1983
Glycine soja	0.236	—	—	0.50	2.48[a]	139	—	23–51	
Hordeum vulgare subsp. vulgare[b]	—	0.068	0.25	0.21	1.39	1[b]	50	28	Nevo et al., 1979a, b
Hordeum vulgare subsp. spontaneum (Israel only)	—	0.098	0.30	0.89	3.75	28	30–50	28	
Hordeum vulgare subsp. vulgare	0.564	—	0.25	1.00	7.00	1358	4	4	Kahler and Allard 1981
Hordeum vulgare subsp. spontaneum	0.514	—	0.59	1.00	7.25	148	4	4	
Hordeum vulgare subsp. vulgare	0.164	0.081	0.32	0.74	2.21	12	20–25	19	Brown and Munday, 1982
Hordeum vulgare subsp. spontaneum (Iran only)	0.218	0.158	0.43	0.68	2.58	13	23–42	19	Nevo et al., 1986a

[a] Alleles per polymorphic locus.
[b] A single composite population synthesized from 6200 cultivated barley genotypes.
[c] Value excluding Cucurbita pepo—C. texana.

Species	H_t	H_e	PLP	PLT	A/L/T	No. Pop.	No. Plants/Pop.	No. Loci	Source
Lactuca sativa	—	0.007	0.02	0.20	1.30	18	25	70	Kesseli and Michelmore, 1986
Lactuca serriola	—	0.018	0.06	0.17	1.20	5	25	70	
Lens culinaris ssp. culinaris	0.103	0.000	0.00	0.32	1.40	29	2.3	25	Hoffman et al., 1986
Lens culinaris ssp. orientalis	0.093	0.000	0.00	0.32	1.64	40	3.4	25	
Lens culinaris ssp. culinaris	0.255	0.030	0.08	0.33	1.60	31	11.0	15	Pinkas et al., 1985
Lens culinaris ssp. orientalis	0.282	0.030	0.10	0.60	1.87	18	15.1	15	
Lycopersicon esculentum var. esculentum	—	—	—	0.50	1.71	178	2–4	14	Rick and Fobes, 1975
Lycopersicon esculentum var. cerasiforme	—	—	—	0.57	1.86	98	8–16	14	
Oryza sativa subsp. japonica	0.12	—	—	—	1.50	144	—	6	Glaszmann et al., 1984
Oryza rufipogon (China)	0.39	—	—	—	2.50	28	—	6	Second, 1985b
Oryza sativa subsp. indica	0.34	—	—	—	2.33	108	—	6	Glaszmann et al., 1984
Oryza rufipogon (S. Asia)	0.42	—	—	—	3.50	82	—	6	Second, 1985b
Oryza glaberrima	0.06	—	—	—	1.50	15–152	—	6	Second, 1982
Oryza breviligulata	0.20	—	—	—	2.00	20	—	6	Second, 1985b
Raphanus sativus—cultivated	0.49	0.34	0.68	0.75	2.63	24	15–45	8	Ellstrand and Marshall, 1985
Raphanus sativus—wild	0.54	0.45	0.94	1.00	3.25	4	15–45	8	
Setaria italica subsp. italica	—	—	—	0.80	2.2	223	—	10	Jusuf and Pernes, 1985
Setaria italica subsp. viridis	—	—	—	0.80	2.4	45	—	10	
Zea mays subsp. mays	0.251	0.182	0.50	0.91	7.1	94	12	23	Doebley et al., 1985
Zea mays var. parviglumis	0.311	0.261	0.69	1.00	6.6	18	12–25	21	Doebley et al., 1984
Zizania palustris (cultivated)	—	0.142	0.35	0.41	1.59	3	12–24	17	Warwick and Aiken, 1986
Zizania palustris var. interior	—	0.139	0.28	0.53	1.76	6	12–24	17	
Mean domesticated	0.193	0.093	0.25	0.49	2.15	—	—	—	—
Mean wild	0.244	0.117	0.32	0.58	2.47	—	—	—	—
% loss from wild to cultivated	25.7	4.5 (23.6)[c]	18.8	11.8	11.4	—	—	—	—

Kahler and Allard (1981) examined variation among four esterase loci for a large sample of cultivated (1358 accessions) and wild (148 accessions) barley. This analysis indicated little differentiation between wild and cultivated barley in that 27 of the 28 alleles found in cultivated barley were also present in wild barley and the most common alleles at each of the four loci were largely the same in the wild and cultivated samples. These data are in agreement with the hypothesis that subsp. *spontaneum* is the progenitor of cultivated barley.

Several studies (Brown et al., 1978; Nevo et al., 1979a, b, 1986a, b, c; Brown and Munday, 1982) have examined the amount and apportionment of isozyme variation in wild and cultivated barley. The data gathered by these authors represent up to 28 loci assayed for 12 barley landraces, two improved populations of cultivated barley, and 41 natural populations of wild barley. These data would be useful for assessing the similarity of wild and cultivated barley if Nei's genetic identity (I) had been calculated between wild and cultivated populations and numerical taxonomic analyses (cluster and principal component analyses) performed.

Chili Peppers

The genus *Capsicum* contains five domesticated taxa and over 20 wild species. These species are native to the warmer regions of the Americas, extending from the southern United States to Chile. Before 1950 some authors hypothesized that all five cultigens arose from a single wild species, but most authors now agree that there were several independent domestications from separate wild species (Eshbaugh et al., 1983). Analysis of isozyme variation in *Capsicum* clearly disproves the former hypothesis by demonstrating that the cultigens *C. annuum* var. *annuum*, *C. baccatum* var. *pendulum*, and *C. pubescens* are far too dissimilar in isozyme constitution to be considered recent domesticated derivatives of a single ancestral species (Jensen et al., 1979; McLeod et al., 1982, 1983a, b). Rather, the isozyme data suggest at least three independent lines of domestication in *Capsicum*. First, isozyme analyses indicate that the wild form *C. baccatum* var. *baccatum* is ancestral to *C. baccatum* var. *pendulum*. Second, the cultigen *C. pubescens* has no clear ancestral form, but shows its closest relationship to the wild species *C. eximum*, which can be considered a potential progenitor. Finally, three cultigens, *C. annuum* var. *annuum*, *C. chinense*, and *C. frutescens*, were remarkably similar to one another in isozyme pattern (Jensen et al., 1979). A single wild form included in the analysis, *C. annuum* var. *aviculare*, is closely associated with all three of these cultigens; however, wild populations of *C. chinense* and *C. frutescens*, which are known (Pickersgill et al., 1979), were not assayed for isozymes. Thus, the isozyme data do not distinguish between two possible alternatives: (a) separate domestication events for each of the three, or (b) a single domestication with subsequent divergence into the three cultigens. Nevertheless, these analyses have considerably clarified the origins of the chili peppers.

Foxtail Millet

Foxtail millet (*Setaria italica* subsp. *italica*) is a small-grained cereal grown throughout much of Eurasia. Archaeological and biogeographical information suggests that this crop originated in central China and spread from there to Europe and India (Rao et al., 1987). The nearest wild relative of foxtail millet is green foxtail (*Setaria italica* subsp. *viridis*), which grows over much of Eurasia. Some weedy foxtails from Europe and North America are much more robust than typical green foxtail. These plants are referred to as giant green foxtail, and deWet has hypothesized that they arose as the hybrid derivatives of the crop and the wild form (see Rao et al., 1987).

Jusuf and Pernes (1985) studied isozyme variation at 10 loci for 223 accessions of *S. italica* subsp. *italica* and 45 accessions of *S. italica* subsp. *viridis*. Their data show a geographic pattern of variation among accessions of the cultigen with European and Asian types at opposite extremes. This pattern was corroborated by a study of variation at two esterase loci in foxtail millets, which also revealed a cline between European and Asian types (Kawase and Sakamoto, 1984). Comparison of the wild and cultivated forms demonstrated that they share the same common alleles at each of the 10 loci analyzed by Jusuf and Pernes (1985). Nei's mean genetic identity between the crop and wild form is high ($I > 0.95$), and in the same range as between other crops and their wild relatives (Table 8.2). These data support the hypothesis that subsp. *viridis* was the progenitor of subsp. *italica* (Rao et al., 1987).

Jusuf and Pernes (1985) characterized subsp. *viridis* from both France and China and examined the relationship of these wild types to the cultigen. They reported that European cultivars most closely resemble the French wild type, and the Asian cultivars resemble the Chinese wild type. Based on this result they suggested two separate origins, one in Europe and one in Asia. An alternative hypothesis, which the isozyme data do not eliminate, would include a single domestication in China, spread of the domesticate to Europe, and then introgression of the domesticate by European subsp. *viridis*. The fact that the European cultivars appear intermediate in allele frequencies between the Asian cultivars and the European wild types supports this scenario.

"Irish" Potato

Solanum is a large pantropical genus with over 1,000 species. Within it, there is a relatively small group of tuber-bearing species belonging to subgenus *Pachystemonum*, section *Tuberarium*, subsection *Hyperbasarthrum*. Within this group, about six wild diploids are taxonomically similar enough to the cultigen to have potentially contributed to the origin of cultivated potatoes (Simmonds, 1979). One commonly mentioned diploid "wild potato" thought to have a role in the evolution of the cultivated forms is *S. sparsipilum*.

The cultivated potatoes include both diploid and tetraploid forms, which have been treated taxonomically in various ways. *Solanum goniocalyx* and *S. stenotomum* are cultivated diploids, and *S. tuberosum* is the cultivated tetraploid. *Solanum tuberosum* includes group Andigena, which is native to South America, and group Tuberosum, which was long thought to have arisen under cultivation in Europe. Two questions surrounding the origin of cultivated potatoes have been: (1) Which diploid(s) gave rise to the cultivated tetraploid (*S. tuberosum*)?; and (2) What is the relationship between the European group Tuberosum and the South American group Andigena?

Oliver and Martinez-Zapater (1984) examined variation among some wild and cultivated potato species at 13 isozyme loci. Formal genetic analyses for these loci have been performed (Martinez-Zapater and Oliver, 1984a, b; Quiros and McHale, 1985). Their data reveal that the tetraploid *Solanum tuberosum* group Andigena has an isozyme constitution that largely represents a combination of alleles found in the diploids *S. stenotomum* and *S. sparsipilum*. This seems to support the hypothesis that *S. tuberosum* group Andigena arose as the tetraploid derivative of these two diploid taxa. However, group Andigena shows a much closer isozymic similarity to *S. stenotomum* ($I = 0.95$) than it does to *S. sparsipilum* ($I = 0.79$), suggesting that the two diploids have not contributed equally to the tetraploid. Thus, the authors suggest that group Andigena had a polyphyletic origin involving allopolyploidy between *S. stenotomum* and *S. sparsipilum*, autopolyploidy of *S. stenotomum*, and subsequent introgression between these taxa. The authors also uncovered some alleles in group Andigena that were not present in either *S. sparsipilum* or *S. stenotomum*, suggesting that other wild diploids may have been involved in the origin of cultivated tetraploid potatoes. Unfortunately, *S. vernei*, previously implicated in the origin of group Andigena, was not included in the study.

As to the origin of group Tuberosum, Oliver and Martinez-Zapater (1984) found that this group shares most of its alleles with group Andigena, although several alleles were apparently derived from strains known only in Chile. These Chilean types also resemble European group Tuberosum in both morphology and ability to flower under long day-lengths. This suggests that group Tuberosum did not arise simply by selection under cultivation in Europe as previously hypothesized, but rather that the European varieties were derived from Chilean types. This agrees well with analysis of cytoplasmic factors, which also implicated Chile as the region for the origin of group Tuberosum (Grun, 1979).

A final contribution made by Oliver and Martinez-Zapater (1984) is that their data show a very high genetic identity ($I = 0.96$) between the two species of cultivated diploid potatoes (*S. stenotomum* and *S. goniocalyx*). The authors suggest that these should be considered a single species.

Lentil

The genus *Lens* includes the cultivated lentil (*L. culinaris* subsp. *culinaris*) and one to three closely related wild species, depending on the taxonomic system followed. Morphological and cytological evidence provided strong support for the hypothesis that *L. culinaris* subsp. *orientalis* was ancestral to the cultivated lentil; however, alternative opinion cast *L. nigricans* in the role of progenitor to the cultigen (Hoffman et al., 1986). Isozymic evidence provided critical support for the hypothesis that *L. culinaris* subsp. *orientalis* is the ancestral taxon (Pinkas et al., 1985; Hoffman et al., 1986). The genetic similarity between *L. culinaris* subsp. *culinaris* and subsp. *orientalis* ($I = 0.81$ to 0.87) is within the range of variation among populations of subsp. *orientalis*, although somewhat lower than that between other crops and their wild ancestors (Table 8.2). On the other hand, *L. nigricans* was surprisingly shown to be the isozymically most distinct member of the genus, having a Nei's genetic identity of only 0.51–0.64 with lentil, arguing convincingly against any possibility that it might be ancestral to the cultigen. *Lens ervoides* and *L. odomensis* also showed low genetic identities (0.72–0.74 and 0.65–0.80, respectively) to the cultigen, indicating that they are too dissimilar to be considered its ancestor.

Lettuce

Isozyme analysis of lettuce and its wild relatives involved a large number of loci (70) although only 30% (22) were polymorphic (Kesseli and Michelmore, 1986). Phenetic analysis of these data revealed substantial heterogeneity among the cultivated populations with different cultivated forms showing affinities to different wild species. Kesseli and Michelmore (1986) interpreted this to mean that either there were multiple independent domestications of lettuce or introgression subsequent to an initial domestication.

Maize

The genus *Zea* contains maize (*Z. mays* subsp. *mays*) and its wild relatives, the teosintes (*Zea* spp.). The teosintes are treated taxonomically as members of four species: *Z. diploperennis*, *Z. perennis*, *Z. luxurians*, and *Z. mays* subsp. *mexicana* and subsp. *parviglumis*. The latter subspecies includes two varieties, var. *parviglumis* and var. *huehuetenangensis* (Iltis and Doebley, 1980). As this taxonomy suggests, subspecies *mexicana* and *parviglumis* are judged to be very closely related to maize. The teosintes are known from Chihuahua in northern Mexico, south to Honduras. They are most common in central and western Mexico.

The origin of maize has long been controversial. Over the past 100 years, a wealth of both information and misinformation has been published on the topic. In spite of this (or because of it), the community of maize evolutionists has been decidedly split into two camps on this issue. One hypothesis (the "wild maize" hypothesis), supported by Mangelsdorf (1974), maintains that maize evolved from a "wild maize," which had all the essential morphological features of modern maize (i.e., paired female spikelets and polystichous ears surrounded by husks). This "wild maize" has never been seen, and thus it is strictly hypothetical. Mangelsdorf considers it to be extinct. The "wild maize" hypothesis gives teosinte no direct role in the origin of maize.

A second hypothesis (the "teosinte hypothesis"), most vigorously supported by Beadle (1972), proposes quite simply that maize is a domesticated form of teosinte. This hypothesis has had greater acceptance with several particularly vocal supporters (Galinat, 1983; Iltis, 1983). The proponents of this hypothesis have tried to explain the evolution of the unusual morphological traits of this obligate cultigen, whereas Mangelsdorf has avoided this question by having the essential characteristics of cultivated maize arise in nature rather than under domestication. The morphological evolution of maize is a complex issue and beyond the scope of this present review.

Maize and teosinte are among the most thoroughly analyzed crop-progenitor complexes for isozyme variation. These analyses involved the survey of over 80 accessions of teosinte (almost all known populations) and several hundred accessions of maize for variation at 21 to 23 loci (Doebley et al., 1983, 1984, 1985, 1986, 1987; Goodman and Stuber, 1983a; Smith et al., 1984, 1985). Genetic control and linkage relationships for the 23 enzyme loci have been thoroughly examined (Goodman and Stuber, 1983b; Stuber and Goodman, 1983). Using the isozyme data, the hypothesis that teosinte is ancestral to maize can be tested. Doebley et al. (1984) demonstrated that maize and one teosinte (*Z. mays* var. *parviglumis*) share a very close allozymic relationship, with the mean I between populations of these two taxa being 0.92 (Table 8.2). This is the same level of allozymic similarity found between other crops and their wild ancestors (Table 8.2). Further, populations of maize and var. *parviglumis* could not be separated by principal component analysis based on isozyme allele frequen-

Table 8.2. Nei's genetic identities (I) between populations of some crop plants and their nearest wild relatives and among populations within these taxa

Species	Nei's I (range)	Source
Capsicum annuum var. *aviculare*		McLeod et al., 1983b
× *C. annuum* var. *annuum*	0.93(n. a.)[a]	
× *C. chinense*	0.94(n. a.)[a]	
× *C. frutescens*	0.94(n. a.)[a]	
Capsicum baccatum var. *pendulum*		McLeod et al., 1983b
× *C. baccatum* var. *baccatum*	0.98(n. a.)[a]	
Capsicum pubescens		McLeod et al., 1983b
× *C. eximum*	0.81(n. a.)[a]	
Chenopodium quinoa subsp. *quinoa*	0.95(0.81–1.00)	Wilson, unpubl. data
Chenopodium quinoa subsp. *milleanum*	0.98(0.91–1.00)	
C. q. quinoa × *C. q. milleanum*	0.96(0.86–1.00)	
Cucumis melo (cultivars) × *C. melo* var. *agrestis*	0.91(n. a.)[a, b]	Perl-Treves et al., 1985
Cucurbita mixta	0.98(0.94–1.00)	Decker, 1986
Cucurbita sororia	0.99(0.94–1.00)	
C. mixta × *C. sororia*	0.89(0.76–0.98)	
Cucurbita pepo	0.83(0.53–1.00)	Decker and Wilson, 1987
Cucurbita texana	0.96(0.90–1.00)	
C. pepo × *C. texana*	0.73(0.49–0.96)	
Glycine max × *G. soya*	0.93(n. a.)[a]	Kiang and Gorman, 1983
Hordeum vulgare subsp. *spontaneum* (Iran only)	0.92(0.86–0.99)	Nevo et al., 1986a
Hordeum vulgare subsp. *spontaneum* (Israel only)	0.89(0.75–0.99)	Nevo et al., 1979b
Lens culinaris ssp. *culinaris*	0.89(0.76–1.00)	Hoffman et al., 1986
Lens culinaris ssp. *orientalis*	0.90(0.76–1.00)	
L. c. culinaris × *L. c. orientalis*	0.87(0.76–1.00)	
Lens culinaris ssp. *culinaris*	0.86(n. a.)	Pinkas et al., 1985
Lens culinaris ssp. *orientalis*	0.83(n. a.)	
L. c. culinaris × *L. c. orientalis*	0.81(n. a.)	
Lactuca sativa	0.93(n. a.)[b]	Kesseli and Michelmore, 1986
Lactuca serriola	0.89(n. a.)[b]	
L. sativa × *L. serriola*	0.90(n. a.)[a, b]	
Oryza glaberrima	n. a.	Second, 1985a
Oryza breviligulata	0.91(0.75–1.00)	
O. glaberrima × *O. breviligulata*	0.91(0.80–1.00)	
Oryza sativa subsp. *indica*	n. a.	Second, 1985a
Oryza rufipogon (South Asia)	0.78(0.54–1.00)	
O. s. indica × *O. rufipogon* (South Asia)	0.80(0.6–0.94)	
Oryza sativa subsp. *japonica*	n. a.	Second, 1985a

[a]Values calculated using mean allele frequencies for the two taxa.
[b]Data taken from a dendrogram published by the authors and should be considered approximations.
n. a. = not available

Species	Nei's I (range)	Source
Oryza rufipogon (China)	0.82(0.57–1.00)	
O. s. japonica × O. rufipogon (China)	0.87(0.69–1.00)	
Setaria italica subsp. italica (China) × S. italica subsp. viridis (China)	0.98(n. a.)[a]	Jusuf and Pernes, 1985
Setaria italica subsp. italica (France) × S. italica subsp. viridis (France)	0.95(n. a.)[a]	Jusuf and Pernes, 1985
Zea mays subsp. mays	0.95(0.87–0.99)	Doebley et al., 1984, 1985
Zea mays var. parviglumis	0.91(0.82–0.98)	
Z. mays mays × Z. mays parviglumis	0.92(n. a.)	
Zizania palustris (cultivated)	0.98(n. a.)	Warwick and Aiken, 1986
Zizania palustris var. interior	0.90(n. a.)	
Z. palustris (cultivated) × Z. p. interior	0.93(n. a.)	

cies (Doebley et al., 1984, 1987). Although these data do not prove that maize is domesticated teosinte, they do add considerable strength to that hypothesis.

Whereas isozyme data implicate Z. mays var. parviglumis in the origin of maize, they also demonstrate that several teosintes could not be ancestral to this cultigen. Zea diploperennis, Z. perennis, and Z. luxurians all have isozyme constitutions far different from maize, as expected from previous morphological and cytological investigations (see Doebley et al., 1984). Curiously, Z. mays subsp. mexicana was also shown to be isozymically distinct from maize. This clashes with morphological data, which suggested to some authors that subsp. mexicana is the most maize-like ("maizoid") teosinte.

Isozyme analyses in Zea have also aided in understanding the relationships among maize landraces. Doebley et al. (1985) demonstrated that maize of Mexico forms three loose racial complexes based on isozyme allele frequencies, and that these complexes correspond well with ecological, geographic, and morphological parameters. Similarly, Goodman and Stuber (1983a) found clear isozymic differentiation between Andean and lowland maize races in Bolivia.

Quinoa

The genus Chenopodium contains two New World cultivated crops: Huauzontle (C. berlandieri subsp. nuttalliae) of Mexico, and Quinoa (C. quinoa subsp. quinoa) of the Andes Mountains in South America. Both are grown for grain (seed) and as leaf vegetables. Quinoa has been thoroughly investigated by both classical taxonomic and isozymic methods (Wilson, 1976, 1978, 1980, 1981, 1988; Wilson et al., 1983). These studies have considerably clarified the relationships among Quinoa and its near relatives, (1) C. hircinum (sensu stricto) of the Argentinean lowlands and eastern Andean slopes, and (2) weedy plants of the Andean highlands known by the natives as Ajara. This latter weedy type, which is sympatric with the cultivated Quinoa, has been variously treated taxonomically as either C. quinoa var. melanospermum, C. quinoa subsp. milleanum, or as part of C. hircinum (sensu lato). I will refer to it as subsp. milleanum or Ajara. Both C. hircinum (sensu stricto) and the weedy Ajara have been suggested as potential progenitors of Quinoa (see Wilson and Heiser, 1979).

Wilson (1988, unpublished data) has examined both morphological and allozymic (21 loci) variation among Quinoa and its wild relatives. These analyses demonstrated clearly that C. hircinum (sensu stricto) and the weedy Ajara of the Andes are distinct both allozymically ($I = 0.83$ between them) and morphologically and deserve separate taxonomic appellations. Similarly, C. hircinum (sensu stricto) and the cultivated Quinoa are both morphologically and allozymically ($I = 0.82$) distinct. This calls into question the hypothesis that this wild taxon could be ancestral to the cultivated Quinoa. Finally, Wilson's data demonstrated that the weedy Ajara (C. quinoa subsp.

milleanum) and the cultivated quinoa (subsp. *quinoa*) are allozymically quite close ($I = 0.96$) and no more divergent than populations within Quinoa ($I = 0.95$) and Ajara ($I = 0.98$). These data leave intact the hypothesis that Ajara could be ancestral to Quinoa. However, Wilson (1988, unpublished data) considers that the weedy Ajara present in South America today may have been altered by introgression from the crop and by selection to adapt it to existence as a weed in cultivated fields. Thus, he feels it may be misleading to think of it as the progenitor of Quinoa and prefers to envision Ajara as a taxon which has "co-evolved" with the cultigen.

Rice

The genus *Oryza* contains about 20 species distributed throughout the warmer regions of Asia, Africa, Australia, and tropical America. Two cultivated taxa have generally been recognized: *O. glaberrima* or African rice, and *O. sativa* or Asiatic rice. The former species is native to west Africa where it has an associated wild relative, *O. breviligulata* (= *O. barthii*). *Oryza breviligulata* has both wild and weedy forms, the latter often invading cultivated fields. *Oryza sativa* is native to a large portion of Asia from India to northern China and Japan. Asiatic rice has traditionally been divided into three subspecies, *japonica, indica,* and *javanica.* Over parts of its range, *O. sativa* has an associated wild and weedy relative called *O. rufipogon* (= *O. perennis*). This relative of *O. sativa* has annual, perennial, and intermediate growth forms.

Various authors have proposed a series of hypotheses concerning the relationships among the various cultivated and wild forms. According to one hypothesis, there was a single domestication with subsequent divergence into the African and Asian cultigens (see Second, 1982, 1985b). Alternatively, it has been proposed that not only do African and Asiatic rice have separate origins, but that the different subspecies of *O. sativa* are also the products of independent domestications.

The isozymes of rice have received careful study (Endo and Morishima, 1983). About 14 enzyme systems and over 40 loci have been resolved. Genetic control and linkage analyses of many of these have been completed. Numerous studies of populational variation in both wild and cultivated forms have also been conducted. Among these, the work of Second (1982, 1984, 1985a, b, c) has addressed many questions surrounding the evolution of cultivated rice. This work involved the analyses of several hundred accessions for variation at 24 to 40 isozyme loci.

Second (1982, 1985b, c), combining information from isozymes, biogeography and hybridization studies, proposed the following model for evolution of cultivated rice. (1) African rice arose under domestication from the wild form of *O. breviligulata*. Populations of these taxa have a high genetic similarity ($I = 0.91$) and could not be separated by multivariate analysis of isozyme data. (2) Asiatic rice was independently domesticated twice: once in southern Asia (subsp. *indica*), and once in China (subsp. *japonica*). Isozyme data show that these two subspecies are quite distinct (see also Endo and Morishima, 1983). Each of these subspecies shows its closest isozymic relationship to populations of *O. rufipogon* from the region in which it is cultivated, i.e. *indica* is most like *O. rufipogon* from southern Asia, and *japonica* most like *O. rufipogon* from China (Taiwan and Kwangsi provinces). (3) *Javanica* types appear somewhat intermediate between the other two subspecies but are closer to *japonica* (see also Glaszmann et al., 1984). They may have arisen in part through crosses between *japonica* and *indica*.

Second (1982, 1985b) further suggested that the evolution of Asiatic rice has continued via introgressive hybridization between the two principal subspecies and between these and the wild and weedy *O. rufipogon*. This seems a reasonable scenario, but exactly how the isozyme data demonstrate introgression from the wild into the cultivated taxa is not fully discussed. Second's (1985b) data did not discriminate between annual, perennial, and intermediate forms of *O. rufipogon,* and thus whether one or all of these made a contribution to cultivated rice could not be resolved. Other authors (Endo and Morishima, 1983) have found isozymic differences between the different seasonal growth forms of *O. rufipogon,* but did not compare these to the cultivated forms. The principal differences between the growth form classes is that the perennials show higher levels of isozymic polymorphism (higher observed heterozygosity, higher proportions of polymorphic loci, and more alleles/locus) than the annuals (Endo and Morishima, 1983). It would be of interest to determine the levels of polymorphism in the cultivars relative to the annual and perennial wild forms.

Nakagahra (1978) examined variation at three esterase loci among 1,317 cultivated varieties of *O. sativa* and found greatest variation in Burma, Laos, northern India, and southwest China, regions where subsp. *indica* predominates. Other lines of evidence (Second, 1982; Endo and Morishima, 1983; Glaszmann et al., 1984) also demonstrate that *indica* is the isozymically more variable of the two subspecies.

Silver-Seeded Squash and Cushaw

The silver-seeded squash and Cushaw are native domesticates of Mexico that on occasion find their way into North American markets. Both belong to a single species, *Cucurbita mixta*. This cultivated species has an associated wild one (*C. sororia*) that ranges throughout the Pacific lowlands of Mexico with some populations known from Veracruz (H. Wilson, personal communication). *Cucurbita sororia* has been considered the progenitor of *C. mixta*. Decker (1986) made a limited isozyme analysis of these taxa, examining 11 accessions of *C. mixta* and 11 accessions of *C. sororia* for only seven loci. Nevertheless, these data show a close genetic relationship between these two taxa ($I = 0.76$ to 0.98 with a mean of 0.89), supporting the hypothesis that *C. sororia* is the progenitor of *C. mixta*.

Squash, Pumpkin, and Ornamental Gourds

Cucurbita pepo, as currently defined, is a morphologically polymorphic species, which includes: (1) pumpkin, (2) zucchini, crookneck, acorn, and other edible squash, and (3) the ornamental gourds, a familiar sight in supermarkets every Halloween. These cultigens are native to Mexico and the U.S. where they played a prominent role in pre-Columbian agriculture. Two wild species, *C. texana* and *C. fraterna*, are judged to be the closest known wild relatives of *C. pepo*. A study of isozyme variation among these taxa (excluding *C. fraterna*) has shed new light on the origin of *C. pepo*; however, many questions remain unanswered (Decker, 1985, 1986; Decker and Wilson, 1987). Perhaps, the most startling revelation of the allozymic analyses was that those entities included under the name *C. pepo* form two divergent groups with a very low Nei's genetic identity (0.51 to 0.70) between them (Decker, 1985). One of these groups includes pumpkin, Mexican calabaza criolla, and marrow squash, and the other includes most ornamental gourds plus crookneck, scallop, acorn, and some other squash types. Given the low genetic identity between the two groups and the clear circumscription of these groups, the possibility that they represent the products of two separate domestications from two separate wild species must be considered. Interestingly, Heiser (1985) and others have proposed that *C. pepo* was twice independently domesticated.

The question of whether or not *C. texana* represented the progenitor of all or part of *C. pepo* has been addressed by Decker and Wilson (1987) using isozyme analysis. Their data show a reasonably close relationship between *C. texana* and some members of *C. pepo*, such as most ornamental gourds and acorn, crookneck, and scallop squash. The Nei's genetic identity between this wild and these cultivated forms averages around 0.80 to 0.85, which is lower than that between most crop-progenitor pairs. Moreover, *C. texana* and *C. pepo* were clearly separated by principal component and cluster analyses based on isozyme data. Finally, there is considerable allelic variation in these cultivated forms, while *C. texana* has much lower levels of allelic polymorphism. Thus, while it is possible that *C. texana* had a role in the origin of *C. pepo*, it is also clear that the present-day populations of *C. texana* analyzed by Decker and Wilson (1987) do not represent a germplasm pool sufficient in scope and character to represent the sole wild ancestor of *C. pepo*. *Cucurbita fraterna*, a native of northeastern Mexico, may also have been involved in the origin of *C. pepo* (Nee and Andres, 1986). This hypothesis should be tested with isozymic or other suitable genetic-biochemical analyses.

Tobacco

The genus *Nicotiana* comprises over 60 species including the cultivated allotetraploid *Nicotiana tabacum*. Tobacco is native to South America, but has no known wild tetraploid relatives that could be cast in the role of progenitor. This suggested that *N. tabacum* arose by polyploidy under domestication from two wild diploid species. Taxonomic, biogeographic, and genetic evidence indicated that one of these two wild diploids was *N. sylvestris*; however, the identity of the second has remained in doubt. Conflicting information suggested that either N. otophora or N. tomentosiformis was the other parental species.

Sheen (1972) examined isozymic variation among a small group of Nicotiana species including those diploids putatively involved in the origin of N. tabacum. Also included in the analysis were F_1 hybrids between these diploids. No genetic analyses were performed, but similarities and differences in banding patterns for eight enzyme systems were recorded and a similarity index computed among the species and hybrids. These analyses revealed that N. tabacum was much more similar to the N. sylvestris × N. tomentosiformis hybrid (86%) than to the N. sylvestris × N. otophora hybrid (64%). Thus, N. tomentosiformis was implicated as the second diploid species in the origin of N. tabacum.

Tomato

The genus Lycopersicon contains nine species that are distributed throughout Central and South America. Rick and his associates (Rick et al., 1974; Rick and Fobes, 1975; Rick, 1983) have conducted extensive analyses of isozyme variation among these species, including studies of breeding systems, genetic control, linkage, and populational variation. Much of their work focused on the three red-fruited species, L. cheesmanii, L. pimpinellifolium, and L. esculentum, which form a natural assemblage within the genus. Lycopersicon cheesmanii is a strictly wild taxon endemic to the Galápagos Islands. Lycopersicon pimpinellifolium is also wild and native to Ecuador and Peru. Lycopersicon esculentum contains two elements: (a) the cultivated tomato (var. esculentum), and (b) a weedy form (var. cerasiforme) found in Central and South America.

Isozyme analysis for 14 loci (Rick and Fobes, 1975) revealed L. cheesmanii to be: (1) highly uniform, consistent with its hypothesized recent origin by dispersal from coastal South America to the Galápagos, and (2) completely homozygous, consistent with its predominantly self-pollinating breeding system. The data also demonstrate that L. cheesmanii is isozymically most similar to L. pimpinellifolium from northwestern Peru, indicating that L. pimpinellifolium (or its ancestral form) gave rise to L. cheesmanii and that the original introduction to the Galápagos came from coastal Peru.

As to the origin of the cultivated tomato, Rick and Fobes (1975) found both L. cheesmanii and L. pimpinellifolium to possess different alleles than the cultigen at several isozyme loci, thus removing these two species as potential wild ancestors of tomato. Lycopersicon cheesmanii would seem an unlikely candidate on biogeographic grounds, it being endemic to the Galápagos Islands. Lycopersicon esculentum var. cerasiforme, on the other hand, possessed the same basic set of alleles as tomato, supporting the view that this weedy form gave rise to the cultigen. Rick and Fobes (1975) argued on the basis of isozymic and other evidence that the weedy var. cerasiforme spread from Ecuador-Peru to Mexico, where it was domesticated. Later, the cultigen spread back to South America where introgression between the cultigen and L. pimpinellifolium took place, as evidenced by the joint occurrence of several alleles in both taxa in Ecuador and Peru (Rick et al., 1974). Thus, whereas most tomato cultivars tend to be completely monomorphic for isozymes, those sympatric with L. pimpinellifolium show some polymorphism. Tomato varieties from Europe and North America share a similar isozyme constitution with the Mexican-Central American types, suggesting the tomato was introduced into North America and Europe from Mexico or Central America.

Wheat

The cultivated wheats compose a group of diploid, tetraploid, and hexaploid cereals within the genus Triticum. There are four primary taxa of cultivated wheat, each of which has its own presumed ancestor: T. monococcum var. monococcum (AA genome) derived from the wild type T. monococcum var. boeoticum (AA); T. turgidum var. dicoccum (AABB) derived from the wild type T. turgidum var. dicoccoides (AABB); T. timopheevii var. timopheevii (AAGG) derived from the wild type T. timopheevii var. araraticum (AAGG); and T. aestivum (AABBDD) derived from a cross of the cultivated T. turgidum (AABB) and the wild T. tauschii (DD). As indicated, each of the cultivated polyploids has a wild ancestral form of the same ploidal level with the exception of T. aestivum, which presumably arose by polyploidy under domestication.

The origins of the A, B, and D genomes of T. aestivum have not been fully resolved. Based on cytological analyses, two of the genomes can be clearly associated with specific wild diploid taxa: (1) *T. monococcum* var. *boeoticum* for the A genome, and (2) *T. tauschii* for the D genome. The donor of the B genome is uncertain; however, several species of *Triticum* have been favored based on cytological, taxonomic, and biochemical data. These include *T. speltoides, T. longissimum,* and *T. bicorne*. The donor of the G genome of *T. timopheevii* is also uncertain, although *T. speltoides* has been suggested by some authors (see Feldman, 1976).

Several authors have attempted to determine the origins of the genomes in the cultivated polyploid wheats by comparison of their isozyme constitutions to those of the various wild diploids. These studies have produced mixed results. In some cases, different authors have reached different conclusions when studying different enzyme systems. In other cases, the results are discordant with results from other biochemical analyses. In part, these problems may result from considerable isozymic heterogeneity within both the wild and cultivated taxa. With these limitations in mind, the contribution of isozyme analysis to unraveling the origin of the cultivated wheats can be briefly summarized.

The origin of the A genome of *T. monococcum* var. *monococcum* from the wild *T. monococcum* var. *boeoticum* has been proposed by several independent lines of evidence. Not surprisingly, several investigations of isozyme variation in wheat have supported this conclusion (Nakai, 1979b; Asins and Carbonell, 1986b). *Triticum tauschii* has generally been regarded as the donor of the D genome to *T. aestivum* based on cytological and taxonomic criteria. Isozyme analyses confirm this result (Jaaska, 1976, 1980, 1981; Nakai, 1979a, 1981; Nishikawa et al., 1980). These studies also provide greater detail on the origin of the D genome by demonstrating that *T. tauschii* subsp. *strangulata* rather than subsp. *tauschii* was involved. Moreover, isozyme studies indicate the origin of the hexaploid wheats probably occurred in the southern Caspian Sea-Transcaucus region where subsp. *strangulata* is found.

The use of isozyme analyses to identify the donor of the B genome has been attempted by several researchers. Jaaska (1976, 1978, 1980, 1981, 1984) has examined electrophoretic variation for aspartate aminotransferase, alcohol dehydrogenase, NAD-dependent aromatic alcohol dehydrogenase, NADP-dependent aromatic alcohol dehydrogenase, and esterase. His comparisons consistently showed that *T. speltoides* was the most likely donor of the B genome. His analyses were also consistent in demonstrating that neither *T. longissimum* nor *T. bicorne* could have donated the B genome to cultivated wheats. Nakai's (1979b) analysis of esterase isozymes suggested that either *T. speltoides* or *T. longissimum* could have contributed the B genome. Asins and Carbonell (1986a, b) analyzed peroxidase and alkaline phosphatase isozymes and reported that *T. longissimum* and *T. speltoides* have high intraspecific variability. These two species also exhibited close isozymic affinity to one another (see also Brody and Mendlinger, 1980). These facts taken together prevented Asins and Carbonell (1986b) from concluding which wild species contributed the B genome to the cultivated polyploids. Thus, despite the consistent implication of Jaaska's analyses that *T. speltoides* donated the B genome, the question does not appear satisfactorily resolved. Jaaska (1978) has suggested that the B genome of the cultigens may represent a hybrid genome of several wild species or that introgression from *T. longissimum* may have occurred. Analysis of chloroplast DNA restriction patterns indicated unequivocally that the cytoplasmic genome of *T. aestivum* (AABBDD) and *T. turgidum* (AABB) came from *T. longissimum* (Tsunewaki and Ogihara, 1983). Minimally, therefore, *T. longissimum* contributed the cytoplasm, and in all probability, made a contribution to the B nuclear genome as well.

The origin of the G genome of *T. timopheevii* var. *timopheevii* has been less thoroughly investigated by isozyme analysis. Asins and Carbonell (1986b) found *T. timopheevi* to differ from *T. turgidum,* suggesting that these two tetraploids have separate origins. Jaaska (1976, 1978) suggested that *T. speltoides* (including *T. aucheri*) may have contributed the G genome to *T. timopheevii*. Analysis of chloroplast DNA restriction site variation supports the hypothesis that *T. aucheri* is the donor of the G genome (Tsunewaki and Ogihara, 1983).

"Wild" Rice

Warwick and Aiken (1986) examined isozyme variation in the genus *Zizania,* which includes the cultivated "wild" rice. This is one of very few crops domesticated by modern plant breeders (i.e., within the past 100 years). The isozyme data support the recognition of two separate species of annual wild rice, Z. aquatica and Z. palustris. Within the latter species, Warwick and Aiken (1986) distinguished vars. palustris and interior. Finally, the isozyme data indicate that the cultivated "wild" rice was derived from Z. palustris var. interior rather than from var. palustris as previously proposed by deWet and Oelke (1978).

Some Other Crops

Questions surrounding the origin and evolution of some other crop plants have been addressed through isozyme studies. Many of these studies have limitations in that either very few loci were analyzed (often only one or two), no genetic analyses were performed, statistical analyses of the data were limited, sampling of the taxa was restricted, or evolutionary questions were not the authors' primary concern. Nevertheless, contributions have been made by isozymic analysis to understanding the origin and evolution of the following crops (see also the chapter by Torres in this volume): Alfalfa (Quiros, 1983), Amaranths (Jain et al., 1980), Avocado (Torres, 1983), Buckwheat (Ohnishi, 1983, 1985), Citrus (Torres et al., 1978, 1982; Scora et al., 1982; Torres, 1983), Cole crops (Arus and Shields, 1983; Arus and Orton, 1983), Cotton (Cherry et al., 1972), Cucumber (Dane, 1983; Perl-Treves et al., 1985), Grape (Loukas et al., 1983), Melons (Perl-Treves et al., 1985), Naranjilla (Whalen and Caruso, 1983), Oats (Price and Kahler, 1983), Pistachio (Loukas and Pontikis, 1979), Radish (Ellstrand and Marshall, 1985), Rye (Jaaska, 1983; Puchalski and Molski, 1981), Soybean (Kiang, 1981; Kiang and Gorman, 1983), Strawberry (Arulsekar and Bringhurst, 1983), and Watermelon (Zamir et al., 1984).

INTROGRESSION BETWEEN CROPS AND THEIR WILD RELATIVES

The occurrence of introgressive hybridization between crops and their wild relatives seems very probable. Crops and their near wild relatives often grow sympatrically, and the latter frequently occur as weeds in cultivated fields. In some cases, as crops spread beyond their cradle of domestication, they came into contact with other closely related wild relatives creating new opportunities for introgression. Crops and their wild relatives often lack genetic barriers to hybridization, and naturally occurring hybrids are frequently reported by crop scientists involved in field studies. This is not to deny the existence of temporal and genetic barriers to gene flow between crops and their wild relatives, but merely to note that such barriers, even in self-pollinating species, are usually imperfect.

While there is some evidence of introgression between crops and their wild relatives, there are reasons why all reports of such introgression should be viewed critically. First, many reports of introgression between crops and their relatives are based solely on examination of morphology without any independent confirmation from genetic or biochemical analysis. Second, because the occurrence of this introgression *seems* so likely, authors have frequently claimed introgression to be the cause of parallel variation between a crop and its wild relatives without considering alternative hypotheses such as convergence. These claims are frequently elevated to "facts" in uncritical reviews.

In this section, I review the very few attempts to document introgression between crops and their wild relatives by isozyme analysis. In some cases, these studies support previous opinion based on morphological observations. In others, isozyme analyses disprove or greatly modify claims of introgression based solely on morphology. The general lesson appears to be that introgression between crops and their wild relatives does occur, but that predictions of introgression based on morphology are not always reliable. Further, whether introgression proceeds most commonly from weed into crop or the reverse cannot be ascertained from the literature despite unsubstantiated claims that it is primarily unidirectional from crop into weed (Ladizinsky, 1985).

The ability to detect introgressive isozyme alleles among crops and their wild relatives is limited by the fact that a crop and its progenitor are apt to possess the same constellation of isozyme alleles. Therefore, the shared presence of an allele may indicate either introgression or joint inheritance from their common ancestor. The best evidence for introgression is most likely to be obtained between crops and more distant wild relatives that have distinct allelic variants fixed (or nearly fixed) at several loci. If, where a crop is sympatric with such a wild relative, it possesses in low frequency alleles typical of the wild taxon (but lacking in the crop elsewhere), then this may be appropriately interpreted as evidence of introgression from the wild form into the cultigen. Introgression from the crop into the wild form could be documented by the reverse situation.

Rick and his colleagues (Rick et al., 1974; Rick and Fobes, 1975) have examined isozyme variation in *Lycopersicon esculentum* var. *esculentum* (tomato), its probable progenitor *L. esculentum* var. *cerasiforme*, and a distinct wild species, *L. pimpinellifolium*. They found that tomato and its progenitor are nearly identical in isozyme constitution and nearly monomorphic throughout their ranges. On the other hand, *Lycopersicon pimpinellifolium* had a distinct isozyme constitution. Where the cultigen is sympatric with *L. pimpinellifolium*, along coastal Peru, certain alleles typical of the latter species ($Prx-1^1$, $Prx-2^1$ and $Prx-4^1$) are found in the cultigen. Rick et al. (1974) concluded that this distribution "suggests that the wild alleles tend to substitute in the cultivated forms as a result of introgression."

A situation similar to that of tomato and *Lycopersicon pimpinellifolium* was reported by Doebley et al. (1987) in maize (*Zea mays* subsp. *mays*), its probable progenitor (*Z. mays* var. *parviglumis*), and a distinct wild form (*Z. mays* subsp. *mexicana*). As in *Lycopersicon*, the crop and its progenitor were nearly identical in isozyme allele constitution, whereas the more distant wild relative (subsp. *mexicana*) possessed distinct alleles at several loci. Where maize is sympatric with subsp. *mexicana*, certain alleles typical of the wild taxon (*Enpl-14* and *Glul-8*) were found in the cultigen. These alleles were not found in maize from other regions. Doebley et al. (1987) concluded that this distribution provides evidence for introgression from the wild form into the cultigen.

Doebley et al. (1984) reported some isozymic evidence for introgression from maize into two of its most distant wild relatives, *Z. luxurians* and *Z. diploperennis*. For the former species, a few plants possessed the allele *Glul-7* that is otherwise totally unknown in this species, although it is a common allele in maize. More convincing evidence exists for introgression from maize into *Z. diploperennis* (Doebley et al., 1984). One plant of this wild species possessed two alleles (*Enpl-8* and *Pgdl-3.8*) that are otherwise unknown in this species, but common in maize. These two loci are tightly linked (3 map units apart) on chromosome 6, suggesting that the segment of this chromosome that carries these two loci was transferred from maize into *Z. diploperennis*.

Zamir et al. (1984) examined isozyme variation in *Citrullus lanatus* (watermelon) and a sympatric wild relative, *C. colocynthis*. These species are fixed for distinct alleles at nine isozyme loci. Watermelon cultivars from three countries were invariant with the exception of one accession from Israel, which was highly polymorphic and contained several alleles characteristic of *C. colocynthis*. Zamir et al. (1984) concluded that "this suggests that there was considerable introgression of the wild species genes into the genetic background of the cultivated watermelon." Curiously, these authors found no evidence for the spread of these introgressive genes beyond this localized population. This suggests that introgression is highly restricted and that "strong disruptive selection" is operating against it.

A study of pollen flow between experimental plantings of *Cucurbita* species and cultivars has been carried out by marking bees and using isozymes to detect successful cross-pollinations (Kirkpatrick and Wilson, 1988). This study documents that cross-pollination and hybridization can occur between the cultivated *C. pepo* and the wild *C. texana*. In another study, Decker and Wilson (1987) reported on isozymic variation in *Cucurbita pepo* (squash, etc.) and *C. texana* (wild Texas gourd). Their data show that some alleles (*Pgi-3o*, *Per-2d*, *Got-1b*, and *Pgm-2v*) typical of the cultigen occur in low frequency in the wild form. Coupled with the experimental study of pollen flow in *Cucurbita*, these data suggest that gene flow from the cultigen into *C. texana* may occur.

A few authors attempted to verify by isozyme analysis cases of introgression hypothesized or claimed on morphological evidence. Doebley et al. (1987) were unable to confirm that *Zea mays* subsp. *mexicana* had been much affected by introgression from maize as hypothesized by some authors (Wilkes, 1967; Galinat, 1973). On the other hand, Second (1982) provided isozymic evidence that a weedy form of *Oryza breviligulata* possesses alleles obtained through introgression from cultivated rice. Because these two examples involve the origin of a weed, they will be discussed more fully in the next section.

ORIGINS OF WEEDY RELATIVES OF CROP PLANTS

Essentially all crop species have wild relatives that grow sympatrically with them over part or all of their range. These wild relatives tend to favor open or routinely disturbed habitats for a simple historical reason. Man, in selecting wild species to convert into annual crops, deliberately or inadvertently chose taxa with weedy propensities. Such plants fit his needs well: (1) they adapted well to soil disturbance typical of agricultural fields; (2) they produced abundant harvestable seed instead of channeling energy into perennating structures such as rhizomes; and (3) they grew rapidly and competed well against other plants. Anderson (1952) termed plants that became the progenitors of our crops "camp followers" as he hypothesized that they frequently grew in the disturbed soil and middens around the encampments of pre-agricultural societies.

In some cases, a near relative of a crop may grow exclusively (or nearly so) in naturally disturbed habitats such as lake shores, stream sides, or other sites that undergo frequent or seasonal natural disturbance. Such plants are generally termed "wild." In other cases, a relative of a crop may grow exclusively (or nearly so) in humanly disturbed habitats, especially agricultural fields. Such plants are generally termed "weedy." Some "weedy" relatives of crops may be restricted to agricultural fields for simple historical reasons. For example, Chalco teosinte (Zea mays subsp. mexicana) occurs almost exclusively in agricultural fields, but where else can it grow as its whole natural range in the Valley of Mexico is now covered with either corn fields or concrete?

There is an apparent continuum from wild to weedy to cultivated that plainly implies a mode of origin for weedy crop relatives. Namely, the weed originated as the hybrid derivative of the crop and wild form. Alternatively, weeds may have originated through the modification of wild species by natural selection in the absence of significant introgressive hybridization, or they may have arisen through a reversion of the crop to natural seed dispersal. Each of these modes of origin leads to a distinct prediction about the degree of isozymic similarity between the wild, weedy, and cultivated forms. In at least three cases, the origin of weedy crop relatives has been investigated electrophoretically. These studies demonstrate that the modes of origin of weeds are diverse and that predictions based on morphology are not always reliable.

African cultivated rice (*Oryza glaberrima*) has associated weedy and wild forms that are both treated taxonomically as *O. breviligulata*. Second (1982) provided isozymic data that help resolve the relationships among these plant forms. It had been suggested by some authors (see Second, 1982) that wild *O. breviligulata* was merely an escape from cultivation. However, Second's data show that the wild form contains far greater isozymic variation than the cultigen, which is inconsistent with this mode of origin. It had also been hypothesized that the weedy form of *O. breviligulata* arose as a hybrid derivative of wild *O. breviligulata* and either *O. glaberrima* and/or *O. sativa*. Second (1982) reported that some of the weedy populations contain alleles typical of *O. sativa* (Pox-C^o and Est-C^2), but unknown in wild *O. breviligulata*. These data suggest that some weedy *O. breviligulata* populations did arise as the hybrid derivatives of the wild form and cultivated *O. sativa*.

Maize (*Z. mays* subsp. *mays*) has several wild and weedy relatives, some of which routinely form F_1 hybrids with maize where they grow sympatrically. One of these relatives, known as Chalco teosinte (*Z. mays* subsp. *mexicana*), is the most weedy of the maize relatives and "appears" most like maize morphologically. Because of the morphological similarity between Chalco teosinte and maize sympatric with it, this teosinte has been presumed to have been selected to mimic maize to evade the eye of the farmers weeding the fields (Wilkes, 1967, 1977). However, as discussed by Doebley (1984), there are good reasons to believe that Chalco teosinte and its sympatric maize have certain morphological resemblances simply because they are both adapted to the relatively cool, high elevations of the Valley of Mexico. Thus, convergence and not mimicry may be the primary evolutionary force.

The morphological similarity between maize and Chalco teosinte in the Valley of Mexico is said to have risen through hybridization of some "pure" teosinte and maize. Thus, the weedy Chalco teosinte is envisioned as the hybrid derivative of a wild form and a crop. Wilkes (1967) proposed that Chalco teosinte may be an introgressive form of Balsas teosinte (*Z. mays* var. *parviglumis*). Galinat (1973) has outlined a genetic mechanism by which gene exchange could take place: "Apparently, the

Mexican teosintes and their maize partners have undergone a co-evolution that permits a constant gene flow between them while maintaining their distinct female spikes through block inheritance of controlling genes" (Galinat, 1973). This, if true, should lead to the homogenization of isozyme allele frequencies in sympatric maize and Chalco teosintes, unless the isozyme loci are tightly linked to the hypothesized blocks of morphological genes. Even if some linkage exists, it is unlikely that all 21 isozyme loci that have been studied (Doebley et al., 1984) should be linked to one of the hypothetical blocks of controlling genes. The 21 loci are known to reside on seven separate chromosomes.

Doebley et al. (1984, 1987) examined isozyme variation in the weedy Chalco teosinte and maize collected from the same fields in the Valley of Mexico, and in more wild forms of annual teosinte from other regions of Mexico. The results were contrary to expectations based on the model for the origin of Chalco teosinte discussed above. Rather than showing a continuum from wild (var. *parviglumis*) to weedy (Chalco) to cultivated types, the data revealed that the weedy Chalco teosinte was the least like maize and the more wild var. *parviglumis* the most like maize. Further, maize and Chalco teosinte collected in the same fields did not show any isozymic similarity, but maintained distinct isozyme constitutions. These data essentially disprove the hypothesis that sympatric maize and Chalco teosinte populations undergo a constant gene exchange and maintain their distinct morphologies only by block inheritance of genes controlling morphology. Similarly, they discredit the idea that Chalco is a Balsas teosinte (var. *parviglumis*) contaminated with maize genes. Rather, the isozyme data suggest that Chalco teosinte is a distinct taxon that has adapted to life as a weed in maize fields. There is no evidence that introgression from maize played a role in this adaptation. However, as discussed in the previous section, there is some evidence of limited gene flow from Chalco teosinte into maize. The results of these isozyme studies agree fully with the cytological and field studies of Kato (1976, 1984) and Doebley (1984), which provided evidence that introgression from maize is not a major factor in the genetic constitution of teosinte populations or in the evolution of the weedy Chalco teosinte.

A final example of a wild relative of a crop whose origin has been investigated electrophoretically comes from the genus *Cucurbita*. In the United States, *Cucurbita* contains one primarily wild taxon (*C. texana*), the cultigen *C. pepo*, and some other non-cultivated populations which have been regarded as escapes from cultivation, and thus, classified as *C. pepo* (Decker and Wilson, 1987). Isozyme data demonstrate that these non-cultivated populations possess several alleles not found in the cultigen and lack other alleles typical of the cultigen (Decker and Wilson, 1987). In principal component analyses, these non-cultivated populations appear intermediate between *C. texana* and *C. pepo*. These populations are apparently not escapes from cultivation as previously thought. They may represent either hybrid derivatives between *C. texana* and *C. pepo* or an evolutionary intermediate between these two species.

In conclusion, the isozymic evidence on the origins of weedy crop relatives indicates that hypotheses based solely on morphology are not always reliable. In the cases of *Zea* and *Cucurbita*, original hypotheses had to be abandoned or greatly modified. Only in *Oryza* was the original hypothesis confirmed. More isozyme studies of the origins of weedy crop relatives are clearly needed.

LOSS OF GENETIC VARIATION IN CROP PLANTS

Little is known about the early stages of crop domestication, but it is generally and reasonably assumed that the first farmers experimented with only a small fraction of the variation present within the species of interest. One expects that during each generation undesirable phenotypes were culled out and the best phenotypes preserved when setting aside seed for planting the following year. Because a single plant may produce dozens or even thousands of seeds, it is theoretically possible that a very small number of individuals (fewer than 10) might contribute the seed for the next year's crop of a single farmer. Over the six to 10 millennia since most crops were first domesticated, one might expect a severe loss of genetic variation. However, there are competing forces such as introgression of new variation from wild relatives, selection against loss of fitness due to inbreeding, and trading of cultivars among farmers over considerable geographic distances. Thus, while some loss of variation in domesticated plants may be expected, the extent of reduction to anticipate is unclear.

The question of whether bottlenecks during domestication have caused a reduction in genetic variation in crops can be addressed with isozymic data. Table 8.1 lists various measures of genetic diversity for some crops and their nearest known wild relative, which in most cases represents their probable progenitor. The averages for domesticated and wild taxa are revealing. Total heterozygosity (H_t) drops from 0.24 in the wild forms to 0.19 in the cultigens, mean expected heterozygosity (H_e) from 0.12 to 0.09, proportion of polymorphic loci per population (PLP) from 0.32 to 0.25, proportion of polymorphic loci per taxon (PLT) from 0.58 to 0.49, and alleles per locus per taxon (A/L/T) from 2.47 to 2.15. If one calculates the average percentage of variation lost in a crop as compared to its wild relative, then H_t drops 26%, H_e drops 5% (24% if Cucurbita pepo—C. texana are excluded), PLP drops 19%, PLT drops 12%, and A/L/T drops 11%. These data suggest that some loss of variation has occurred; however, caution must be exercised in interpreting these numbers because the sampling of the crop and wild type may have differed radically. For example, wild taxa may be represented by seed collected directly in the field, whereas crops may be represented by lines maintained in germplasm banks for 20 years or more. Further, in most of the studies, the crops were more thoroughly sampled than their wild relatives. Nevertheless, the general conclusion that some loss in variation has occurred agrees with results of previous authors based on smaller numbers of taxa (Brown, 1978; Ellstrand and Marshall, 1985). As noted by several previous authors, this result indicates that wild progenitors represent a pool of new genetic variation for crop improvement.

EFFECTS OF DOMESTICATION ON THE APPORTIONMENT OF GENETIC VARIATION

The extent to which isozyme variation in a crop species tends to be localized within landraces and cultivars or distributed more equally among them provides a picture of the genetic structure of the crop (Levin, 1976; Ellstrand and Marshall, 1985). This structure may be compared to that of its progenitor to determine if domestication has altered this structure. If any change has occurred, it might be toward either greater concentration of variation within categories or greater dispersion of variation among categories, depending on the operative forces. Strong artificial selection, genetic bottlenecks, mode of maintenance of the crops (selfing vs. outcrossing), trading practices among farmers, and introgression from wild forms could all come into play.

Table 8.3 lists measures of genetic variation within and among populations, cultivars, or landraces of four crop species and their probable progenitors. For one of the species listed (Lens culinaris), the amount and distribution of variation in the wild and cultivated forms are nearly identical. The remaining three species show a reduction in the amount of variation within the crop (lower H_e and H_t) and a tendency for the variation present to be among rather than within cultivars or landraces (higher D_{st} and G_{st}). These data are too limited and the sampling problems in each of these studies too severe to ascribe much meaning to these data. As more studies are performed, it may be possible to determine whether the genetic structure of a crop species generally differs from that of its progenitor. Unravelling the causes of any restructuring of genetic variation in landraces and cultivars of crops as compared to their progenitors will be more difficult because of the many variables involved.

Table 8.3. Measures of genetic diversity within and among populations, cultivars or landraces of some crop species and their probable progenitors. H_e is mean expected heterozygosity within populations, H_t is total heterozygosity, D_{st} is H_t-H_e and G_{st} is D_{st}/H_t. See Table 1 for source of data

Taxon	H_e	H_t	D_{st}	G_{st}
Hordeum vulgare subsp. vulgare	0.08	0.16	0.08	0.51
Hordeum vulgare subsp. spontaneum	0.16	0.22	0.06	0.28
Lens culinaris subsp. culinaris	0.03	0.26	0.22	0.88
Lens culinaris subsp. orientalis	0.03	0.28	0.25	0.89
Raphanus sativus—cultivated	0.34	0.49	0.15	0.31
Raphanus sativus—wild	0.45	0.54	0.08	0.15
Zea mays subsp. mays	0.18	0.25	0.07	0.28
Zea mays var. parviglumis	0.26	0.31	0.05	0.16

CENTERS OF GENETIC DIVERSITY

Based on the premise that the degree of diversity in a crop from a given area should be correlated to the length of time it has been cultivated there, Vavilov (1950) proposed that crop species have centers of diversity that coincide with their centers of origin. Further, these centers should typically coincide with the distribution of the crop's nearest wild relatives. Vavilov (1950) recognized that, in addition to primary centers, crops might have secondary centers of diversity that could arise as the result of hybridization between crops and their wild relatives.

Harlan (1971, 1975) challenged the idea that centers of origin can be established for all crops and suggested that some crops may have "non-centric" origins. Other authors have also criticized Vavilov's concept and noted that trading patterns, farming practices, and environmental diversity can all influence the degree of diversity in a crop from a particular region. The length of time of cultivation of a crop in a given area is only one of many interacting factors. Vavilov, too, was aware of the limitations of his system.

Vavilov's determinations of centers of diversity for various crop species were based on his extensive field studies and observations of morphology and other phenotypic traits. Brücher (1969) posed the question, "do gene-centers exist?" Pickersgill (1977) noted that despite the availability of an appropriate methodology (namely isozymes), few broad scale studies of genetic diversity within crop species have been undertaken. While the situation has improved over the past 10 years, genetic diversity has been quantified for fewer than 10 crop species over their ranges. Data for a few other crops probably exist, but are not published in a form suitable for discerning the existence of gene centers.

Rick and Fobes (1975) examined diversity for four enzyme systems (14 loci) among 178 tomato cultivars from Latin America, Europe, and the United States. Their data show that the center for allelic diversity in tomato is in Peru. Yet, Mexico is the presumed center of origin for the tomato based on isozymic and other sources of evidence. As noted by Rick and Fobes (1975), the cause of the greater allelic diversity for tomato cultivars from Peru appears to be introgression from a wild relative of tomato, *Lycopersicon pimpinellifolium*.

Kahler and Allard (1981) examined variation at four esterase loci among 1,358 accessions of barley from throughout the world. Their data show that the highest levels of polymorphism are found in accessions from middle south Asia. Yet, based on archaeological evidence and the distribution of wild barley, it appears that barley was domesticated in the Fertile Crescent of the Middle East. Again, it appears that the center of origin and the center of isozymic diversity do not coincide.

Ohnishi (1983, 1985) examined isozyme variation in buckwheat from Japan, Europe, Nepal, and India. His data show that the highest levels of polymorphism are found in Nepal with the other regions sampled showing various degrees of reduced polymorphism. Whereas region of origin of buckwheat is not known, wild buckwheat is found in Nepal and southern China, but not in Japan and Europe.

Nakagahra (1978) examined variation for three esterase loci in 1,317 varieties of Asiatic rice. His data demonstrate that the center of genetic diversity for these loci exists from Assam, India, through Burma, Laos, and Yunnan, China, with considerable reductions in diversity in north China, Japan, and southern India. The region of diversity coincides with the region in which both of the principal varieties of rice, subsp. *japonica* and subsp. *indica* are grown.

Goodman and Stuber (1983a) and Doebley et al. (1983, 1985, 1986, 1988) examined isozyme variation in maize landraces from North America, Mexico, and Bolivia. The sampling of maize races in Mexico and Bolivia was very similar, both studies including approximately 100 accessions representing nearly 30 landraces. Despite the equality of sampling, Mexican maize was found to contain far greater genetic variation than Bolivian maize (163 vs. 115 alleles at 23 loci, $H_t = 0.251$ vs. 0.198). Similarly, maize landraces from the U.S.A. and Canada showed a reduction in variation compared to those from Mexico (Doebley et al., 1986). Mexico is both the center of origin for maize and the region in which its closest wild relatives are native.

Centers of genetic (isozyme) diversity can be defined for crop species that have been analyzed. In some cases, such as tomato and barley, these centers do not correspond to the center of origin. In other cases, such as maize and buckwheat, there does appear to be correspondence between center of isozyme diversity and center of origin. As predicted by the critics of Vavilov, and, to some extent, by Vavilov himself, centers of diversity should not be equated with centers of origin. To determine which crops show a "non-centric" pattern of genetic diversity (i.e., equivalent levels of diversity over a large portion of the range), more detailed studies of other crops are needed.

MORPHOLOGICAL AND ISOZYMIC VARIATION IN CROPS AND THEIR WILD RELATIVES

Crop plants often show wide morphological departures from their nearest wild relatives, and within crop species one cultivated variety may differ from another by equally dramatic morphological variations. This situation is perhaps most startling in crop species such as *Brassica oleracea*, *Cucurbita pepo*, and *Zea mays*. Taxonomists previously responded to this situation by placing crops and their nearest wild relatives in separate species or even separate genera, and by treating different cultivars of a single crop as separate taxonomic species. Conversely, wild species within a genus containing a cultigen may be morphologically so similar that only an expert can readily distinguish them. Within these wild species one typically finds relatively little morphological variation compared to that found within crop species. This situation seems paradoxical as it is known that crop species are apt to be much younger than their wild counterparts.

The tables are turned in this relationship if isozymic rather than morphological variation is considered. By this measure cultigens typically show close isozymic similarity to their progenitor species, and varieties within a single crop species are not apt to show large isozymic differences. Congeneric wild species, which are difficult for all but the experts to distinguish, may show substantial degrees of isozymic divergence. The lack of correlation between morphological and isozymic divergence among crops and their relatives bespeaks the extreme effects that strong artificial selection may have on morphology and aptly demonstrates that extreme morphological divergences can be achieved in relatively short periods of time with little accompanying genetic variation (Gottlieb, 1984). The uncoupling of morphological and isozymic variation is nicely documented in *Lycopersicon* (Rick et al., 1974; Rick and Fobes, 1975; Rick, 1983), *Amaranthus* (Jain et al., 1980), *Zea* (Doebley, 1984, 1985, 1987), *Cucurbita* (Decker, 1985; Decker and Wilson, 1986), and *Capsicum* (McLeod et al., 1982, 1983a, b), although not all of these authors have discussed their data from this perspective.

CONCLUSIONS

Isozyme analysis has contributed in a variety of ways to our understanding of the origin and evolution of crop plants. Most notably, it has provided a measure of relatedness between crops and their wild relatives that is independent of morphology. This is critical because crop species have typically undergone such extreme morphological departures from their wild relatives that morphology becomes an uncertain measure of evolutionary affinity. By isozyme analysis, one can easily test two expectations that will generally be met if the wild form in question is ancestral to the cultigen: (1) the crop falls within the range of variation of its presumed progenitor; and (2) the crop possesses a subset of the allelic diversity found within its progenitor. Isozyme analysis has revealed that, despite major morphological differences, teosinte (*Zea mays* subsp. *parviglumis*) meets the expectations necessary to be considered the progenitor of maize. Similarly, isozyme analysis revealed at least three separate lines of domestication among the chili peppers, in agreement with recent taxonomic work on the genus. Within the cultigen *Cucurbita pepo*, isozyme analysis has shown that there are two major and quite distinct groups of cultivars, raising the question: does *C. pepo*, as currently defined, actually include the products of two separate domestication events from two separate ancestral species? Finally, isozyme analysis provided evidence that Asiatic rice was independently domesticated twice and that the products of these two domestications have remained largely distinct with little gene exchange between them.

Isozyme analysis has also contributed to our understanding of the origin of polyploid crops. For tobacco and wheat, isozyme analysis has helped identify the diploid ancestors of the polyploid cultigens. For banana and potato, isozyme analysis has shown that these polyploid crops are probably polyphyletic in origin.

Another area in which isozyme analyses have made an important contribution relates to the degree of reciprocal gene flow between crops and their wild relatives. Isozyme studies have provided the best evidence that introgression between crops and their relatives exists. At the same time, these analyses show that the occurrence and direction of gene flow has not always been accurately predicted by morphological analyses. In *Zea*, where introgression has been proposed as the mode of origin of a weedy form, isozyme analysis has failed to support this claim. However, the hybrid origin of a weedy intermediate between wild African rice and cultivated Asiatic rice does find support from isozyme study.

Isozymes can probably contribute more than any other form of analysis to quantifying the amount and distribution of genetic variation within a crop. Isozyme studies have shown that geographic centers of genetic diversity can be demonstrated for some crops; however, these centers do not necessarily coincide with the crops' centers of origin. Isozyme data provide fairly strong evidence that crop species have reduced levels of variation compared to their probable progenitors. This reduction may be the legacy of the domestication process. Isozyme studies also provide preliminary indications that genetic variation may be differently apportioned in some crops compared to their wild relatives. Specifically, three of the four crops that have been examined have relatively less variation within landraces or cultivars and relatively more variation among these categories when compared to populations of their progenitors. Whether such reapportionment is a feature of domestication or has some other cause remains to be determined.

Isozyme analysis has substantially advanced our knowledge of both the process and the consequences of plant domestication. As further studies are completed, they will enable us to understand better both the evolution of specific crops and the general features of the domestication process itself.

ACKNOWLEDGMENTS

I extend my thanks to Charles Heiser and Jonathan Wendel for the helpful comments they provided on this manuscript. The preparation of this manuscript was supported in part by grants from the U.S.D.A. (86-CRCR-1-2161) and Pioneer Hi-Bred International of Johnston, Iowa.

LITERATURE CITED

Anderson, E. 1952. *Plants, man and life*. University of California Press, Berkeley.

Arulsekar, S., and R. S. Bringhurst. 1983. Strawberry. *In* S. D. Tanksley and T. J. Orton [eds.], *Isozymes in plant genetics and breeding*. Part B, 391–400. Elsevier, Amsterdam.

Arus, P., and T. J. Orton. 1983. Inheritance and linkage relationships of isozyme loci in *Brassica oleracea*. J. Heredity 74: 405–412.

_____, and C. R. Shields. 1983. Cole crops (*Brassica oleracea* L.). *In* S. D. Tanksley and T. J. Orton [eds.], *Isozymes in plant genetics and breeding*, Part B, 339–350. Elsevier, Amsterdam.

Asins, M. J., and E. A. Carbonell. 1986a. A comparative study on variability and phylogeny of *Triticum* species. I. Intraspecific variability. Theor. Appl. Genet. 72: 551–558.

_____, and _____. 1986b. A comparative study on variability and phylogeny of *Triticum* species. 2. Interspecific relationships. Theor. Appl. Genet. 72: 559–568.

Beadle, G. W. 1972. The mystery of maize. Field Mus. Nat. Hist. Bull. 43: 2–11.

_____. 1980. The ancestry of corn. Sci. Amer. 242: 112–119.

Brody, T., and S. Mendlinger. 1980. Species relationships and genetic variation in the diploid wheats (*Triticum, Aegilops*) as revealed by starch gel electrophoresis. Pl. Syst. Evol. 136: 247–258.

Brown, A. H. D. 1978. Isozymes, plant population genetic structure and genetic conservation. Theor. Appl. Genet. 52: 145–157.

_____. 1983. Barley. *In* S. D. Tanksley and T. J. Orton [eds.], *Isozymes in plant genetics and breeding,* Part B, 57–77. Elsevier, Amsterdam.

_____, and J. Munday. 1982. Population-genetic structure and optimal sampling of land races of barley from Iran. *Genetica* 58: 85–96.

_____, E. Nevo, D. Zohary, and O. Dagan. 1978. Genetic variation in natural populations of wild barley (*Hordeum spontaneum*). *Genetica* 49: 97–108.

Brücher, H. 1969. Gibt es Gen-Zentren? *Naturwissenschaften* 56: 77–84.

Cherry, J. P., F. R. Katterman, and J. E. Endrizzi. 1972. Seed esterases, leucine aminopeptidase and catalases of species of *Gossypium*. *Theor. Appl. Genet.* 42: 218–226.

Dane, F. 1983. Cucurbits. *In* S. D. Tanksley and T. J. Orton [eds.], *Isozymes in plant genetics and breeding,* Part B, 369–390. Elsevier, Amsterdam.

Decker, D. S. 1985. Numerical analysis of allozyme variation in *Cucurbita pepo*. *Econ. Bot.* 39: 300–309.

_____. 1986. *A biosystematic study of* Cucurbita pepo. Ph.D. thesis, Texas A&M Univ., College Station, Texas.

_____, and Hugh D. Wilson. 1987. Allozyme variation in the *Cucurbita pepo* complex: *C. pepo* var. *ovifera* vs. *C. texana*. *Syst. Bot.* 12: 263–273.

DeWet, J. J., and E. A. Oelke. 1978. Domestication of American wild rice (*Zizania aquatica* L. Gramineae). *Journ. d'Agric. et de Bota. Appl.* 25: 67–84.

Doebley, J. F. 1984. Maize introgression into teosinte—a reappraisal. *Ann. Missouri Bot. Gard.* 71: 1100–1113.

_____, M. M. Goodman, and C. W. Stuber. 1983. Isozyme variation in maize from the Southwestern United States: taxonomic and anthropological implications. *Maydica* 28: 97–120.

_____, _____, and _____. 1984. Isoenzymatic variation in Zea (Gramineae). *Syst. Bot.* 9: 203–218.

_____, _____, and _____. 1985. Isozyme variation in the races of maize from Mexico. *Amer. J. Bot.* 72: 629–639.

_____, _____, and _____. 1986. Exceptional genetic divergence of Northern Flint Corn. *Amer. J. Bot.* 73: 64–69.

_____, _____, and _____. 1987. Patterns of isozyme variation between maize and Mexican annual teosinte. *Econ. Bot.* 41: 234–246.

_____, J. Wendel, J. S. C. Smith, C. W. Stuber, and M. M. Goodman. 1988. The origin of cornbelt maize: the isozyme evidence. *Econ. Bot.* 42: 120–131.

Ellstrand, N. C., and D. L. Marshall. 1985. The impact of domesticaton on distribution of allozyme variation within and among cultivars of radish, *Raphanus sativus* L. *Theor. Appl. Genet.* 69: 393–398.

Endo, T., and H. Morishma. 1983. Rice. *In* S. D. Tanksley and T. J. Orton [eds.], *Isozymes in plant genetics and breeding,* Part B, 129–146. Elsevier, Amsterdam.

Eshbaugh, W. H., S. I. Guttman, and M. J. McLeod. 1983. The origin and evolution of domesticated *Capsicum* species. *J. Ethnobiol.* 3: 49–54.

Feldman, M. 1976. Wheats. *Triticum* spp. (Gramineae-Triticinae). *In* N. W. Simmonds [ed.], *Evolution of crop plants,* 120–127. Longman, New York.

Galinat, W. C. 1973. Preserve Guatemalan teosinte, a relict link in corn's evolution. *Science* 180: 323.

_____. 1983. The origin of maize as shown by key morphological traits of its ancestor, teosinte. *Maydica* 28: 121–138.

Glaszmann, J. C., H. Benoit, and M. Arnaud. 1984. Classification des riz cultives (*Oryza sativa* L.): utilisation de la variabilite isoenzymatique. *L'Agronomie Tropicale* 39: 51–66.

Goodman, M. M., and C. W. Stuber. 1983a. Races of maize. VI. Isozyme variation among races of maize in Bolivia. *Maydica* 28: 169–187.

_____, and _____. 1983b. Maize. *In* S. D. Tanksley and T. J. Orton [eds.], *Isozymes in plant genetics and breeding,* Part B., 1–33. Elsevier, Amsterdam.

Gottlieb, L. D. 1981. Electrophoretic evidence and plant populations. *Prog. Phytochem.* 7: 1–46.

_____. 1984. Genetics and morphological evolution in plants. *Amer. Naturalist* 123: 681–709.

Grun, P. 1979. Evolution of the cultivated potato: a cytoplasmic analysis. *In* J. G. Hawkes, R. N.

Lester, and A. D. Skelding [eds.], *The biology and taxonomy of the* Solanaceae, 655–666. Academic Press, New York.

Harlan, J. R. 1971. Agricultural origins: centers and noncenters. *Science* 174: 468–473.

———. 1975. Geographic patterns of variation in some cultivated plants. *J. Heredity* 66: 182–191.

Heiser, C. B. 1985. *Of plants and people*. University of Oklahoma Press, Norman.

Hoffman, D. L., D. E. Soltis, F. J. Muehlbauer, and G. Ladizinsky. 1986. Isozyme polymorphism in *Lens* (Leguminosae). *Syst. Bot.* 11: 392–402.

Iltis, H. H. 1983. From teosinte to maize: the catastrophic sexual transmutation. *Science* 222: 886–894.

———, and J. F. Doebley. 1980. Taxonomy of *Zea* (Gramineae). II. Subspecific categories in the *Zea mays* complex and a genetic synopsis. *Amer. J. Bot.* 67: 994–1004.

Jaaska, V. 1976. Aspartate aminotransferase isoenzymes in the polyploid wheats and their diploid relatives. On the origin of tetraploid wheats. *Biochem. Physiol. Pflanzen.* 170: 159–171.

———. 1978. NADP-dependent aromatic alcohol dehydrogenase in polyploid wheats and their diploid relatives. On the origin and phylogeny of polyploid wheats. *Theor. Appl. Genet.* 53: 209–217.

———. 1980. Electrophoretic survey of seedling esterases in wheats in relation to their phylogeny. *Theor. Appl. Genet.* 56: 273–284.

———. 1981. Aspartate aminotransferase and alcohol dehydrogenase isoenzymes: intraspecific differentiation in *Aegilops tauschii* and the origin of the D genome polyploids in the wheat group. *Pl. Syst. Evol.* 137: 259–273.

———. 1983. *Secale* and *Triticale*. *In* S. D. Tanksley and T. J. Orton [eds.], *Isozymes in plant genetics and breeding*, Part B, 79–101. Elsevier, Amsterdam.

———. 1984. NAD-dependent aromatic alcohol dehydrogenase in wheats (*Triticum* L.) and goatgrasses (*Aegilops* L.): evolutionary genetics. *Theor. Appl. Genet.* 67: 535–540.

Jain, S. K., L. Wu, and K. R. Vaidya. 1980. Levels of morphological and allozyme variation in Indian amaranths: a striking contrast. *J. Heredity* 71: 283–285.

Jarret, R. L., and R. E. Litz. 1986. Enzyme polymorphism in *Musa acuminata* Colla. *J. Heredity* 77: 183–188.

Jensen, R. J., M. J. McLeod, W. H. Eshbaugh, and S. I. Guttman. 1979. Numerical taxonomic analyses of allozymic variation in *Capsicum* (Solanaceae). *Taxon* 28: 315–327.

Jorgensen, R. B. 1986. Relationships in the barley genus (*Hordeum*): an electrophoretic examination of proteins. *Hereditas* 104: 273–291.

Jusuf, M., and J. Pernes. 1985. Genetic variability of foxtail millet (*Setaria italica* P. Beauv.). *Theor. Appl. Genet.* 71: 385–391.

Kahler, A. L., and R. W. Allard. 1981. Worldwide patterns of genetic variation among four esterase loci in barley (*Hordeum vulgare* L.). *Theor. Appl. Genet.* 59: 101–111.

Kato, T. A. 1976. *Cytological studies of maize*. Massachusetts Agric. Exper. Sta. Bull. 635.

———. 1984. Chromosome morphology and the origin of maize and its races. *Evol. Biol.* 17: 219–253.

Kawase, M., and S. Sakamoto. 1984. Variation, geographical distribution and genetical analysis of esterase isozymes in foxtail millet, *Setaria italica* (L.) P. Beauv. *Theor. Appl. Genet.* 67: 529–533.

Kesseli, R. V., and R. W. Michelmore. 1986. Genetic variation and phylogenies detected from isozyme markers in species of *Lactuca*. *J. Heredity* 77: 324–331.

Kiang, Y. T. 1981. Inheritance and variation of amylase in cultivated and wild soybeans and their wild relatives. *J. Heredity* 72: 382–386.

———, and M. B. Gorman. 1983. Soybean. *In* S. D. Tanksley and T. J. Orton [eds.], *Isozymes in plant genetics and breeding*, Part B, 295–328. Elsevier, Amsterdam.

Kirkpatrick, K. J., and H. D. Wilson. 1988. Interspecific gene flow in *Cucurbita*: *C. texana* vs. *C. pepo*. *Amer. J. Bot.* 75: 519–527.

Ladizinsky, G. 1985. Founder effect in crop-plant evolution. *Econ. Bot.* 39: 191–199.

Levin, D. A. 1976. Consequences of long-term artificial selection, inbreeding and isolation in *Phlox*. II. The organization of allozymic variability. *Evolution* 30: 463–472.

Loukas, M., and C. A. Pontikis. 1979. Pollen isozyme polymorphism in types of *Pistacia vera* and related species as an aid in taxonomy. *Jour. Hort. Sci.* 54: 95–102.

———, M. N. Stavrakakis, and C. B. Krimbas. 1983. Inheritance of polymorphic isoenzymes in grape cultivars. *J. Heredity* 74: 181–183.

Mangelsdorf, P. C. 1974. *Corn: its origin, evolution and improvement.* Harvard University Press, Cambridge, MA.

Martinez-Zapater, J. M., and J. L. Oliver. 1984a. Identification of potato varieties: an isozyme approach. In D'Arcy, W. G. [ed.], *Solanaceae: biology and systematics,* 457–467. Columbia University Press, New York.

———, and ———. 1984b. Genetic analysis of isozyme loci in tetraploid potatoes (*Solanum tuberosum* L.). *Genetics* 108: 669–679.

McLeod, M. J., S. I. Guttman, and W. H. Eshbaugh. 1982. Early evolution of chili peppers (*Capsicum*). *Econ. Bot.* 36: 361–368.

———, ———, and ———. 1983a. Peppers (*Capsicum*). In S. D. Tanksley and T. J. Orton [eds.]., *Isozymes in plant genetics and breeding,* Part B, 189–201. Elsevier, Amsterdam.

———, ———, ———, and R. E. Rayle. 1983b. An electrophoretic study of evolution in *Capsicum* (Solanaceae). *Evolution* 37: 562–574.

Nakagahra, M. 1978. The differentiation, classification and center of genetic diversity of cultivated rice (*Oryza sativa* L.) by isozyme analysis. *Trop. Agr. Res.* 11: 77–82.

Nakai, Y. 1979a. Isozyme variation in *Aegilops* and *Triticum.* IV. The origin of the common wheats revealed from the study on esterase isozymes in synthesized hexaploid wheats. *Jap. J. Genet.* 54: 175–189.

———. 1979b. The origin of the tetraploid wheats revealed from the study of esterase isozymes. *Proc. 5th Inter. Wheat Genet. Symp.* 1: 108–119.

———. 1981. D genome donors for *Aegilops cylindrica* (CCDD) and *Triticum aestivum* (AABBDD) deduced from esterase isozyme analysis. *Theor. Appl. Genet.* 60: 11–16.

Nee, M., and T. Andres. 1986. Search for the wild ancestors of the cultivated species of *Cucurbita.* Abstracts of Symposium and Contributed Papers of the 27th Annual Meeting of the Society for Economic Botany, Bronx, New York.

Nei, M. 1972. Genetic distance between populations. *Amer. Naturalist* 106: 283–292.

Nevo, E., D. Zohary, A. H. D. Brown, and M. Haber. 1979a. Genetic diversity and environmental associations of wild barley, *Hordeum spontaneum,* in Israel. *Evolution* 33: 815–883.

———, A. H. D. Brown, and D. Zohary. 1979b. Genetic diversity in the wild progenitor of barley in Israel. *Experientia.* 35: 1027–1029.

———, A. Beiles, D. Kaplan, N. Storch, and D. Zohary. 1986a. Genetic diversity and environmental associations of wild barley, *Hordeum spontaneum* (Poaceae), in Iran. *Pl. Syst. Evol.* 153: 141–164.

———, ———, and D. Zohary. 1986b. Genetic resources of wild barley in the near east: structure, evolution and application in breeding. *Biol. J. Linn. Soc.* 27: 355–380.

———, D. Zohary, A. Beiles, D. Kaplan, and N. Storch. 1986c. Genetic diversity and environmental association of wild barley, *Hordeum spontaneum,* in Turkey. *Genetica* 68: 203–213.

Nishikawa, K., Y. Furuta, and T. Wada. 1980. Genetic studies on alpha-amylase isozymes in wheat. III. Intraspecific variation in *Aegilops squarrosa* and birthplace of hexaploid wheat. *Jap. J. Genet.* 55: 325–336.

Ohnishi, O. 1983. Isozyme variation in common buckwheat, *Fagopyrum esculentum* Moench, and its related species. *Proc. 2nd Intl. Symp. Buckwheat Miyazaki,* Buckwheat Research, 39–50.

———. 1985. Population genetics of cultivated common buckwheat, *Fagopyrum esculentum* Moench. IV. Allozyme variability in Nepali and Kashmirian populations. *Jap. J. Genet.* 60: 293–305.

Oliver, J. L., and J. M. Martinez-Zapater. 1984. Allozyme variability and phylogenetic relationships in the cultivated potato (*Solanum tuberosum*) and related species. *Pl. Syst. Evol.* 148: 1–18.

Perl-Treves, R., D. Zamir, N. Navot, and E. Galun. 1985. Phylogeny of *Cucumis* based on isozyme variability and its comparison with plastome phylogeny. *Theor. Appl. Genet.* 71: 430–436.

Pickersgill, B. 1977. Taxonomy and the origin and evolution of cultivated plants in the New World.

Nature 268: 591–595.

———, C. B. Heiser, Jr., and J. McNeill. 1979. Numerical taxonomic studies on variation and domestication in some species of *Capsicum*. In J. G. Hawkes, R. N. Lester, and A. D. Skelding [ed.], *The biology and taxonomy of the Solanaceae*, 679–700. Academic Press, New York.

Pinkas, R., D. Zamir, and G. Ladizinsky. 1985. Allozyme divergence and evolution in the genus *Lens*. *Pl. Syst. Evol.* 151: 131–140.

Price, S., and A. L. Kahler. 1983. Oats (*Avena* spp.). In S. D. Tanksley and T. J. Orton [eds.], *Isozymes in plant genetics and breeding*, Part B, 103–127. Elsevier, Amsterdam.

Puchalski, J., and B. Molski. 1981. Isoenzyme variation within the wild *Secale* L. species. *Kulturpflanzen* 29: 391–399.

Quiros, C. F. 1983. Alfalfa, Luzerne (*Medicago sativa* L.) In S. D. Tanksley and T. J. Orton [eds.], *Isozymes in plant genetics and breeding*, Part B, 253–294. Elsevier, Amsterdam.

Quiros, C. F., and N. McHale. 1985. Genetic analysis of isozyme variants in diploid and tetraploid potatoes. *Genetics* 111: 131–145.

Rao, K. E. P., J. M. J. DeWet, D. E. Brink, and M. H. Mengesha. 1987. Infraspecific variation and systematics of cultivated *Setaria italica*, Foxtail millet (Poaceae). *Econ. Bot.* 41: 108–116.

Rick, C. M. 1983. Tomato (*Lycopersicon*). In S. D. Tanksley and T. J. Orton [eds.], *Isozymes in plant genetics and breeding*, Part B, 147–165. Elsevier, Amsterdam.

———, and J. F. Fobes. 1975. Allozyme variation in the cultivated tomato and closely related species. *Bull. Torrey Bot. Club.* 102: 376–384.

———, R. W. Zobel, and J. F. Fobes. 1974. Four peroxidase loci in red-fruited tomato species: genetics and geographic distribution. *Proc. Natl. Acad. Sci. USA* 71: 835–839.

Scora, R. W., J. Kumamoto, R. K. Soost, and E. M. Nauer. 1982. Contribution to the origin of the grapefruit, *Citrus paradisi* (Rutaceae). *Syst. Bot.* 7: 170–177.

Second, G. 1982. Origin of the genic diversity of cultivated rice (*Oryza* spp.): study of the polymorphism scored at 40 isozyme loci. *Jap. J. Genet.* 57: 25–57.

———. 1984. Different rates of genome divergence presumed between two species groups in the genus *Oryza*. *The Nucleus* 27: 44–48.

———. 1985a. A new insight into the genome differentiation in *Oryza* L. through isozymic studies. In A. K. Sharma and A. Sharma [eds.], *Advances in chromosome and cell genetics*, 45–78. Oxford and IBH.

———. 1985b. Evolutionary relationships in the *Sativa* group of *Oryza* based on isozyme data. *Genet. Sel. Evol.* 17: 89–114.

———. 1985c. Geographic origins, genetic diversity and the molecular clock hypothesis in the Oryzeae. In P. Jacquard, G. Heim and J. Antonovics [eds.], *Genetic differentiation and dispersal in plants*, 42–56. Springer-Verlag, Berlin.

Sheen, S. J. 1972. Isozymic evidence bearing on the origin of *Nicotiana tabacum* L. *Evolution* 26: 143–154.

Simmonds, N. W. 1979. Potatoes: *Solanum tuberosum* (Solanaceae). In N. W. Simmonds [ed.], *Evolution of crop plants*, 279–283. Longman, New York.

Smith, J. S. C., M. M. Goodman, and C. W. Stuber. 1984. Variation within teosinte. III. Numerical analysis of allozyme data. *Econ. Bot.* 38: 97–113.

———, ———, and ———. 1985. Relationships between maize and teosinte of Mexico and Guatemala: numerical analysis of allozyme data. *Econ. Bot.* 39: 12–24.

Stuber, C. W., and M. M. Goodman. 1983. *Allozyme genotypes for popular and historically important inbred lines of corn, Zea mays L.* U.S.D.A. Agric. Res. Results, Southern Series 16, New Orleans, LA.

Torres, A. M. 1983. Fruit trees. In S. D. Tanksley and T. J. Orton [eds.], *Isozymes in plant genetics and breeding*, Part B, 401–421. Elsevier, Amsterdam.

———, R. K. Soost, and U. Diedenhofen. 1978. Leaf isozymes as genetic markers in citrus. *Amer. J. Bot.* 65: 869–881.

———, ———, and T. Mau-Lastovicka. 1982. Citrus isozymes. Genetics and distinguishing nucellar from zygotic seedlings. *J. Heredity* 73: 335–339.

Tsunewaki, K., and Y. Ogihara. 1983. The molecular basis of genetic diversity among cytoplasms of *Triticum* and *Aegilops* species. II. On the origin of polyploid wheat cytoplasms as suggested by chloroplast DNA restriction fragment patterns. *Genetics* 104: 155–171.

Vavilov, N. I. 1950. The origin, variation and immunity and breeding of cultivated plants. *Chron. Bot.* 13: 1–366.

Warwick, S. L., and S. G. Aiken. 1986. Electrophoretic evidence for the recognition of two species in annual wild rice (*Zizania*, Poaceae). *Syst. Bot.* 11: 464–473.

Whalen, M. D., and E. E. Caruso. 1983. Phylogeny in *Solanum* sect. *Lasiocarpa* (Solanaceae): congruence of morphological and molecular data. *Syst. Bot.* 8: 369–380.

Wilkes, H. G. 1967. *Teosinte: the closest relative of maize.* Bussey Inst., Harvard University, Cambridge, MA.

———. 1977. Hybridization of maize and teosinte, in Mexico and Guatemala and the improvement of maize. *Econ. Bot.* 31: 254–293.

Wilson, H. D. 1976. Genetic control and distribution of leucine aminopeptidase in the cultivated chenopods (*Chenopodium*) and related weed taxa. *Biochem. Genet.* 14: 913–919.

———. 1978. *Chenopodium quinoa* Willd.: variation and relationships in southern South America. *Natl. Geographic Soc. Res. Reports* 711–721.

———. 1980. Artificial hybridization among species of *Chenopodium* section *Chenopodium*. *Syst. Bot.* 5: 253–263.

———. 1981. Genetic variation among South American populations of tetraploid *Chenopodium* sect. *Chenopodium* subsect. *Cellulata*. *Syst. Bot.* 6: 380–398.

———. 1988. Allozyme variation and phenetic relationships of *Chenopodium hircinum* Schrader (s. lat.) *Syst. Bot.* 13: 215–228.

———, S. C. Barber, and T. Walters. 1983. Loss of duplicate gene expression in tetraploid *Chenopodium*. *Biochem. Syst. Ecol.* 11: 7–13.

———, and C. B. Heiser, Jr. 1979. The origin and evolutionary relationships of "Huauzontle" (*Chenopodium nuttalliae* Safford), a domesticated chenopod of Mexico. *Amer. J. Bot.* 66: 198–206.

Zamir, D., N. Navot, and J. Rudich. 1984. Enzyme polymorphism in *Citrullus lanatus* and *C. colocynthis* in Israel and Sinai. *Pl. Syst. Evol.* 146: 163–170.

Zohary, D. 1969. The progenitors of wheat and barley in relation to domestication and agricultural dispersal in the Old World. *In* P. J. Ucko, and G. W. Dimbleby [eds.], The domestication and exploitation of plants and animals, 47–66. Gerald Duckworth and Co. Ltd., London.

CHAPTER 9

Isozyme Analysis of Tree Fruits

Andrew M. Torres

Department of Botany
University of Kansas
Lawrence, Kansas 66045

The advent of isozyme genetics has revolutionized the study of tree fruit crops. Each crop has had its own set of problems, sometimes unique, that remained intractable until this versatile tool became available and was applied. This chapter summarizes the work done on those fruit crops whose isozymes have been studied sufficiently to constitute a major advance. It is interesting and instructive that this breakthrough resulted in many cases from the collaboration of researchers trained in horticulture with those trained in more basic plant genetics.

The six major fruit crops based on tonnage of production are, in order, grape, citrus, banana, apple, plantain, and mango (Samson, 1980). The next six in order are pineapple, date, fig, avocado, papaya, and cashew. This review includes citrus, banana and plantain, apple, mango, dates, fig, avocado, and papaya. In addition, reports were found dealing with the pear, peach, mulberry, cherimoya, and olive. Grape, pineapple, and cashew are not included in this review, nor are the numerous isozyme studies that are primarily biochemical or concern isozyme changes incidental to morphogenesis or pathological conditions.

More complete discussions of the temperate and subtropical tree fruit crops treated here may be found in the excellent compendium edited by Janick and Moore (1975). Each entry includes information on the origin, history, breeding system, and cytology for the crops treated and includes extensive bibliographies. A companion volume edited by Moore and Janick (1983) focuses on breeding methods employed with fruit crops. Although fruit chemistry is emphasized, the equivalent work for tropical and subtropical fruits is that of Nagy and Shaw (1980). The small volume by Samson (1980) is also very useful.

Despite the great variation among the fruit trees, all have several characteristics in common. Each cultivated variety (cv) contains a unique gene combination which results in the expression of a genetic program that produces a desirable fruit. This combination must ordinarily be kept intact by cloning via some means of asexual propagation such as budding or grafting. Fruit of predictably consistent quality can only rarely be obtained from seedlings because seed embryos generally develop from reshuffled gene combinations carried through the sperm and egg. Exceptions to this general pattern are discussed below. Since most of our present-day cvs were selected over very long periods of time, even millennia, new gene combinations will most commonly result, not surprisingly, in fruit and other plant characteristics that are inferior to those already at hand. The rarely discovered recombinant of greater value is, of course, again vegetatively propagated.

One trait that makes fruit trees such poor experimental genetic organisms is their long period of juvenility, the time from germination to fruit set. This period can vary considerably, but five to seven years is most common. Consequently, any breeding program normally requires a long-term commitment. Techniques used to shorten the seed to seed generation time to accelerate breeding efforts include embryo culture, artificial stratification of seeds followed by germination in flats, girdling, and grafting to dwarfing rootstocks. More complete discussions of these topics may be found in Sherman and Lyrene (1983) and in Hesse (1975) relative to peaches, but applicable to other crops as well.

Many horticultural traits of fruit crops are due to multiple genes, and hence genetic studies are complex as well as time-consuming. However, until recently, even analyses of single-gene characters in tree crops were not common in comparison with many annual crops. The best known fruit tree genetically is apparently the peach for which more than 25 single-gene characters have been analyzed (Hesse, 1975) and some linkage studies have been carried out. Only within the past three years have the first efforts at producing genetic maps for any tree crop using isozyme markers appeared, but as yet none are comparable to those for maize, wheat, or tomato.

In general, mapping studies become more difficult as chromosome numbers increase. However, this is not a serious problem with most fruit trees because they typically have low chromosome numbers. Most are $n = 8$ to 12; the apple and pear are $n = 17$ and the date palm is $n = 18$.

USES OF ISOZYMES IN TREE FRUITS

There are at least two reasons why isozyme techniques have resulted in significant and exciting strides in fruit tree genetics. One is that codominance of the alleles which specify allozymes permits genetic analyses to be carried out with F_1s. A second is that many F_1, backcross, and other generations have been produced by breeders, and many of these populations are presently available for analysis. As examples, 10,000 seedlings from controlled hybridizations have been grown to fruiting in a nectarine (fuzzless peach) breeding program (Okie et al., 1985), and Storey (1975) based his review of figs on more than 30,000 seedlings from over 300 crosses. Similarly, there are hundreds or even thousands of seedlings of citrus, avocados, date palms, peaches, pears, apples, and olives. Their existence has not escaped notice, and isozyme geneticists with the requisite skills have started to utilize these invaluable resources.

For those crops without extant progeny, the time required for genetic analysis of isozyme markers could be relatively short if the earliest possible recombinant tissues are used. If two parents had been examined for a polymorphic enzyme system for which an hypothesis of its genetic control had been advanced, then crossed, the first available recombinant tissue would be found in the resulting seed, in embryo and cotyledons. Next, generally in a matter of weeks or months, the seedlings with their leaves, stems, and roots could be used. Thus, experimentation with different tissues could be rewarded with considerable data in a relatively short time. If the same isozymes in different tissues are to be examined, studies are required to assure that the products of the same genes are being compared.

Various tissues have been used for isozyme studies, but leaves are most commonly employed. Some tissues (avocado mesocarp, citrus leaves) require no special treatment and have simply been crushed directly onto wicks whereas others (apple bark) have required more elaborate techniques. Because the leaves of most fruit trees contain varying amounts of phenolic compounds that usually interfere with enzyme activity, appropriate extraction techniques generally must be developed to obtain satisfactory results (see the chapter on methodology by Wendel and Weeden in this volume). In general, the simpler the methods for sampling, the more efficient the studies. Pollen can be especially useful in addressing problems of subunit structure and organelle localization of isozymes (Makinen and Brewbaker, 1967; Makinen and MacDonald, 1968; Weeden and Gottlieb, 1979, 1980; Zamir, 1983; Weeden, 1986). Because pollen is haploid it produces simplified zymograms compared with those of diploid sporophytic tissue.

Of what value are isozymes as markers in horticulture? An early and common use of isozymes has been the characterization and identification of cultivars. Other studies have determined the genetic origin of seedlings, distinguished progeny from selfing versus crossing, documented the parentage of cvs, and examined similarities of cvs (Peirce and Brewbaker, 1973). Sizable capital investments in horticultural enterprises require that nurserymen, growers, and breeders have confidence in the identification of their material—opportunities for errors in labeling and handling are numerous. Morphological traits have traditionally been used for identification, but in many cases a morphological assessment of flowering and fruiting material is not possible because of season, immaturity of material, or because the material of interest is a rootstock. Clearly, isozymes offer a possible solution to such problems (Torres, 1983). Isozymes are most useful when the pattern of bands is fully understood in terms of the primary factors that produce it; the number of genes and alleles which code for the enzyme and the subunit structure of the enzyme being studied.

Optimism compels a search for correlations between isozyme markers and traits of horticultural significance; for example, disease resistance or the sex of a young plant in the case of dioecious cvs. The odds against finding such associations are great, and any that are found are most likely fortuitous. Markers correlated with horticultural traits are simply correlations and do not indicate a cause-effect relationship unless it is shown how the isozyme relates to the appropriate environmental factor to produce the observed effect. Nevertheless, more correlations will be found as further studies are conducted.

In the treatments that follow, genes and alleles will be indicated by italics and will be abbreviated with the symbols used by the authors. Commonly mentioned enzymes will be abbreviated as follows: acid phosphatase, ACP; alcohol dehydrogenase, ADH; esterase, EST; 6-phosphogluconate dehydrogenase, 6PGD; glutamate oxaloacetate transaminase, GOT, also called aspartate aminotransferase, AAT; hexokinase, HK; isocitrate dehydrogenase, IDH; leucine aminopeptidase, LAP; malate dehydrogenase, MDH; malic enzyme, ME; peroxidase, PX or POX; phosphoglucose isomerase, PGI, also called glucose phosphate isomerase, GPI; phosphoglucose mutase, PGM; and triose phosphate isomerase, TPI.

The treatment of each species or cv includes general notes on floral traits, reproductive system, methods of propagation, special problems, and a brief account of how isozyme markers have been used to address them.

EXAMPLES OF ISOZYME ANALYSIS IN TREE FRUITS

Apple

Apple (*Malus pumila/ M. domestica*; $n = 17$) flowers are perfect, but most cvs are self-incompatible (Brown, 1975) and crosses are readily made. Cvs are propagated by grafting. Among the rosaceous fruit species, there are chromosome numbers of $n = 8$, 9, and 17. The possible relationship of these numbers is obvious, and isozymes are starting to provide valuable evidence for genome duplication and for the origin of the apple.

The earliest attempt to assess the similarities of isozyme patterns of the apple and related taxa was that of Veidenberg et al., (1977). They examined PX, diphenol oxidase, MDH, and EST isozymes from leaves and pollen in several apple species and cvs. Although no genetic control mechanisms were postulated, it was found that the zymograms of related taxa were similar.

Chyi and Weeden (1984) used isozymes to determine whether the male or female parent contributed the unreduced, diploid gamete to triploid cvs. As had been shown in several other studies, mostly with endosperm, some enzyme systems reflect a dosage effect when specified by three alleles. If two of the three alleles in a triploid are identical (allele a), and the third is different (allele b), the two doses of a may result in a more intensely staining band relative to that of allele b. In the case of dimers, the homodimer specified by the allele in double dose will be more intense than the allozyme specified by the single allele given that the dimers are equally active, are produced in equal numbers per allele and, for heterodimer activity, that hetero- and homodimerization occur equally. The theoretical relative intensities in such a case are 4:4:1 (homodimer from the allele in two doses:heterodimer:homodimer from the allele in single dose). Of several enzyme systems examined, 6PGD provided the clearest evidence that unreduced eggs of 'Red Spy' and 'Golden Delicious' were involved in the production of 'Spigold' and 'Jonagold', respectively.

Chevreau et al. (1985) used pollen to analyze seven isozyme-specifying genes in over 200 progeny from controlled crosses. Several genes gave the expected results, but EST-specifying genes were exceptional in that they apparently produced double-banded patterns. 'Golden Delicious' was considered homozygous so that all of its pollen would carry the same allele but a pollen sample produced two bands. 'Reinette Clochard' was heterozygous and half of its pollen would carry one allele and the other half the other. Pollen extracts from this mixture produced a four-banded pattern. Half of the 'Golden Delicious' × 'Reinette Clochard' hybrids produced a two-banded pattern and the other half a four-banded pattern. The overall conclusion was that bigenic disomic inheritance, which would result from a diploidized tetraploid with homoeologous genomes (Allard, 1960; Sanford, 1983), provided the best explanation for their data.

Statistically significant single-gene segregation distortion was noted for a few of the markers. As will be seen, this phenomenon is so common in tree fruit cvs (and other woody plants) that it raises the question of its possible significance and of its possible occurrence more commonly among particular genes.

Weeden and Lamb (1985) examined leaf and floral tissue for several enzyme systems of more than 50 apple cvs. They determined the genetic control for two 6PGD-specifying genes, two for AAT (GOT), two for IDH, and one each for GPI, TPI, and diaphorase. Many of the cvs they used could be identified from their unique combinations of enzyme phenotypes, and some of the long-standing problems concerning the origin of certain cvs were addressed.

Menendez et al. (1986a) used bark protein extracts and polyacrylamide gel electrophoresis (PAGE) to produce patterns of proteins and isozymes for cv identification. Gross protein patterns, as expected, were not useful. Isozymes of PX, indoleacetic acid oxidase, EST, and ACP were resolved and produced clear banding patterns. In subsequent studies (Menendez et al., 1986b), the isozyme patterns of each enzyme system were systematically analyzed and classified using densitometry. Each of the 34 clonal rootstocks was assigned a pattern type, and combinations of patterns were used for identification. Apple scion clones were similarly analyzed (Menendez et al., 1986c). The banding patterns of some enzymes of cvs in tissue culture were different from those of the original explant sources, and the ACP patterns of virus-tested clones were different from original contaminated material. The nature and relationships of the individual bands in the patterns were not determined from genetic studies.

Weeden and Lamb (1987) studied isozyme polymorphism in nine enzyme systems and identified four linkage groups. No correlations between isozyme markers and disease resistance were found. The high incidence of duplicate genes observed was taken as evidence of the assumed ancient tetraploidy of the apple.

Very tight linkage between a *Got* gene and a gene for an extremely important horticultural trait, incompatibility, was detected by Manganaris and Alston (1987). The cis/trans relationships between the alleles involved were also determined. Clearly, this finding will be of significance in breeding attempts and in planting strategies as incompatibilities between cvs can now be predicted with a simple test for *Got* genotype.

Pear

The pear (*Pyrus* spp. and *P. calleryana*; $n = 17$) flower is perfect but the cvs are generally self-incompatible; propagation is by grafting. Several species of *Pyrus* and their hybrids provide edible fruit, but the most widely used species in the Western world is *P. communis* L., the common pear. The Callery pear has inedible fruit, but it has horticultural value as an ornamental and as a source of genetic resistance to fire blight, the most serious pear disease in North America (Layne and Quamme, 1975). There are several cvs of the Callery pear but perhaps the best known is the profusely flowering 'Bradford'.

Santamour and Demuth (1980) examined PX zymograms from leaf and cambium extracts in six cvs of the Callery pear. In all, 10 anodal and five cathodal bands were detected, but the patterns of the bands in the two tissues were quite different in each cv. There was sufficient polymorphism in the zymograms to distinguish all six cvs. A few seedlings were examined for the segregation ratio of two bands. The PX zymograms were also used as evidence that blighted nursery plants being sold as 'Bradford' which is resistant to blight, were probably not 'Bradford' itself but rather seedlings of 'Bradford'. A seedling is a propagule from a cv, generally from an embryo derived from the union of sperm and egg and thus recombinant. The 'Bradford' seedlings apparently lacked the resistance to blight.

Patterns of PX, EST, and ACP were analyzed with gradient PAGE (Menendez and Daley, 1986).

Identification of many cvs of edible pears and accessions of the Callery pear could be made from the patterns or combinations of patterns.

Peach

Peach (*Prunus persica*; $n = 8$) flowers are perfect, and most cvs are self-incompatible. Crosses are readily made, and propagation is by grafting and budding.

Crude protein extracts from stem tissue were examined by isoelectric focusing (Carter and Brock, 1980). The number of protein bands in the five cvs tested varied from 12 to 15, and the profile of each cv was distinct.

The genetic segregation of peach MDH was analyzed and the mitochondrial isozyme was identified from leaf fractionation extracts (Arulsekar et al., 1986). The complex zymogram was only partially interpreted.

Hybrids of plum (*P. salicina*) and peach may provide improved rootstocks but a means to verify the hybrids during the seedling stage is desirable to avoid culturing unwanted plum × plum or peach × peach progeny. Parfitt et al. (1985) addressed this problem using isozymes of GPI (PGI) and PGM. The authors propose that two genes (*Pgm-1*, *Pgm-2*) specify PGM, a monomeric enzyme. Yet, one of the genes, *Pgm-1*, apparently produces three enzyme bands in plums. An alternative interpretation which appears consistent with the PGM banding patterns is that plum and peach each have three PGM-coding genes due to gene duplication. All three genes are presumably homozygous in the peach cvs tested, whereas two in the plums, those specifying the slowest and the fastest migrating sets, are variable. Examination of additional cvs and perhaps pollen extracts will no doubt clarify the situation. Regardless of the genetics, peach and plum PGM isozymes were sufficiently distinct to allow the identification of interspecific hybrids.

The slower migrating set of GPI bands (probably the cytoplasmic isozymes) was variable in plum cvs and appeared to represent the typical dimeric system with the gene having two alleles. The corresponding peach gene was apparently homozygous for a different allele in the cvs examined, and consequently, hybrids could be identified. Especially interesting is the indication (Figs. 1 and 2 of Parfitt et al., 1985) that the presumed plastid isozymes encoded by *Gpi-1* are different in plum and peach. With suitable controls in the gels, it seems that *Gpi-1* by itself could be used to distinguish selfs from hybrids.

Mulberry

Several species of mulberry (*Morus spp.*; $n = 14, 21$) are of economic importance, some for their fruit and others for their leaves as food for the silkworm. Mulberries are either monoecious or dioecious; controlled pollinations are therefore rather straightforward. Cvs are propagated by cuttings.

Hirano (1977) examined PX isozymes from leaves of 240 mulberry cvs belonging to three species: *M. bomysis*, *M. alba*, and *M. latifolia*. Isoelectric focusing in thin layer acrylamide gels yielded a total of 10 isozymes which produced five different patterns among the cvs. The patterns were not affected by the age of the tree or leaf nor by the location of planting. Cv affinities based on morphology generally agreed with zymogram similarity. Controlled matings were made, and parents and progeny were analyzed for segregation of the bands in the PX zymograms (Hirano and Naganuma, 1979). In all, three patterns were found, and each pattern consisted of either three (one major and two minor) or four (two major and two minor) bands. It was hypothesized that the two three-banded patterns resulted from the two homozygotes of the gene *Px-1*—either genotype 1/1 or 2/2. The four-banded pattern resulted from the 1/2 heterozygote. Crosses between all six possible two-by-two combinations of plants with the three phenotypes were made and the progeny analyzed. In all instances, observed and expected segregation ratios were in agreement. Correlations were sought between specific PX isozyme patterns and phylotaxis, leaf shape, tree vigor, and resistance to powdery mildew, but none were found.

Hirano et al. (1980) searched for correlations between peroxidase genotypes and amino acid content. It was suggested that the weak correlations noted may have been due to pleiotropism or linkage of the peroxidase genes with genes controlling amino acid content and that selection for amino acid composition could be based on isozyme genotypes.

Polyembryony (two to several embryos in the same seed) is uncommon in flowering plants, but it does occur in the mulberry. Embryos of nucellar origin and subsequent seedlings would be genetically identical to the maternal parent. Ogure (1979) used isozymes of PX and indoleacetic acid oxidase to determine whether twin seedlings from the same seed were identical to each other and to the female parent. The PX zymograms indicated that nucellar seedlings were not formed, and differences in patterns between some of the twin seedlings were attributed to fertilization of two or more eggs, or to the fertilization of one egg and a synergid or antipodal cell rather than to some asexual process.

Fig

Figs (Ficus carica; n = 13) are effectively dioecious. Some plants are strictly female and produce male-sterile, longistyle, female flowers exclusively whereas others are monoecious producing both male-sterile, brevistyle, pistillate flowers and male-fertile flowers. Protogyny of the female flowers of the monoecious trees precludes self-pollination. The strict females produce the edible domestic figs called simply figs, and the others are the caprifigs which provide pollen for the figs and shelter the pollinating wasp (Storey, 1975). Sex is determined either by a pair of tightly linked genes (Storey, 1975) or by one gene with two alleles (Valizadeh, 1977). In either case, there is a 1:1 segregation ratio of the two types of trees, fig and caprifig. Cloning of the many cultivars is by rooted cuttings.

Genetic studies of isozymes were based on some 400 progeny from five caprifig cvs crossed with four fig cvs (Valizadeh, 1977, 1978). Isozyme extracts were from small syconia and/or leaves. The enzyme systems studied and used for cv identification included EST, ACP, GOT, and POX.

Breeding programs aimed at selecting better edible figs must include the cultivation until fruiting of unwanted caprifigs, the effective males, which comprise 50% of any seedling population. Tests for linkage between tree type and enzyme genes indicated that *Pox-C* was 15 recombination units from the factor(s) for sex determination (Valizadeh, 1978). This discovery should enhance the efficiency of future fig breeding programs.

Olives

The olive (Olea europaea; n = 23) flower is perfect but most cvs are self-incompatible so that cross-pollination is required. Olive cvs are propagated asexually.

Pollen was used to study isozymes of 16 resolvable enzyme systems (out of 22 examined) in 27 olive cvs (Pontikis et al., 1980). There was considerable variation in the complexity of the zymograms, and genetic interpretations were postulated on variations in staining intensities and rates of migration, but progeny tests were not conducted. All 27 cvs could be identified by a combination of the EST and ME zymograms, and 20 cvs had unique zymograms for at least one enzyme system.

Citrus

The flowers of citrus (Citrus spp.; n = 9) are perfect, and most cvs are capable of selfing (Soost and Cameron, 1975), but controlled pollinations following emasculation are routinely made with good results.

Formal genetic studies of citrus leaf isozymes were first carried out using many cvs of known parentage, as well as several F_1 populations from controlled crosses (Torres et al., 1978).

A wide range of problems with citrus has been examined with isozymes (reviewed by Soost and Torres, 1981). A major problem is polyembryony. The embryos can be of two different genetic origins; nucellar, in which case they are genetically identical to the seed parent, or zygotic, from the fusion of sperm and egg. Seven of the nine cultivated species of Citrus (Swingle, 1943) are polyembryonic. When seeds are germinated, the genetic origins of the seedlings frequently cannot be determined from morphology. Nucellar seedlings, when identifiable, would normally be discarded since the seed parent genotype can easily be propagated by budding which also shortens the period of juvenility. Unambiguous identification based on morphology often must await fruiting which occurs five to 10 years later. The space required and the costs of caring for trees of questionable value for such a long time limit the number of crosses that could be attempted.

The rationale for using isozyme markers to distinguish nucellar from zygotic seedlings is simple: because of codominance, the presence of a male contribution to the embryo or seedling can easily be documented. For example, assume a seed parent is S/S for enzyme A, and the pollen parent is F/F. A nucellar seedling will be S/S, but a zygotic seedling must be F/S. Not all possible interspecific pairwise crosses in citrus are this straightforward, but all or nearly all nucellar and zygotic seedlings can be distinguished in the majority of the crosses. Using PGI and PGM isozymes, Soost et al. (1980) demonstrated the power of isozymes to address this problem. An analysis of the efficiency of the technique using available markers showed that 86% of the zygotics from selfing and more than 99% of the zygotics from all interspecific pairwise crossing combinations should be distinguishable from nucellars (Torres et al., 1982).

Iglesias et al. (1974) first used citrus isozymes to distinguish zygotic from nucellar seedlings. PX and EST isozymes from leaf extracts were separated by PAGE. No attempt was made to provide a genetic explanation for the isozymes, but zygotic seedlings were identified by the presence of a paternal isozyme not possessed by the seed parent. Some cases, however, were doubtful because the progeny had the same isozymes as the seed parent yet could have been recombinant. This again illustrates the need for an understanding of the underlying genetic basis of the isozymes. A later report (Iglesias and Lima, 1979) included amylases run under various electrophoretic conditions in combination with PX isozymes. Amylases were considered more effective than PX as markers because there were more isozymes with greater mobility. Button et al. (1976) and Manzocchi et al. (1981) distinguished nucellar from zygotic seedlings using PX isozyme patterns from root extracts.

Despite the problems it causes, nucellar embryony can be useful. It is used in the citrus industry to propagate rootstocks from seed with desirable traits. Morphological off-types are simply rogued. Using isozyme markers, Moore and Castle (1988) evaluated the efficiency of the practice with 15 rootstock cvs and found that high percentages of zygotic seedlings could be detected and that the frequency of their production varied considerably with the cv.

PX isozyme patterns from citrus leaves of known cvs and their hybrids were compared by Ueno and Nishiura (1976). Zymograms of nucellar offspring were identical with their seed parents whereas hybrids had different patterns. Ueno (1976) also found that cvs and species could generally be identified from their PX zymograms.

Esen and Scora (1977) examined amylase patterns from leaf extracts and compared zymograms to infer species relationships. Isozyme complementation was considered evidence for the hybrid origin of certain of the species. Scora and Kumamoto (1981) used isozyme evidence to support the previously proposed hypothesis of the hybrid origin of the grapefruit from pummelo × sweet orange.

Ben-Hayyim et al. (1982) analyzed zymograms of 10 enzyme systems in calli and roots of nine taxa to determine which systems could be used to identify protoplast fusion products. Six systems were found useful for specific cv fusion combinations. Using PX and EST isozyme phenotypes, Yamashita (1983) proposed a highly unusual origin of a cv as a chimera originating from the junction of a rootstock of one species with the scion of another.

The possible linkage of 12 genes specifying isozyme markers was examined in a population of seedlings from cvs known to be monoembryonic (Torres et al., 1985). Nearly 400 progeny in nine families from two crossing combinations were sampled. Segregation data and the 37 pairwise gene combinations were analyzed with the LINKAGE-1 computer program (Suiter et al., 1983). There was considerable single-gene segregation distortion, and it seems apparent that this phenomenon is common in woody plants. The genes $Got-1$ and $Mdh-1$ were linked as were $Mdh-2$, $MeO1$, and $MeO2$. Cis/trans arrangements of the alleles were deduced from the data. Each of the nine citrus chromosomes has at least one isozyme marker.

Isozymes provided insight into the origin of cvs within species. Examination of the isozyme genotypes of the several to many cvs of each species (Torres et al., 1978) indicates that each species has a unique combination of genotypes, and most of the cvs within a species have identical combinations of genotypes. This strongly suggests that the cvs in each species arose from some common progenitor by asexual means. As one example, all seven cvs of lime are identically heterozygous for seven of the 10 genes studied and identically homozygous for the other three. It therefore seems likely that a single hybridization event produced the first cv, from which the other six cvs were derived asexually. This general pattern holds for all the species, with most exceptions found in the mandarins, the most variable group morphologically and the group which includes cvs known to be of hybrid origin. For each species there apparently arose a prototype, possibly through hybridization, and from it the other cvs arose by mutation. The distinctive mutants were then propagated asexually. Isozymes provide ample evidence to force a reconsideration of the taxonomy of cultivated citrus (Hirai and Kozaki, 1981).

Avocado

The avocado (*Persea americana*; $n = 12$) flower is perfect but is at any one time functionally either male or female. The breeding system is complementary synchronous dichogamy (Stout, 1924; Samson, 1980), a breeding mechanism that promotes, but does not assure, cross-pollination. Successful hand pollinations for controlled crosses are exceedingly difficult to achieve principally because of fruitlet abscission after pollination. The odds that a given pollinated flower will result in a fruit are estimated conservatively at about 1 to 1000. The pollen parent of any seed is frequently uncertain even when precautions are taken (Bergh, 1975).

The mesocarp of the avocado is the edible portion of the fruit and is of purely maternal origin so that, unlike the embryo and cotyledons, all mesocarps on a tree and on all the trees of the same cv are genetically identical. The mesocarp contains many active enzymes involved in the pathways for carbohydrate, protein, and fat metabolism (Biale and Young, 1971), requires no protectants, and is a very good tissue for isozyme analysis. Cotyledons are the first available recombinant tissue and can also be used for isozyme assays.

As with citrus, two approaches were used to carry out formal isozyme genetic studies on avocados, initially with mesocarp (Torres et al., 1978) and later with leaves (Torres and Bergh, 1980) and cotyledons (Torres, 1984). One approach was to analyze a series of individual progeny and their known parents when the progeny were established cvs. The second approach was to analyze available F_1 selfs and hybrids and other generation populations of nearly certain parentage. PGI had been a difficult system to analyze because of the complexity of its zymogram due to apparent gene duplication. Goldring et al. (1985) provided an insightful interpretation based in part on comparisons of pollen and somatic tissue zymograms.

Perhaps the most useful collection for genetic studies has been a population of progeny from hand pollination and embryo rescue (Skene and Barlass, 1983). This group formed the basis for linkage studies (Torres et al., 1986) among nine genes specifying seven enzymes. Only *Got1* and *Got2* were found to be linked, and the other seven genes each mark a chromosome; hence, eight of the 12 avocado chromosomes have an isozyme marker. Vrecenar-Gadus and Ellstrand (1984) had earlier reported independent assortment among four of the genes examined. Single-gene segregation distortion was again found.

The early detection of ADH in mesocarp was unexpected because its possible function there is uncertain. It is widely held that ADH in the seed helps provide energy during periods of anoxia, but its role in the mesocarp is unknown. Perhaps it produces ethanol to make the fruit more attractive to dispersal agents such as small mammals (B. O. Bergh, R. J. Knight, and S. Gazit, personal communication).

Torres and Bergh (1978a) examined a large number of 'Pinkerton' seedlings to measure the extent of outcrossing versus selfing. Sixty-two percent were outcrossed progeny from a 1973 planting and 98% in a 1976 planting. Because a map of the plantings was available and the locations of the parental 'Pinkertons' relative to other cvs was known, it was possible to identify with considerable confidence the precise pollen parent. Because only selfs were needed for this project, the breeder could discard the known hybrids. The same methods could be used to select for hybrids from outcrossing as demonstrated by Degani and Gazit (1984). Outcrossing rates were also measured in groves of different planting designs to provide information as to planting patterns for optimal yields (Vrecenar-Gadus and Ellstrand, 1985).

Isozymes were used to address the problem that arose when two collections of 'Duke' were found to have different genotypes for *Adh2* and other enzymes (Torres and Bergh, 1978b). The oldest surviving 'Duke' and trees from its budwood were examined and it was found that the real 'Duke' was F/S for Adh2 and that the S/S types were 'Duke' seedlings mostly from unknown pollen parents. One such seedling was 'Duke 7', an important cv because its resistance to root rot is even greater than that of 'Duke'. Thus, it was of interest to identify the pollen parent of 'Duke 7' since the male parent had apparently contributed some of the resistance. The profiles of about 100 cvs were examined in the search for the one that had the right combination of genotypes. Three cvs were found but only one seemed likely from its morphology. Thus, it was suggested that 'Clifton' be examined as a possible source of resistance.

The rarity of the S/S genotype of Lap2 is apparent from an examination of over 100 cv genotypes (Torres and Bergh, 1980). Degani et al. (1986) found that progeny of this genotype from selfing of an F/S cv tended to occur in expected proportions (1FF:2FS:1SS) in young fruits, but became deficient as fruit maturity progressed. It is doubtful that the isozymes themselves are responsible for the differential mortality, but more likely, they are linked to other factors. The data actually show a significant excess of heterozygotes relative to both homozygotes (see the chapter by Mitton in this volume).

Date Palm

The date palm (Phoenix dactylifera; $n = 18$) is dioecious with separate male and female plants. Because the date is a monocotyledon, albeit arborescent, it lacks a vascular cambium and cannot therefore be propagated by budding or grafting. The male and female cvs must be laboriously propagated from offshoots which form at the base of the parent tree (Popenoe, 1973; Nixon and Carpenter, 1978). Tissue culture techniques for propagation have been developed to alleviate this problem.

Date palm leaflets are so tough that the only practical way to crush them was with a hammer (Torres and Tisserat, 1980). Presumably, the same methods would be appropriate for other palms and other plants with equally fibrous leaves. Seven genes specifying five enzyme systems were analyzed in 24 female and 20 male cvs and in over 100 progeny of known parents. Seven genes, each with two alleles, can produce 21 genotypes. Of the 21, 15 were observed in females and 14 were observed in males. Of the 42 possible crossing combinations (six for each of the seven genes) 23 were analyzed in 10 of the 26 F_1 populations available. It was found that most cvs could be identified by their isozyme profiles. Among the 24 females, 'Amir Hijj', 'Zahidi', 'Badragah', and 'Horra' shared the same profiles. The other 20 cvs had unique profiles.

The detection of ADH in date palm leaves was unexpected because the date palm is clearly a plant of very dry habitats where ADH induction by flooding is not obvious. Why then, is ADH in leaves? Possibly the deep root system is flooded from time to time causing anaerobiosis which induces root ADH followed by the transport of ethanol to the leaves where it is converted to acetate by leaf ADH.

In dioecious crops, it would be very useful to find a marker associated with the gene or genes for sex determination. This would allow the breeder to discard young males (or females) to save time and space because first flowering can take from four to seven years (Carpenter and Ream, 1976). No such associations have been found for the date palm.

Mango

Mangos (Mangifera indica; $n = 20$) bear numerous inflorescences of thousands of flowers which are either staminate or perfect. Because bagged inflorescences can produce fruit, at least some cvs are self-compatible (Samson, 1980). Some cvs are monoembryonic, others are polyembryonic.

As with most cultivated plants, there is considerable confusion in the naming of mango 'types'. Identical clones have been given different names, and different 'clones' have been given the same name. The situation is described as chaotic, and Gan et al. (1981) used leaf isozymes of EST, GOT, ACP, and alkaline phosphatase to study the possible variation among members of so-called clones and varieties. A total of 244 individuals from six 'clones' was studied. Since there was variation in all six clones for all four enzymes, it was concluded that the various samples were not members of the same clone, and it was suggested that the term variety be used instead.

Cherimoya

The cherimoya (Annona cherimola; n = 7) is dichogamous but can be selfed by hand pollination. The fruit is compound, developing from many pistils. The sops are not well known in temperate climes, but certain cvs are said to produce fruit of superb flavor. Plants are propagated by budding and grafting.

Isozyme markers were developed to address some of the horticultural problems with cherimoya, especially that of cv identification (Ellstrand and Lee, 1987a, b). Polymorphism was examined in eight enzyme systems specified by 15 genes in 15 cvs and an 'atemoya' (A. cherimola × A. squamosa). It was found that each cv could be identified by its combination of zymograms.

Selfing five cvs that were polymorphic for 13 suitable isozyme systems, Lee and Ellstrand (1987) used leaf and seed tissue to analyze single-gene segregation and linkage of 46 gene pairs. Segregation ratios were distorted for two genes, and two linkage groups were mapped, one with five of the loci tested and the other with four. The remaining four genes provide a marker each for four more chromosomes. Thus, six of the seven cherimoya chromosomes have one or more isozyme markers.

Papaya

The papaya (Carica papaya; n = 9) is a short-lived perennial and is either dioecious or hermaphroditic. It is generally propagated from seed but can be proliferated from explants in tissue culture.

Isozyme polymorphism was examined to determine the genetic nature of somatic embryos. The embryos originated from ovules from crossing attempts between normally incompatible C. papaya and C. cauliflora, which carries resistance to a ringspot virus (Moore and Litz, 1984). Although the genetics of the markers were not reported and some of the results seemed unusual, banding patterns suggested that the embryos may be of zygotic rather than of asexual origin. Plants from such embryos may represent a way to transfer resistance to the papaya.

Banana

The bananas (Musa spp.; x = 11) and the starchier, usually cooked, plantains, are giant herbs rather than trees. They are included because they are a significant source of food for millions. The flowers are hermaphroditic or unisexual, and fruits in commercial cvs develop parthenocarpically. Propagation is typically by suckers but plantlets produced in tissue culture are being more widely used.

Bananas comprise a genomically complex group of cvs that occur as diploids, triploids, and tetraploids. The A genome derives from Musa acuminata whereas the B genome originates with M. paradisiaca and M. balbisiana. Cvs of commerce are typically triploid, most commonly AAA, but also AAB and ABB. The genomic origin of many cvs has been a long-standing problem. Jarret and Litz (1986a) described the zymograms of eight enzyme systems in 35 clones to elucidate taxonomic relationships. Five markers were used to characterize 24 clones of various ploidal levels (Jarret and Litz, 1986b). Certain isozymes were specific to B-genome taxa, and these would likely be useful in examining problems of cv origin. For further discussion see the chapter by Doebley in this volume.

CONCLUSIONS

The list of tree fruit crops that have had problems addressed with isozymes has grown impressively since the late 1970s. The particular problem has varied with the crop, but every major crop is now represented. Problems have included the genetic origin of seedlings in polyembryonic cultivated varieties (cvs) of citrus, identification of cvs and of protoplast fusion products, documentation of parentage of cvs, and the source of unreduced gametes in triploids. The significance of single-gene segregation distortion so commonly noted in fruit crops remains a mystery. Significant progress has been made toward developing linkage maps in several crops, and these efforts will no doubt continue.

The next decade will see more fully developed linkage maps, and isozymes will be included in protoplast studies, cell fusion programs, gene transfer projects, and even patent applications. Cell culture and genetic engineering techniques equivalent to those being applied to herbs will be applied to woody crops. Much challenging work remains to be done in the years ahead but in all these efforts, isozymes will play a key role.

ACKNOWLEDGMENTS

Thanks to Drs. Ricardo Menendez, Gloria Moore, Chemda Degani, Norman Ellstrand, Janet Lee, and Norman Weeden for preprints of papers in press. The editors of this volume and Norman Weeden offered many useful suggestions for improving the manuscript. I am very grateful to Drs. R. J. Knight, USDA, Miami, Florida; B. O. Bergh, R. K. Soost, and W. Storey, University of California, Riverside; S. Gazit, Bet Dagan, Israel; B. Tisserat, USDA, Pasadena, California; V. Vithanage, Commonwealth Scientific and Industrial Research Organization, Brisbane, Australia, and M. Sedgley, The Waite Institute, University of Adelaide, Australia, for their generosity in introducing me to their crops and for teaching me about the problems that could be approached with isozymes. Ours has been a most fruitful collaboration.

LITERATURE CITED

Allard, R. W. 1960. *Principles of plant breeding.* John Wiley and Sons, New York.

Arulsekar, S., D. E. Parfitt, W. Beres, and P. E. Hansch. 1986. Genetics of malate dehydrogenase isozymes in the peach. *J. Heredity* 77: 49–51.

Ben-Hayyim, G., A. Shani, and A. Vardi. 1982. Evaluation of isozyme systems in *Citrus* to facilitate identification of fusion products. *Theor. Appl. Genet.* 64: 1–5.

Bergh, B. O. 1975. Avocados. *In* J. Janick and J. N. Moore [eds.], *Advances in fruit breeding,* 541–567. Purdue University Press, West Lafayette, IN.

Biale, J. B., and R. E. Young. 1971. The avocado pear. *In* A. C. Hulme [ed.], *The biochemistry of fruits and their products,* 2–63. Acedemic Press, London.

Brown, A. G. 1975. Apples. *In* J. Janick and J. N. Moore [eds.], *Advances in fruit breeding,* 3–37. Purdue University Press, West Lafayette, IN.

Button, J., A. Vardi, and P. Speigel-Roy. 1976. Root peroxidase isoenzymes as an aid in *Citrus* breeding and taxonomy. *Theor. Appl. Genet.* 47: 119–123.

Carpenter, J. B., and C. L. Ream. 1976. *Date palm breeding, a review.* Date Growers' Inst. Rep. 53: 25–33.

Carter, G. E., and M. M. Brock. 1980. Identification of peach cultivars through protein analysis. *HortSci.* 15: 292–293.

Chevreau, E., Y. Lespinasse, and M. Gallet. 1985. Inheritance of pollen enzymes and polyploid origin of apple (*Malus* × *domestica* Borkh.). *Theor. Appl. Gen.* 71: 268–277.

Chyi, Y. S., and N. F. Weeden. 1984. Relative isozyme band intensities permit the identification of the 2N gamete parent for triploid apple cultivars. *HortSci.* 19: 818–819.

Degani, Ch., and S. Gazit. 1984. Selfed and crossed proportions of avocado progenies produced by caged pairs of complementary cultivars. *HortSci.* 19: 258–260.

———, A. Goldring, S. Gazit, and U. Lavi. 1986. Genetic selection during the abscission of avocado fruitlets. *HortSci.* 21: 1187–1188.

Ellstrand, N. C., and J. M. Lee. 1987a. Cherimoya cultivar identification; Part 1. A tale of two 'Pierces'. *The Fruit Gardener* 10: 5–7.

———, and ———. 1987b. Cultivar identification of cherimoya (*Annona cherimola* Mill.) using isozyme markers. *Scientia Horticulturae* 32: 25–31.

Esen, A., and R. W. Scora. 1977. Amylase polymorphism in *Citrus* and some related genera. *Amer. J. Bot.* 64: 305–309.

Gan, Y. Y., S. Zaini, and A. Idris. 1981. Genetic variation in the grafted vegetatively propagated mango (*Mangifera indica*). *Pertainida* 4: 53–62.

Goldring, A., D. Zamir, and Ch. Degani. 1985. Duplicated phosphoglucose isomerase genes in avocado. *Theor. Appl. Genet.* 71: 491–494.

Hesse, C. O. 1975. Peaches. In J. Janick and J. N. Moore [eds.], *Advances in fruit breeding*, 285–335. Purdue University Press, West Lafayette, IN.

Hirai, M., and I. Kozaki. 1981. Isozymes of citrus leaves. In K. Matsumoto [ed.], *Proc. Intern. Soc. Citriculture*, 9–12. International Citrus Congress, November 9–12, 1981, Tokyo, Japan.

Hirano, H. 1977. Varietal affinities in mulberry (*Morus spp.*) assessed by peroxidase isozymes. *Jap. J. Breed.* 27: 350–358.

———, and K. Naganuma. 1979. Inheritance of peroxidase isozymes in mulberry (*Morus spp.*). *Euphytica*, 28: 73–79.

———, T. Inokuchi, and T. Nakajima. 1980. Relationships between amino acid contents and peroxidase isozymes in leaf blades of mulberry (*Morus spp.*). *Euphytica* 29: 145–153.

Iglesias, L., and H. Lima. 1979. Diferenciación de posturas nucelares y cigóticas en citrus utilizando isoenzimas. Analisis comparativo de amilasas y peroxidasas. *Agrotecnia de Cuba* 11: 135–145.

———, ———, and J. P. Simon. 1974. Isozyme identification of zygotic and nucellar seedlings in *Citrus*. *J. Heredity* 65: 81–84.

Janick, J., and J. N. Moore, [eds.]. 1975. *Advances in fruit breeding*. Purdue University Press, West Lafayette, IN.

Jarret, R. L., and R. E. Litz. 1986a. Isozymes as genetic markers in bananas and plantains. *Euphytica* 35: 539–549.

———, and ———. 1986b. Enzyme polymorphism in *Musa acuminata* Colla. *J. Heredity* 77: 183–188.

Layne, R. E. C., and H. A. Quamme. 1975. Pears. In J. Janick and J. N. Moore [eds.], *Advances in fruit breeding*, 38–70. Purdue University Press, West Lafayette, IN.

Lee, J. M., and N. C. Ellstrand. 1987. Inheritance and linkage of isozymes in the cherimoya (*Annona cherimola* Mill., Annonaceae). *J. Heredity* 78: 383–387.

Makinen, Y., and J. L. Brewbaker. 1967. Isoenzyme polymorphism in flowering plants. I. Diffusion of enzymes out of intact pollen grains. *Physiol. Pl.* 20: 447–482.

———, and T. MacDonald. 1968. Isoenzyme polymorphism in flowering plants. II. Pollen enzymes and isoenzymes. *Physiol. Pl.* 21: 477–486.

Manganaris, A. G., and F. H. Alston. 1987. Inheritance and linkage relationships of glutamate oxaloacetate transaminase isoenzymes in apple. I. The gene *Got-1*, a marker for the S incompatibility locus. *Theor. Appl. Genet.* 74: 154–161.

Manzocchi, L. A., N. Tusa, and G. Geraci. 1981. Peroxidase isoenzymes as genetic markers in *Citrus*. (abstract). *Genetica Agraria (Italy)* 35: 73.

Menendez, R., and L. S. Daley. 1986. Characterization of *Pyrus* species and cultivars using gradient polyacrylamide gel electrophoresis. *J. Environ. Hort.* 4: 56–60.

———, F. E. Larsen, and R. Fritts, Jr. 1986a. Protein and isozyme electrophoresis and isoelectric focusing for the characterization of apple clones. *Scientia Horticulturae* 29: 211–220.

———, ———, and ———. 1986b. Identification of apple rootstock cultivars by isozyme analysis. *J. Amer. Soc. Hort. Sci.* 111: 933–937.

———, ———, and ———. 1986c. Fingerprinting apple cultivars by electrophoretic isozyme banding patterns. *J. Environ. Hort.* 4: 101–107.

Moore, G. A., and R. E. Litz. 1984. Biochemical markers for *Carica papaya*, *Carica cauliflora*, and plants from somatic embryos of their hybrid. *J. Amer. Soc. Hort. Sci.* 109: 213–218.

———, and W. S. Castle. 1988. Morphological and isozymic analysis of open-pollinated citrus rootstock populations. *J. Heredity* 79: 59–63.

Moore, J. N., and J. Janick, [eds.]. 1983. *Methods in fruit breeding*. Purdue University Press, West Lafayette, IN.

Nagy, S., and P. E. Shaw, [eds.]. 1980. *Tropical and subtropical fruits*. AVI Publishing, Inc., Westport, CT.

Nixon, R. W., and J. B. Carpenter. 1978. *Growing dates in the United States*. USDA Agric. Info. Bull. 207.

Ogure, M. 1979. Growth and isozyme pattern of seedlings from polyembryonic seed in the mulberry. (Engl. abstract). *Nippon Sanshigaku Zasshi Tokyo* 48: 433–438.

Okie, W. R., D. W. Ramming, and R. Scorza. 1985. Peach and nectarine and other stone fruit breeding by the USDA in the last two decades. *HortSci.* 20: 633.

Parfitt, D. E., S. Arulsekar, and D. W. Ramming. 1985. Identification of plum × peach hybrids by isoenzyme analysis. *HortSci.* 20: 246–248.

Peirce, L. C., and J. L. Brewbaker. 1973. Applications of isozyme analysis in horticultural science. *HortSci.* 8: 17–22.

Pontikis, C. A., M. Loukas, and C. Kousounis. 1980. The use of biochemical markers to distinguish olive cultivars. *J. Hort. Sci.* 55: 333–343.

Popenoe, P. B. 1973. *The date palm.* Field Research Projects, Coconut Grove, Miami.

Samson, J. A. 1980. *Tropical fruits.* Longman, London.

Sanford, J. C. 1983. Ploidy manipulations. In J. N. Moore and J. Janick [eds.], *Methods in fruit breeding.* 100–123. Purdue University Press, West Lafayette, IN.

Santamour, F. S., and P. Demuth. 1980. Identification of Callery pear cultivars by peroxidase isozyme patterns. *J. Heredity* 71: 447–448.

Scora, R., and J. Kumamoto. 1981. On the origin of the grapefruit. *XIII International Botanical Congress, Abstracts,* 333.

Sherman, W. B., and P. M. Lyrene. 1983. Handling seedling populations. In J. N. Moore and J. Janick [eds.], *Methods in fruit breeding,* 66–73. Purdue University Press, West Lafayette, IN.

Skene, K. G. M., and M. Barlass. 1983. In vitro culture of abscissed immature avocado embryos. *Ann. Bot.* 52: 667–672.

Soost, R. K., and J. W. Cameron. 1975. Citrus. In J. Janick and J. N. Moore [eds.], *Advances in fruit breeding,* 507–540. Purdue University Press, West Lafayette, IN.

———, and A. M. Torres. 1981. Leaf isozymes as genetic markers in citrus. In K. Matsumoto [ed.], *Proc. Intern. Soc. Citriculture,* 7–10. International Citrus Congress, November 9–12, 1981, Tokyo, Japan.

———, T. E. Williams, and A. M. Torres. 1980. Identification of nucellar and zygotic seedings of *Citrus* with leaf isozymes. *HortSci.* 15: 728–729.

Storey, W. B. 1975. Figs. In J. Janick and J. N. Moore [eds.], *Advances in fruit breeding,* 568–590. Purdue University Press, West Lafayette, IN.

Stout, A. B. 1924. The flower mechanism of avocados with reference to pollination and the production of fruit. *J. New York Bot. Gard.* 25: 1–7.

Suiter, K. S., J. F. Wendel, and J. S. Case. 1983. LINKAGE-1: A Pascal computer program for the detection and analysis of genetic linkage. *J. Heredity* 74:203–204.

Swingle, W. T. 1943. The botany of *Citrus* and its wild relatives of the orange subfamily (family Rutaceae, subfamily Aurantioideae). In H. J. Webber and L. D. Batchelor [eds.], *The citrus industry,* Vol. 1, 190–422. Div. Agric. Sci., Univ. Calif., Berkeley.

Torres, A. M. 1983. Fruit trees. In S. D. Tanksley and T. J. Orton [eds.], *Isozymes in plant genetics and breeding,* Part B, 401–421. Elsevier, Amsterdam.

———. 1984. Isozymes from avocado cotyledons. *J. Heredity* 75: 300–302.

———, and B. O. Bergh. 1978a. Isozymes as indicators of outcrossing among 'Pinkerton' seedlings. *Calif. Avocado Soc. Yrbk.* 62: 103–110.

———, and ———. 1978b. Isozymes of 'Duke' and its derivatives. *Calif. Avocado Soc. Yrbk.* 62: 111–117.

———, and ———. 1980. Fruit and leaf isozymes as genetic markers in avocado. *J. Amer. Soc. Hort. Sci.* 105: 614–619.

———, U. Diedenhofen, B. G. Bergh, and R. J. Knight. 1978. Enzyme polymorphisms as genetic markers in the avocado. *Amer. J. Bot.* 65: 134–139.

———, T. Mau-Lastovicka, T. E. Williams, and R. K. Soost. 1985. Segregation distortion and linkage of *Citrus* and *Poncirus* isozyme genes. *J. Heredity* 76: 289–294.

———, ———, V. Vithanage, and M. Sedgely. 1986. Segregation distortion and linkage analysis of hand pollinated avocados. *J. Heredity* 77: 445–450.

———, R. K. Soost, and U. Diedenhofen. 1978. Leaf isozymes as genetic markers in citrus. *Amer. J. Bot.* 65: 869–881.

———, ———, and T. Mau-Lastovicka. 1982. Citrus isozymes: genetics and distinguishing nucellar from zygotic seedlings. *J. Heredity* 73: 335–339.

———, and B. Tisserat. 1980. Leaf isozymes as genetic markers in date palms. *Amer. J. Bot.* 67: 162–167.

Ueno, I. 1976. Application of zymography to citrus breeding. II. Variations in peroxidase isozymes for species, varieties and strains of citrus and its relatives. *Bull. Fruit Tree Res. Stn.* (Japan) Series B 3: 9–24.

———, and M. Nishiura. 1976. Application of zymography to citrus breeding. I. Identification of hybrid and nucellar seedlings in citrus by peroxidase isozyme electrophoresis. *Bull. Fruit Tree Res. Stn.* (Japan) Series B 3: 1–8.

Valizadeh, M. 1977. Esterase and acid phosphatase polymorphism in the fig tree (*Ficus carica* L.). *Biochem. Genet.* 15: 1037–1048.

———. 1978. *Aspects genetiques, ecologiques et agronomiques de l'étude de la variabilité des proteines chez les plantes superieures cas de Ficus carica L.* Doctoral Thesis, Academie de Montpellier, Univ. des Sci. et Tech. du Languedoc (France).

Veidenberg, A. E., N. R. Gaziyan, and Kh. E. Yanes. 1977. Genetic specificity of isozymes in apple. (Engl. abstract). *Referativnye Zhur.* 5: 256.

Vrecenar-Gadus, M., and N. C. Ellstrand. 1984. Independent assortment of four isozyme loci in the Bacon avocado (*Persea americana* Mill.). *Calif. Avocado Soc. Yrbk.* 68: 173–177.

———, and ———. 1985. The effect of planting design on outcrossing rate and yield in the Hass avocado. *Scientia Horticulturae* 27: 215–221.

Weeden, N. F. 1986. Identification of duplicate loci and evidence for post-meiotic gene expression in pollen. In D. L. Mulcahy, G. Bergamini, and E. Ottaviano [eds.], *Biotechnology and ecology of pollen,* 9–14. Springer-Verlag, New York.

———, and L. D. Gottlieb. 1979. Distinguishing allozymes and isozymes of phosphoglucoisomerase by electrophoretic comparisons of pollen and somatic tissues. *Biochem. Genet.* 17: 287–296.

———, and ———. 1980. Isolation of cytoplasmic enzymes from pollen. *Pl. Physiol.* 66: 400–403.

———, and R. C. Lamb. 1985. Identification of apple cultivars by isozyme phenotypes. *J. Amer. Soc. Hort. Sci.* 110: 509–515.

———, and ———. 1987. Genetics and linkage analysis of 19 isozyme loci in apple. *J. Amer. Soc. Hort. Sci.* 112: 865–872.

Yamashita, K. 1983. Chimerism of Kobayashi-mikan (*Citrus natsudaidai* × *unshiu*) judged from isozyme patterns in organs and tissues. *J. Jap. Soc. Hort. Sci.* 52: 223–230.

Zamir, D. 1983. Pollen gene expression and selection: applications in plant breeding. In S. D. Tanksley and T. J. Orton [eds.], *Isozymes in plant genetics and breeding,* Part A, 313–330. Elsevier, Amsterdam.

CHAPTER 10

Isozymes as Markers for Studying and Manipulating Quantitative Traits

Charles W. Stuber

U.S. Department of Agriculture, Agricultural Research Service
and
North Carolina State University
Raleigh, North Carolina, 27695-7614

Many plant characteristics (e.g., grain and forage yield, time of flowering, stress tolerance) show continuous variation. This usually implies that the inheritance of such traits is complex (quantitative) and probably involves the collective effects of numerous genetic factors. Classically, the activity of these factors has been characterized en masse, using biometrical procedures, and it has usually not been possible to isolate and measure the individual and interactive parameters of single factors (genes) or segments of chromosomes. Identification, mapping, and examination of individual genes affecting quantitative traits should provide knowledge concerning the organization of genomes and insight into the relative contribution of "major" and "minor" genes to such complexly inherited traits. With a better understanding of the inheritance of such traits, it should be possible to develop new methods for enhancing plant improvement.

A powerful approach for studying the inheritance of quantitatively inherited traits involves the use of mapped genetic markers. Consider, for example, the segregating generation produced from selfing the single cross of two homozygous lines. If the transmission of each chromosomal segment could be followed by identifiable genetic markers, the entire genome could be assayed, segment by segment, for genes associated with the variation of any desired quantitative trait. Thus, the effects contributed by individual chromosomal regions could be identified. Theoretically, a detailed map of all major genes associated with the quantitative trait could be constructed, describing the chromosomal locations of the genes and their individual and interactive effects.

The concept of using mapped, monogenic markers in the study and evaluation of quantitatively inherited traits is not new. Early reports of associations of quantitative traits and major genes included those of Sax (1923), Rasmusson (1933), and Everson and Schaller (1955). Studies on the use of genetic markers for the investigation of polygenic characters in *Drosophila* have been reported by Breese and Mather (1957, 1960), Thoday (1961), and Spickett and Thoday (1966). Law (1967) used intervarietal chromosome substitution lines in wheat (*Triticum aestivum*) to investigate effects associated with four morphological marker loci. Factors influencing four quantitative traits (grain weight, grain number, plant height, and tiller number) were identified and mapped with respect to the marker loci on chromosome 7B.

The reports noted above predated the use of isozymes as markers, so the researchers were restricted to the use of morphological genetic markers. Thus, the studies were limited in scope and utility for practical applications such as in plant or animal improvement. Molecular markers, such as isozymes, have a number of inherent properties that allow the theoretical approaches pioneered by these earlier scientists to be used very effectively for dissecting and manipulating quantitative variation.

Tanksley (1983) stressed that molecular markers are superior to morphological markers for use in studies of quantitative traits for several reasons. (1) Alleles at most molecular marker loci are usually codominant, thus all possible genotypes can be distinguished in segregating populations. For loci encoding morphological traits, dominant-recessive interactions usually preclude distinguishing all genotypes. (2) Genotypes of molecular markers can usually be determined at the whole plant, tissue, or cellular levels. For most morphological markers, genotypes can be ascertained only at the whole plant level. (3) In many plants a number of naturally occurring alleles is available at most molecular marker loci. Frequently, distinguishable alleles must be induced at loci encoding morphological markers through mutagenesis. (4) Deleterious effects only very rarely are associated with specific alleles at molecular marker loci. Alleles at loci affecting morphological traits frequently are responsible for deleterious phenotypic effects. (5) Unfavorable interactions among loci encoding morphological traits severely limit the number of segregating markers that can be used in the same segregating population or generation. With molecular markers, the number that can be monitored in a single population is restricted to those polymorphic markers for which assay techniques are available or by the number that can be handled with the facilities and resources available to the investigator.

The effectiveness of molecular genetic markers for identifying, mapping, and manipulating quantitative trait loci (QTLs) is dependent upon several characteristics of the linkage relationships between the marker loci and the associated QTLs. These characteristics include: (1) number of marker loci and average spacing between them; (2) distribution or uniformity of spacing of the marker loci; and (3) level of linkage disequilibrium. If only a few marker loci are available, a population such as the F_2 derived from the cross of two inbred lines is desirable because linkage disequilibrium is maximized in this generation. However, as shown by Hanson (1959), large genomic regions would be represented by specific marker loci in this generation. Although an F_2 is advantageous for detecting QTLs with a minimum number of markers, the probability that genotypic classes at a marker locus may reflect the effects of multiple QTLs is obviously quite high. An ideal population for using marker loci to investigate quantitative traits would include closely spaced, uniformly distributed, marker loci (i.e., about every 10 centimorgans) in a population that has experienced sufficient recombination so that the size of the genomic region associated with each marker allele has been reduced significantly from that found in an F_2 population.

Extensive sets of mapped isozyme marker loci have been compiled from studies of electrophoretic variability for several plant taxa including maize (*Zea mays*) (Goodman et al., 1980; Goodman and Stuber, 1983; Wendel et al., 1985, 1986, 1988), tomato (*Lycopersicon esculentum*) (Tanksley and Rick, 1980; Tanksley, 1983), wheat (*Triticum aestivum*) (Hart, 1983), and pine (*Pinus*) (Conkle, 1981). Allozyme (isozyme) genotypes at 22 loci for 406 publicly available inbred lines of maize (Stuber and Goodman, 1983) provide an extensive information base for selecting lines with different marker-loci genotypes in this species.

ASSOCIATIONS OF ISOZYME MARKER LOCI WITH QUANTITATIVELY INHERITED TRAITS

Isozyme markers were used to study variation and allelic frequency changes at four esterase loci in Composite Cross V, a widely divergent barley (*Hordeum vulgare*) population (Allard et al., 1972). The population, which originated from intercrossing 30 diverse barley varieties, was grown for 25 years at Davis, California, without any conscious selection. This study showed that balancing selection, with the direction of selection differing for different alleles, was responsible for changes in allelic frequencies over generations and for maintenance of variability within the populations.

Clegg and Allard (1972) used isozyme markers to analyze genetic variability among populations of slender wild oats (*Avena barbata*) in California and described two complementary five-locus allozyme complexes whose distributions were closely associated with environment. Hamrick and Allard (1975) conducted further studies in these populations to separate environmental effects from genetic effects for several quantitative characters (two measures of maturity, plant height, number of tillers, and number of seeds). They reported that plants representing these two five-locus enzyme genotypes differed genetically with respect to four of the five quantitative characters studied.

In maize, several studies evaluated associations of isozyme marker loci with various quantitatively inherited traits. Earlier investigations involved monitoring allelic frequency changes at a large number of enzyme loci in different cycles of recurrent selection experiments. In several long-term recurrent selection experiments in North Carolina, changes of allelic frequencies at eight isozyme loci were statistically significant and greater than would be expected from genetic drift acting alone. Also, these changes were highly correlated with changes in the populations due to selection for increased grain yield (Stuber and Moll, 1972; Stuber et al., 1980). Associations between isozyme marker loci and several agronomic traits in maize have also been reported for several selection experiments by Kahler (1985).

From the results of earlier studies in maize, it was hypothesized that manipulation of allelic frequencies at appropriate isozyme loci might produce responses in the correlated quantitative traits. Experiments to test this hypothesis indicated that selections based solely on manipulations of allelic frequencies at seven enzyme loci in an open-pollinated maize population significantly increased grain yield and the highly correlated trait, ear number (Stuber et al., 1982). Results from a similar study conducted in a population generated from a composite of elite inbred lines showed marker-facilitated selection responses approximately equal to the responses found for phenotypic selection conducted in the same population (Frei et al., 1986b). The large number of generations of random mating in most of the maize populations used in these studies would have reduced the level of linkage disequilibrium between marker loci and QTLs. This undoubtedly limited the effectiveness of the seven marker loci for manipulating the associated quantitative traits. The limited positive results did demonstrate, nevertheless, the utility of using molecular markers for studying and manipulating quantitatively inherited traits.

Positive results in several studies involving the associations of isozyme marker loci with quantitative traits in interspecific crosses of tomato (Lycopersicon spp.) have been reported. Tanksley et al. (1982) used 12 isozyme loci to locate factors influencing four quantitatively inherited characteristics in a backcross population of 400 plants derived from L. esculentum and L. pennellii. Twenty-seven of the 48 possible comparisons of marker loci with quantitative trait expression showed significance, and a minimum of five QTLs was detected per trait. In another study involving 11 isozyme loci in an interspecific cross in Lycopersicon, Vallejos and Tanksley (1983) reported linkages between segregating enzyme loci and genetic factors responsible for cold tolerance. A minimum of three QTLs responsible for growth at low temperatures was detected. Marker-locus associations with 18 quantitative traits were examined in an interspecific F_2 population involving L. pimpinellifolium and L. esculentum (Weller, 1983). Four of the 10 marker loci employed were isozyme loci, and 83 of the 180 possible comparisons among markers and quantitative traits were significant. These studies of tomato demonstrated quite conclusively the effectiveness of markers for identifying and locating QTLs for several traits.

ASSOCIATIONS OF MARKER-LOCUS GENOTYPES WITH HYBRID PERFORMANCE

Detection of inbred lines with superior combining ability (i.e., superior heterosis) in hybrid combinations is very expensive and time-consuming. Several studies have been conducted in which isozyme genotypes have been evaluated as possible predictors of hybrid performance in maize. For example, Hunter and Kannenberg (1971) reported a correlation of only 0.09 between hybrid yield and isozyme diversity among lines in a study involving 15 inbred lines and 11 enzyme loci. Heidrich-Sobrinho and Cordeiro (1975) found a correlation of 0.23 between specific combining ability and isozyme diversity and a correlation of 0.72 between general combining ability and diversity among eight inbred lines using eight isozyme marker loci. In an investigation of seven inbred lines of maize and four isozyme systems, Gonella and Peterson (1978) found a correlation of −0.42 between yield and percentage relatedness. Correlations between diversity of lines and hybrid yield were low and varied with environments in a study of 26 inbred lines of maize and 15 isozyme systems reported by Hadjinov et al. (1982).

In an investigation of 114 hybrids to assess the value of isozyme markers for predicting single-cross hybrid yield performance among maize inbreds (Frei et al., 1986a), line pairs for specific hybrids were classified into similar and dissimilar isozyme groups using 21 marker loci. These isozyme diversity groups were further subdivided into similar and dissimilar pedigree classes according to commonality of pedigree background between line pairs. Grain yield in the group with dissimilar isozymes was significantly higher (10%) than in the similar isozyme group. The dissimilar pedigree class, however, yielded about 37% more than the similar pedigree class. Frei et al. (1986a) concluded that although isozyme marker dissimilarity was significantly associated with higher grain yield, the predictive value of these markers was limited largely to lines with similar pedigrees.

Lamkey et al. (1987) evaluated single-cross hybrids among 24 high-yielding and 21 low-yielding lines selected from a group of 247 inbred lines derived by single-seed descent from the Iowa Stiff Stalk Synthetic maize population. The objective of their evaluation was to determine whether allelic differences at 11 isozyme loci could be used to predict hybrid performance in maize. Comparisons of high × high, high × low, and low × low hybrids of the high- and low-yielding lines indicated that allelic differences at enzyme loci did not reflect the performance of hybrids produced from crosses among unselected lines derived from the Stiff Stalk Synthetic population.

Although these studies suggested a limited utility of isozyme genotypes for predicting hybrid performance in maize, several confounding factors might have caused the somewhat negative outcomes of these investigations. Because so few enzyme loci were assayed in most of the investigations, only a small fraction of the genome was effectively marked. If QTLs are distributed throughout the genome, as is frequently hypothesized, these few isozyme loci surveyed would effectively sample only a small proportion of the genetic factors affecting the quantitative traits measured. More importantly, alleles at most isozyme loci probably do not directly affect the phenotypic expression of the quantitative traits evaluated; rather they may serve only as markers for adjacent segments of the chromosome. Thus, the effects associated with specific alleles present at the linked QTLs must be known. If the lines are derived from a randomly-mated population, or compose some subset of the publicly available inbred lines, then the alleles present at the linked QTLs would likely be somewhat random, i.e., near linkage equilibrium. For marker-facilitated techniques to be successful for predictive or selective purposes, the genome should be well saturated with markers and/or strong linkage disequilibrium is required. Likewise, information on the effects of the specific alleles at the linked QTLs is necessary. Studies involving F_2 populations, which are reported in some detail in the following sections of this chapter, meet these requirements.

THEORETICAL BASIS FOR USING ISOZYME MARKER LOCI FOR IDENTIFICATION AND MANIPULATION OF QUANTITATIVE TRAIT LOCI

Theoretical tools for using genetic markers, such as isozyme loci, for the identification, location, and manipulation of quantitative trait loci (QTLs) have been provided by several investigators (Jayakar, 1970; Mather and Jinks, 1971; McMillan and Robertson, 1974; Soller and Plotkin-Hazan, 1977; Tanksley et al., 1982; Soller and Beckmann, 1983; Edwards et al., 1987; Lebowitz et al., 1987). The theory is based on the fact that the marker locus identifies or "marks" the chromosomal segment in its vicinity and allows that segment to be followed in inheritance. Alternative homologous chromosomal segments, characterized by alternative alleles at the marker locus, can be replicated repeatedly in different individuals and compared for effects on quantitative trait expression whereas other chromosomal regions in the same individuals and the environmental factors affecting them are allowed to vary randomly. With sufficient markers distributed relatively uniformly throughout the genome, all chromosomal regions can be evaluated for their effects on any number of quantitative traits.

Relationships between marker-locus genotypes and quantitative trait expression can be very effectively examined in F_2 populations derived from the cross of two homozygous inbred lines (Edwards et al., 1987). This relationship can be examined by first considering the F_1 derived from two homozygous lines that have different codominant alleles, M_1 and M_2, at a marker locus. Assume that the lines have different alleles, Q_1 and Q_2, respectively, at a QTL linked to the marker locus with some recombinant frequency, r (Table 10.1a). F_2 progeny derived from the selfed F_1 will segregate into nine genotypic classes with respect to these two loci (Table 10.1b). Relative frequencies of the nine genotypes are functions of r, and expressions (genotypic values) of these genotypes for a specific quantitative trait are assigned based upon the genotype at the QTL. Because genotypes at the QTL cannot be discriminated directly, their effects must be inferred through association with the three genotypes at the linked marker locus.

Table 10.1. *Theoretical basis for interpreting the expression of marker-locus genotypes in an F_2 population in terms of additive (a) and dominance (d) effects at a quantitative trait locus linked to the marker locus with recombination frequency, r. M_1, M_2 and Q_1, Q_2 are alleles at the marker and quantitative trait loci, respectively*

a. Hypothetical F_1 genotype:

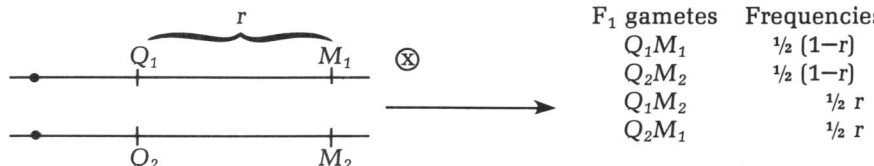

	F_1 gametes	Frequencies
	Q_1M_1	½ (1−r)
	Q_2M_2	½ (1−r)
	Q_1M_2	½ r
	Q_2M_1	½ r

b. Generates F_2 array:

Genotypes	Frequency	Value
$Q_1Q_1M_1M_1$	¼ (1−r)²	+a
$Q_1Q_2M_1M_1$	½ (r−r²)	d
$Q_2Q_2M_1M_1$	¼ r²	−a
$Q_1Q_1M_1M_2$	½ (r−r²)	+a
$Q_1Q_2M_1M_2$	½ [r²+(1−r)²]	d
$Q_2Q_2M_1M_2$	½ (r−r²)	−a
$Q_1Q_1M_2M_2$	¼ r²	+a
$Q_1Q_2M_2M_2$	½ (r−r²)	d
$Q_2Q_2M_2M_2$	¼ (1−r)²	−a

c. Marker-locus class means (frequency-adjusted):

Marker Class	Mean Expression
M_1M_1	$(1-2r)a + 2r(1-r)d$
M_1M_2	$[(1-r)^2 + r^2]d$
M_2M_2	$(1-2r)(-a) + 2r(1-r)d$

d. Expressions to resolve additive and dominance effects:

Additivity: $(M_1M_1 - M_2M_2)/2 = a(1-2r)$
Dominance: $M_1M_2 - (M_1M_1 + M_2M_2)/2 = d(1-2r)^2$

e. Dominance/Additive Ratio:

$$\frac{M_1M_2 - (M_1M_1 + M_2M_2)/2}{(M_1M_1 - M_2M_2)/2} = (1-2r)\, d/a$$

For each marker-locus genotype, sums of frequency multiplied by its associated value produce the expected mean expressions for each of the three marker-locus genotypes in terms of r and the genotypic values due to the QTL (Table 10.1c). Mean expressions in this table have been adjusted from their relative frequencies of ¼, ½, and ¼ to unity to allow comparisons among them. If $r = 0$ (i.e., the marker locus itself is responsible for the detected quantitative effects), expressions for M_1M_1, M_1M_2, and M_2M_2 reduce to the assigned values of the QTL genotypes (a, d, and −a, respectively). Obviously, when the marker is tightly linked to the QTL (i.e., $r \cong 0$) the expressions are similar. If $r = 0.5$ (i.e., the marker locus segregates independently from the QTL), all three marker locus classes are expected to have the same mean value, $d/2$, which is the mean expression for the F_2 population for the quantitative trait of interest. When $0 < r < 0.5$, the expressions for the marker-locus classes are complex functions of r and both additive and dominance effects. Contrasts involving both the heterozygous and homozygous class means provide functions that are simpler expressions of additive or dominance gene effects (Table 10.1d). The ratio of the "dominance" to "additive" contrasts can be used to approximate the apparent degree of dominance at the QTL (Table 10.1e). Although this expression is progressively biased in the direction of underestimating the "true" d/a ratio at the QTL as r approaches 0.5, the magnitude of this bias is small, particularly with small values of r, for which differences between the expression of the marker-locus classes are most likely to be significant (Edwards et al., 1987).

It should be stressed that, for a specific quantitative trait, failure to detect a QTL linked to a particular marker locus does not necessarily imply that there is no QTL in the vicinity of the marker locus. If the two parental lines have identical alleles at the linked QTL, this locus will not be detected. Alternatively, the parents may have different alleles at the QTL that show similar expressions for the particular trait. Even if the genome is saturated with uniformly spaced marker loci, the number of genomic regions exhibiting significant marker-linked effects in an F_2 population must, therefore, represent fewer than the "true" number of QTLs that might be involved in expression of the trait.

INVESTIGATIONS OF QUANTITATIVE TRAIT LOCI IN F_2 MAIZE POPULATIONS USING ISOZYME MARKER LOCI

Two large F_2 maize populations which were designed to identify and locate quantitative trait loci (QTLs) associated with a large number of quantitative traits have been studied by Edwards et al. (1987) and Stuber et al. (1987). In these studies, significant associations between marker-locus genotypes and quantitative trait expressions were interpreted to mean that the marker loci were linked to QTLs that influenced the expression of the traits. Types and magnitudes of gene action expressed by these QTLs were also evaluated.

Populations for these studies were developed by self-pollinating F_1 hybrid plants from two crosses of inbred lines, CO159 with Tx303 and T232 with CM37. Inbred parents for these hybrids were chosen to maximize both the number of segregating marker (isozyme) loci and the segregation for agronomic and morphological characteristics in the F_2 populations. The F_2 populations of CO159 × Tx303 (COTX) and T232 × CM37 (CMT) showed segregation at 15 and 18 isozyme loci respectively, plus loci encoding two easily scorable morphological traits. These isozyme and morphological marker loci are distributed on nine of the 10 maize chromosomes and are located within about 20 cM of 40 to 45% of the genome (Fig. 10.1). Totals of 1776 and 1930 plants were evaluated in COTX and CMT, respectively.

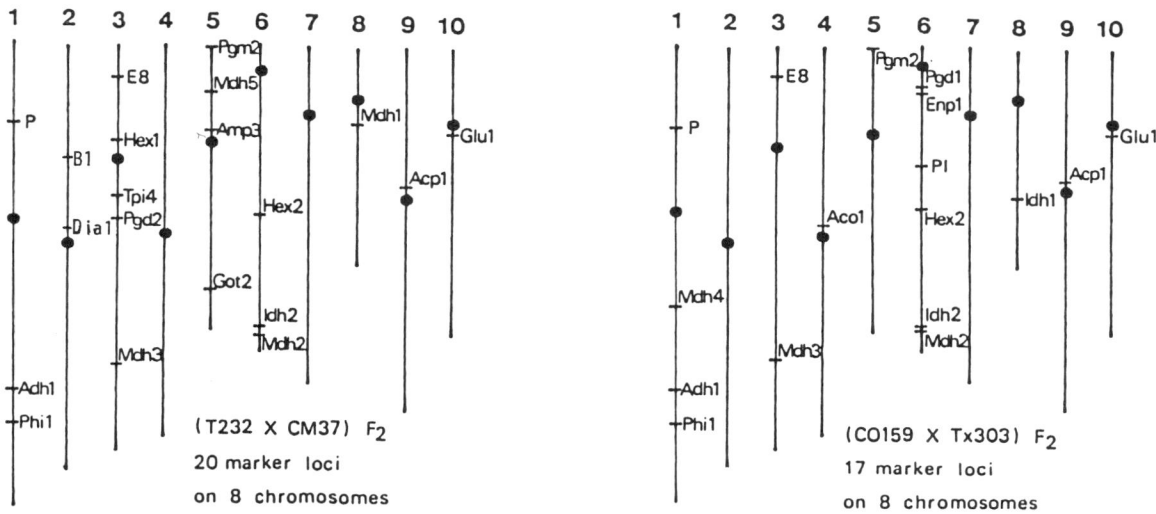

Fig. 10.1. Chromosomal locations of the 20 and 17 marker loci segregating in (T232 × CM37)F_2 and (CO159 × Tx303)F_2.

Each F_2 plant was genotyped for the marker loci and then transplanted into the field where approximately 40 characteristics were measured on each plant. Measurements included weights, dimensions, and counts of vegetative and reproductive plant parts, as well as silking and pollen shedding dates. These measurements were used to construct 82 quantitative variables, many of which were functions of individual measurements (e.g., harvest index = grain weight/total above ground plant weight). Experimental procedures and data analyses are outlined by Edwards et al. (1987). A subset of 25 yield and yield-related traits is discussed by Stuber et al. (1987).

For the 82 quantitative traits evaluated, trait expressions were found to be significantly ($P = 0.05$) associated with marker-locus genotypes in 830 of 1394 comparisons (60%) in COTX and in 1079 of 1640 comparisons (66%) in CMT. About two-thirds of these significant associations were highly significant ($P = 0.001$). Some significant associations were found for each of the 82 variables in each population indicating that QTLs were identified that influenced every trait measured in each population. The number of marker loci significantly associated with factors influencing the expression of each trait averaged 10.2 (of 17) and 13.8 (of 20) in COTX and CMT, respectively.

For a subset of 14 of the 82 quantitative traits evaluated in the two populations (Table 10.2), numbers of significant associations with marker loci ranged from low values of seven of the 17 markers for plant height in COTX, to 19 of the 20 marker loci for stalk weight and several yield-related traits in CMT. In general, yield and most yield-related traits were affected by factors associated with most of the marker loci. Some of the marker loci are located in the same chromosomal arm (Fig. 10.1) and, in some cases, may reflect the effects of the same QTL(s). Members of tightly linked pairs on chromosome 1 (*Adh1—Phi1*) and chromosome 6 (*Pgd1—Enp1* and *Idh2—Mdh2*) likely reflect associations with the same QTLs.

Table 10.2. Numbers of segregating marker loci that showed significant (P<0.05) associations, percent of phenotypic variation ($R^2 \times 100$) accounted for by individual marker loci, and percent of phenotypic variation ($R^2 \times 100$) explained by a regression model including simple effects at all marker loci for 14 of the quantitative traits evaluated in (CO159 × Tx303)F_2 and (T232 × CM37)F_2

Trait	(CO159 × Tx303)F_2 − COTX				(T232 × CM37)F_2 − CMT			
	No.[a]	$R^2 \times 100$[b] Min.	Max.	Model R^2 (× 100)[c]	No.[a]	$R^2 \times 100$[b] Min.	Max.	Model R^2 (× 100)[c]
Whole Plant:								
Grain weight	13	0.61	3.50	14.23	18	0.30	5.08	29.96
Stalk weight	8	0.54	14.82	25.94	19	0.42	7.42	31.77
Harvest index	9	0.50	7.06	16.50	11	0.23	1.88	9.95
Ear number	10	0.35	5.24	12.33	9	0.23	4.13	9.94
Ear height	9	0.39	15.08	26.98	17	0.47	6.81	35.02
Plant height	7	0.85	13.94	28.08	14	0.38	9.72	39.81
Leaf Number	12	0.34	16.27	26.40	15	0.36	10.41	23.61
Top Ear:								
Grain weight	13	0.72	4.86	13.42	19	0.24	8.62	32.33
Circumference	16	0.39	3.47	13.73	19	0.36	11.30	29.74
Length	10	0.40	4.27	13.78	19	0.32	4.24	22.07
Kernel row no.	12	0.47	3.88	12.44	17	0.25	5.49	27.33
100-kernel wt.	8	0.36	1.91	7.66	13	0.38	3.84	14.43
Kernel depth	17	0.53	2.51	13.54	15	0.46	9.88	22.55
Days to silk	11	0.39	15.63	36.73	13	0.36	12.99	39.33

[a] Number of marker loci showing significant associations with quantitative trait expression.
[b] Minimum and maximum percent of the phenotypic variation explained by marker-locus genotypic classes (for those loci that showed significant associations with quantitative trait expression).
[c] Percent of phenotypic variation explained by a regression model that included simple effects at all marker loci.

Because so many plants were measured in each population, factors contributing as little as 0.2% of the phenotypic variation could be detected for some traits. It should be noted, however, that effects detected by marker loci diminish relative to the true effect of the QTL as the distance (recombination frequency) between the marker locus and the QTL increases (Edwards et al., 1987). Therefore, small estimates of variation detected by marker loci may reflect either QTLs with small effects or QTLs with larger effects but that are loosely linked to the marker loci.

In the CMT populations, *Dia1* explained 5.1% of the total phenotypic variation for grain weight (Table 10.3). Genotypic classes at several marker loci (including *Pgm2*, *Mdh5*, and *Amp3* on chromosome 5S) accounted for 6% to more than 11% of the variation of several yield-related traits including grain weight of the uppermost ear, kernel depth, and ear circumference (Table 10.2). In COTX, markers were less effective for predicting trait expression for yield and yield-related traits than in CMT. For several vegetative traits, however, such as stalk weight, plant and ear height, and leaf number, markers were more effective in COTX than CMT for predicting trait expression. From 14 to 16% of the phenotypic variation of these traits was accounted for by individual marker loci in COTX (Table 10.2). Although more than 60% of the associations of marker loci with the 82 quantitative traits were significant, in more than three-fourths of these significant associations individual marker loci accounted for less than 2.0% of the phenotypic variation of these traits.

Table 10.3. *Mean values of homozygous and heterozygous marker classes and percent of variation ($R^2 \times 100$) associated with four unlinked marker loci for two traits, grain weight and ear number, in $(T232 \times CM37)F_2$. Marker loci are Dia1 (chromosome 2S), Pgd2 (chromosome 3L), Amp3 (chromosome 5S), and Mdh1 (chromosome 8L)*

		Marker Classes			
Trait	Marker locus	CM37 Homozygote	Heterozygote	T232 Homozygote	R^2 ($\times 100$)
Whole plant:					
Grain weight	*Dia1*	110.7	133.0	131.3	5.08
(g)	*Pgd2*	113.3	131.3	132.8	3.78
	Amp3	113.8	130.3	135.2	4.04
	Mdh1	114.7	132.6	135.0	4.50
Ear number	*Dia1*	1.53	1.73	1.78	4.13
	Pgd2	1.63	1.70	1.74	0.78
	Amp3	1.73	1.69	1.68	0.17
	Mdh1	1.66	1.71	1.72	0.33

Although individual marker loci accounted for relatively small proportions of the phenotypic variation for most of the yield-related traits studied, differences between mean phenotypic values of homozygous classes at some marker loci occasionally were greater than 16% of the population mean (Tables 10.3 and 10.4). In CMT, for example, grain weight differences between the two homozygous classes for four unlinked loci (*Dia1*, *Pgd2*, *Amp3*, and *Mdh1*) were each slightly more than 20 grams per plant (16% of the mean grain yield in this F_2 population). However, these loci individually accounted for only about 4 to 5% of the total phenotypic variation. It can be noted that for all four of these marker loci, yield was greatest in the homozygous class containing the allele contributed by the inbred T232.

In the COTX population, the grain yield difference associated with the two homozygous classes at *Adh1* was 20 grams per plant (16% of the COTX mean). For two other unlinked loci, *Mdh3* and *Idh1*, differences between the two homozygous classes averaged about 10% of the population mean (Table 10.4). For the *Adh1* locus, the CO159 type homozygous class yielded higher than the Tx303 type homozygous class; however, for the *Mdh3* and *Idh1* marker loci, the Tx303 homozygous class yielded higher than the CO159 homozygous class. The three loci (*Adh1*, *Mdh3*, and *Idh1*) accounted for only 3.50, 1.59, and 1.75% of the total phenotypic variation for grain yield, respectively.

Table 10.4. Mean values of homozygous and heterozygous marker classes and percent of variation ($R^2 \times 100$) associated with three unlinked marker loci for two traits, grain weight and ear number, in (CO159 × Tx303)F_2. Marker loci are Adh1 (chromosome 1L), Mdh3 (chromosome 3L), and Idh1 (chromosome 8L)

		Marker Classes			
Trait	Marker locus	CO159 Homozygote	Heterozygote	Tx303 Homozygote	R^2 (× 100)
Whole plant:					
Grain weight	Adh1	133.2	132.7	112.8	3.50
(g)	Mdh3	118.0	131.4	131.2	1.59
	Idh1	117.7	132.2	130.0	1.75
Ear number	Adh1	1.37	1.39	1.43	0.17
	Mdh3	1.37	1.40	1.42	0.15
	Idh1	1.23	1.42	1.53	5.24

The proportion of the total phenotypic variation for each trait that could be explained by the cumulative, simple effects of all marker loci was determined by calculating a multilocus R^2 for each trait in each population. For the 14 traits shown in Table 10.2, the multilocus model accounted for 8 to 37% of the phenotypic variation in COTX, and for 10 to 40% of the variation in CMT. Four vegetative traits (stalk weight, ear height, plant height, and leaf number) were explained quite well in both populations with 24 to 40% of the phenotypic variation accounted for by the multilocus model. For grain weight, marker-locus genotypes in CMT accounted for 30% of the variation, whereas the multilocus model accounted for only 14% of the grain weight variation in COTX.

The design of these F_2 studies facilitated an assessment of the types of gene action expressed by those genetic factors contributing to the variation of the quantitative traits measured (Edwards et al., 1987). Contrasts among mean values of marker-locus genotypic classes were used to determine whether the variation could be attributed to additive or dominance effects for specific QTLs or chromosomal regions associated with the marker loci. Both F_2 populations showed mostly dominance or overdominance for grain yield and ear length. Additive gene action was largely implicated for several yield-related traits such as ear number, kernel row number, and second ear weight (Stuber et al., 1987). Analyses of interlocus interactions showed little evidence that digenic epistasis was an important source of variation for the traits evaluated in these F_2 populations (Edwards et al., 1987).

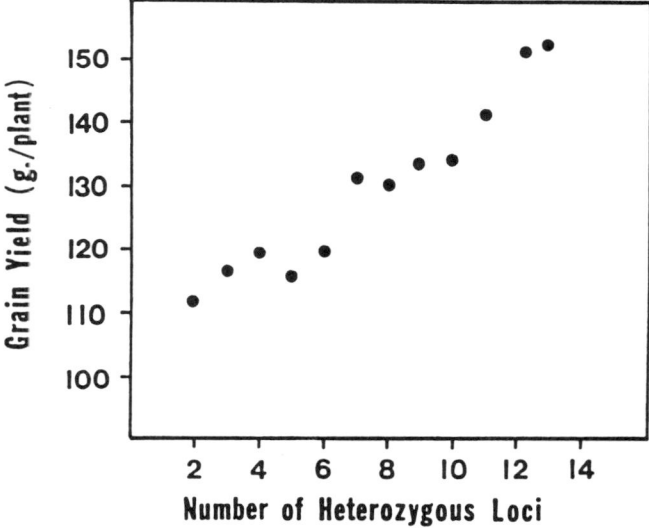

Fig. 10.2. Relationship of grain yield to number of heterozygous loci in (CO159 × Tx303)F_2.

Contributions of heterozygosity to trait expression were obtained by comparing the mean level of trait performance for plants with varying numbers of heterozygous marker loci. As shown in Fig. 10.2, heterozygosity is highly associated with the expression of grain yield, as expected because the gene action was largely in the dominance and overdominance classes. Kahler (1985) showed a similar relationship in an F_2 population derived from Wf9 and Pa405.

Segregating isozyme loci in the F_2 populations, COTX and CMT, therefore were effective in identifying chromosomal regions that affect a wide array of phenotypic characteristics in corn. Factors contributing as little as 0.2% of the phenotypic variation in quantitative traits were detected, and some individual marker loci explained as much as 17% of the phenotypic variation. Cumulative effects of marker loci explained 8 to 40% of phenotypic variation. The results showed that QTLs are distributed throughout the genome, and certain regions affected a greater number of traits than others. All of the markers were significantly associated with some of the traits measured. Types of gene action varied for the different traits evaluated, and the number of marker loci in a heterozygous state was highly associated with the expression of traits such as grain yield.

GENOTYPIC SELECTION FOR IMPROVEMENT OF QUANTITATIVE TRAITS IN MAIZE USING ISOZYME MARKER LOCI

The merits of using marker loci to manipulate quantitative traits, such as grain yield in maize, were reported by Stuber and Edwards (1986). In their investigations, results from the studies discussed in the previous section of this chapter were used as the basis for marker-locus facilitated selection. In one study, selections were made among the F_2 plants grown in the earlier studies, and evaluations were made on progeny of these open-pollinated plants. In a second study, selfed progeny from a different sample of F_2 plants from the same two populations were evaluated.

In both of these marker-facilitated selection studies, a breeding value was determined for each marker for each plant and for each trait being manipulated. This value was equivalent to one-half the difference between the means of the homozygous classes for that locus. This value could be either positive or negative for a specific homozygous class depending on whether that class contained the favorable or unfavorable allele for the quantitative trait of interest. A value of zero was assigned to each heterozygote because its progeny would segregate and, thus, would not contribute to a change in gene frequency in the population. For each trait, breeding values were totaled over marker loci to arrive at a composite breeding score for each plant.

Using these composite scores, individual plants were chosen to provide divergent selection classes (positive and negative) for several traits and combinations of traits, including grain yield, ear height, grain moisture, and ear number. Kernels from selected plants were bulked for each selection criterion to provide each entry in the evaluation studies. In the open-pollinated F_2 population studies, 37 plants were selected for each criterion to give a selection intensity of approximately 2%. Fewer plants were available for the selfed-progeny studies, and the selection intensity was 17% in COTX and 30% in CMT. Mass selection, based solely on the phenotypic expression of each plant, was also conducted as part of the open-pollinated F_2 studies for comparison with the marker-facilitated selection responses.

Evaluations of progenies from plants selected in the open-pollinated F_2 studies were made at three locations in North Carolina with 10 replications for each entry at each location. Evaluations in the selfed-progeny studies were also made at the same three locations with five replications for each entry.

Results of marker-facilitated (genotypic) and mass (phenotypic) divergent selections are presented (Tables 10.5 and 10.6) for two of the single trait selection criteria, grain yield and ear height, in the open-pollinated F_2 populations. In the COTX population, differences between means of entries selected for increased yield and means of entries selected for decreased yield were significant for both genotypic and phenotypic selection (Table 10.5). For genotypic selection, the mean for the increased-yield entry was about 12% greater than the mean for the decreased-yield entry. For phenotypic selection, the same comparison showed a 16% difference. Correlated changes for ear height and ear number were in the same direction as yield, and percentage differences were similar in magnitude.

Table 10.5. Means for three traits following divergent genotypic (marker-facilitated) and phenotypic (mass) selection in (CO159 × Tx303)F$_2$ randomly mated populations

Selection Criterion	Traits		
	Grain yield g/plant	Ear height cm	Ear number
Grain yield:			
Genotypic-increase	138.9	75.0	1.40
Genotypic-decrease	123.9	61.9	1.24
Phenotypic-increase	143.2	79.8	1.43
Phenotypic-decrease	123.0	69.7	1.25
Ear height:			
Genotypic-increase	135.4	83.5	1.44
Genotypic-decrease	117.9	53.4	1.13
Phenotypic-increase	144.1	93.9	1.45
Phenotypic-decrease	120.5	55.5	1.18
Checks—no selection:			
(CO159 × Tx303)F$_2$	137.9	66.2	1.26
(CO159 × Tx303)F$_2$ —randomly mated	133.0	70.0	1.31
S.E.$_{\bar{d}}$[†]	5.81	1.96	0.04

[†]Standard error of mean difference.

Table 10.6. Means for three traits following divergent genotypic (marker-facilitated) and phenotypic (mass) selection in (T232 × CM37)F$_2$ randomly mated populations

Selection Criterion	Traits		
	Grain yield g/plant	Ear height cm	Ear number
Grain yield:			
Genotypic-increase	151.2	73.5	1.48
Genotypic-decrease	107.7	47.1	1.20
Phenotypic-increase	151.7	68.5	1.43
Phenotypic-decrease	122.4	57.8	1.28
Ear height:			
Genotypic-increase	142.6	73.1	1.43
Genotypic-decrease	111.3	46.8	1.25
Phenotypic-increase	143.0	75.2	1.40
Phenotypic-decrease	111.1	48.0	1.28
Ear number:			
Genotypic-increase	132.8	67.8	1.44
Genotypic-decrease	118.2	55.0	1.22
Checks—no selection:			
(T232 × CM37)F$_2$	120.1	55.1	1.30
(T232 × CM37)F$_2$ —randomly mated	127.2	59.2	1.35
S.E.$_{\bar{d}}$[†]	6.36	2.16	0.04

[†]Standard error of mean difference.

Both genotypic and phenotypic selection based on ear height in COTX were considerably more effective for this trait than selection based on yield (Table 10.5). In fact, means of entries selected for increased ear height were about 60% greater for this trait than those for which decreased ear height was selected. Although selection was based on ear height, the correlated responses of yield and ear number tended to be slightly greater than when selection was based solely on yield.

Selection based solely on grain yield was more effective in CMT than in COTX for the primary trait (Table 10.6). This was expected because the marker-locus genotypes accounted for two times as much variation for grain yield in CMT as in COTX. For genotypic selection in CMT, the mean for the entry for which selection was expected to increase yield was 40% greater than the mean for the decreased yield entry. Also, selection for increased yield resulted in nearly 20% gain over the mean of the control. Phenotypic selection for increased yield was as effective as genotypic selection, but phenotypic selection for decreased yield did not differ from the check mean. It was also noted that phenotypic selection for increased yield resulted in about 24% more grain than selection for decreased yield. Correlated responses for ear height and ear number in CMT were similar to those found for COTX.

When selection was based solely on ear height, responses to genotypic selection in CMT were no more effective for changing ear height than selection based solely on yield. Changes in the correlated traits, yield and ear number, were less than when selection was based solely on yield. Responses to phenotypic selection for ear height were very similar to responses to genotypic selection. However, changes in ear height were greater than when phenotypic selection was applied solely to yield.

Genotypic selection in CMT based solely on ear number was no more effective for changing this trait than when selection was practiced for either yield or ear height. Correlated responses from selection on ear number were much less effective for yield than the correlated responses from selection made solely on ear height.

In the selfed progeny evaluation studies, divergent selection for grain yield was also more effective in CMT than in COTX (Table 10.7). The entry associated with selection for increased yield in CMT resulted in about 31% more grain than selection for decreased yield. The same contrast was only about 17% in COTX. Correlated responses for ear height and ear number were similar in magnitude to those found in progeny from the open-pollinated populations.

Table 10.7. *Means for three traits following divergent genotypic (marker-facilitated) selection on grain yield in (CO159 × Tx303)F$_2$ and (T232 × CM37)F$_2$ selfed populations*

Population (selfed) and Selection Criterion	Traits		
	Grain yield g/plant	Ear height cm	Ear number
(CO159 × Tx303)F$_2$:			
Positive	100.4	79.5	1.33
Negative	86.0	67.6	1.28
(T232 × CM37)F$_2$:			
Positive	105.5	67.8	1.48
Negative	80.3	51.1	1.20
S.E.$_{\bar{d}}$†	6.4	1.1	0.05

†Standard error of mean difference.

Because the marker genotypes were known for each selected plant, marker-locus allelic frequencies could be monitored in each of the selected populations (M. D. Edwards and C. W. Stuber, unpublished data). In the CMT population resulting from marker-facilitated selection for increased grain yield, frequencies of the alleles at 15 marker loci averaged about 0.38 greater than in the population derived from selection for decreased grain yield. Differences for individual loci ranged from 0.02 to 0.73. Loci with the greatest differences were those that accounted for the largest proportion of the phenotypic variation for yield. In the populations resulting from phenotypic (mass) selection for grain yield, frequency differences between the populations selected for increased and decreased yield averaged 0.13, only about one-third the 0.38 found for genotypic selection. These results imply that the selection responses for grain yield, which were similar for genotypic and phenotypic selection, resulted from three-fold greater changes for loci linked to marker loci in the populations derived from genotypic selection than in those derived from phenotypic selection. Thus, loci in unmarked regions of the genome, which would receive no selection pressure from genotypic selection, undoubtedly contributed to a major part of the response from phenotypic selection. Results in the COTX populations were similar to those for CMT; however, magnitudes of frequency changes were less for COTX.

The maize selection studies indicated that marker-facilitated (genotypic) selection was effective for manipulating quantitatively inherited traits in these two populations. Selection based on marker loci representing 30 to 40% of the genome was as effective as phenotypic selection which, presumably, involves the entire genome. This implies that increasing the number of markers so that the entire genome is effectively marked with uniformly distributed loci would significantly increase the effectiveness of marker-facilitated selection. These results also suggest that marker-facilitated techniques should be effective in transferring desired factors at quantitative trait loci for improvement of superior lines by using marker-facilitated backcrossing procedures.

CONCLUSIONS

The concept of using mapped, monogenic markers for genetics and breeding research was reported more than 60 years ago. Until the advent of molecular markers, such as isozyme loci, researchers were limited generally to the use of morphological markers, which lack many of the advantages of molecular markers. Although isozyme markers provide the basis for a relatively simple tool for genetic and plant breeding research, it is unlikely that a sufficient number of usable isozyme loci will be found in most species to saturate the genome completely and uniformly. In addition, marker loci are useful only if different alleles are segregating in the population of interest. Frequently, many loci are monomorphic. Thus, restriction fragment length polymorphisms (RFLPs) will likely be used to complement isozyme markers in many genetic or breeding situations (Rivin et al., 1983; Soller and Beckmann, 1983; Helentjaris et al., 1985).

The associations of isozyme marker alleles with specific QTL variants found in maize and tomato may be unique to the populations studied. Further investigations are required to determine whether the findings can be generalized to other populations and to other crop species. It is anticipated, however, that additional studies will identify certain chromosomal segments that tend to be associated with specific quantitative traits. Thus geneticists/breeders may be able to restrict their efforts to particular parts of the chromosomal complement rather than search through the entire genome for useful markers in each new genetic or breeding program.

ACKNOWLEDGMENTS

This investigation represents a joint contribution from the United States Department of Agriculture, Agricultural Research Service and the North Carolina Agricultural Research Service, North Carolina State University, Raleigh, NC, USA 27695-7614. The investigation was supported in part by United States Department of Agriculture Competitive Grant No. 83-CRCR-1-1273 and in part by National Institutes of Health Research Grant No. GM 11546 for the National Institute of General Medical Sciences of the USA. Paper No. 11373 of the Journal Series of the North Carolina Agricultural Research Service, Raleigh, NC 27695-7601.

LITERATURE CITED

Allard, R. W., A. L. Kahler, and B. S. Weir. 1972. The effect of selection on esterase allozymes in a barley population. *Genetics* 72: 489–503.

Breese, E. L., and K. Mather. 1957. The organization of polygenic activity within a chromosome in *Drosophila*. I. Hair characters. *Heredity* 11: 373–395.

_____, and _____. 1960. The organization of polygenic activity within a chromosome in *Drosophila*. II. Viability. *Heredity* 14: 375–400.

Clegg, M. T., and R. W. Allard. 1972. Patterns of genetic differentiation in the slender wild oat species *Avena barbata*. *Proc. Natl. Acad. Sci. USA* 69: 1820–1824.

Conkle, M. T. 1981. Isozyme variation and linkage in six conifer species. In M. T. Conkle [ed.], *Isozymes of North American forest trees and forest insects*, PSW-48, 11–17. USDA Gen. Tech. Rep.

Edwards, M. D., C. W. Stuber, and J. F. Wendel. 1987. Molecular marker-facilitated investigations of quantitative trait loci in maize. I. Numbers, distribution, and types of gene action. *Genetics* 116: 113–125.

Everson, E. H., and C. W. Schaller. 1955. The genetics of yield differences associated with awn barbing in the barley hybrid (Lion × Atlas[10]) × Atlas. *Agron. J.* 47: 276–280.

Frei, O. M., C. W. Stuber, and M. M. Goodman. 1986a. Use of allozymes as genetic markers for predicting performance in maize single cross hybrids. *Crop Sci.* 26: 37–42.

_____, _____, and _____. 1986b. Yield manipulation from selection on allozyme genotypes in a composite of elite corn lines. *Crop Sci.* 26: 917–921.

Gonella, J. A., and P. A. Peterson. 1978. Isozyme relatedness of inbred lines of maize and performance of their hybrids. *Maydica* 23: 55–61.

Goodman, M. M., and C. W. Stuber. 1983. Maize. In S. D. Tanksley and T. J. Orton [eds.], *Isozymes in plant genetics and breeding*, Part B, 1–33. Elsevier, Amsterdam.

_____, _____, K. Newton, and H. H. Weissinger. 1980. Linkage relationships of 19 isozyme loci in maize. *Genetics* 96: 697–710.

Hadjinov, M. I., V. S. Scherbak, N. I. Benko, V. P. Gusev, T. B. Sukhorzheuskaya, and L. P. Voronova. 1982. Interrelationships between isozyme diversity and combining ability in maize lines. *Maydica* 27: 135–149.

Hamrick, J. L., and R. W. Allard. 1975. Correlations between quantitative characters and enzyme genotypes in *Avena barbata*. *Evolution* 29: 438–442.

Hanson, W. D. 1959. The theoretical distribution of lengths of parental gene blocks in the gametes of an F_1 individual. *Genetics* 44: 197–209.

Hart, G. E. 1983. Hexaploid wheat (*Triticum aestivum* L. em Thell) In S. D. Tanksley and T. J. Orton [eds.], *Isozymes in plant genetics and breeding*, Part B, 35–56. Elsevier, Amsterdam.

Heidrich-Sobrinho, E., and A. R. Cordeiro. 1975. Codominant isoenzymic alleles as markers of genetic diversity correlated with heterosis in maize (*Zea mays* L.). *Theor. Appl. Genet.* 46: 197–199.

Helentjaris, T., G. King, M. Slocum, C. Siedenstang, and S. Wegman. 1985. Restriction fragment polymorphisms as probes for plant diversity and their development as tools for applied plant breeding. *Pl. Mol. Biol.* 5: 109–118.

Hunter, R. B., and L. W. Kannenberg. 1971. Isozyme characterization of corn (*Zea mays* L.) inbreds and its relation to single cross hybrid performance. *Canad. J. Genet. Cytol.* 13: 649–655.

Jayakar, S. D. 1970. On the detection and estimation of linkage between a locus influencing a quantitative character and a marker locus. *Biometrics* 26: 451–464.

Kahler, A. L. 1985. Associations between enzyme marker loci and agronomic traits in maize. *Proceedings of 40th Annual Corn and Sorghum Research Conference*, American Seed Trade Association 40: 66–89.

Lamkey, K. R., A. R. Hallauer, and A. L. Kahler. 1987. Allelic differences at enzyme loci and hybrid performance in maize. *J. Heredity* 78: 231–234.

Law, C. N. 1967. The location of factors controlling a number of quantitative characters in wheat. *Genetics* 56: 445–461.

Lebowitz, R. J., M. Soller, and J. S. Beckmann. 1987. Trait-based analyses for the detection of linkage between marker loci and quantitative trait loci in crosses between inbred lines. *Theor. Appl. Genet.* 73: 556–562.

Mather, K., and J. L. Jinks. 1971. *Biometrical genetics.* Cornell University Press, Ithaca, New York.

McMillan, I., and A. Robertson. 1974. The power of methods for detection of major genes affecting quantitative characters. *Heredity* 32: 349–356.

Rasmusson, J. M. 1933. A contribution to the theory of quantitative character inheritance. *Hereditas* 18: 245–261.

Rivin, C. J., E. A. Zimmer, C. A. Cullis, V. Walbot, T. Huynh, and R. W. Davis. 1983. Evaluation of genomic variability at the nucleic acid level. *Plant Mol. Biol. Reporter* 1: 9–16.

Sax, K. 1923. The association of size differences with seed coat pattern and pigmentation in *Phaseolus vulgaris. Genetics* 8: 552–560.

Soller, M., and J. S. Beckmann. 1983. Genetic polymorphism in varietal identification and genetic improvement. *Theor. Appl. Genet.* 67: 25–33.

_____, and J. Plotkin-Hazan. 1977. The use of marker alleles for the introgression of linked quantitative alleles. *Theor. Appl. Genet.* 51: 133–137.

Spickett, S. G., and J. M. Thoday. 1966. Regular responses to selection 3: interaction between located polygenes. *Genet. Res.* 7: 96–121.

Stuber, C. W., and M. D. Edwards. 1986. Genotypic selection for improvement of quantitative traits in corn using molecular marker loci. *Proceedings of 41st Annual Corn and Sorghum Research Conference,* American Seed Trade Assoc. 41: 70–83.

_____, _____, and J. F. Wendel. 1987. Molecular marker-facilitated investigations of quantitative loci in maize: II. Factors influencing yield and its component traits. *Crop Sci.* 27: 639–648.

_____, and M. M. Goodman. 1983. *Allozyme genotypes for popular and historically important inbred lines of corn.* U.S. Dept. of Agriculture. Agric. Res. Serv., Southern Series No. 16.

_____, _____, and R. H. Moll. 1982. Improvement of yield and ear number resulting from selection at allozyme loci in a maize population. *Crop Sci.* 22: 737–740.

_____, and R. H. Moll. 1972. Frequency changes of isozyme alleles in a selection experiment for grain yield in maize (Zea mays L.). *Crop Sci.* 12: 337–340.

_____, _____, M. M. Goodman, H. E. Schaffer, and B. S. Weir. 1980. Allozyme frequency changes associated with selection for increased grain yield in maize (Zea mays L.). *Genetics* 95: 225–236.

Tanksley, S. D. 1983. Molecular markers in plant breeding. *Plant Mol. Biol. Reporter* 1: 3–8.

_____, H. Medina-Filho, and C. M. Rick. 1982. Use of naturally-occurring enzyme variation to detect and map genes controlling quantitative traits in an interspecific backcross of tomato. *Heredity* 49: 11–25.

_____, and C. M. Rick. 1980. Isozyme linkage map of the tomato: applications in genetics and breeding. *Theor. Appl. Genet.* 57: 161–170.

Thoday, J. M. 1961. Location of polygenes. *Nature* 191: 368–370.

Vallejos, C. E., and S. D. Tanksley. 1983. Segregation of isozyme markers and cold tolerance in an interspecific backcross of tomato. *Theor. Appl. Genet.* 66: 241–247.

Weller, J. 1983. *Linkage relationships between quantitative trait loci in an interspecific cross of tomato (L. pimpinellifolium × L. esculentum).* Ph.D. thesis (English summary) Hebrew University of Jerusalem, Israel.

Wendel, J. F., M. M. Goodman, and C. W. Stuber. 1985. Mapping data for 34 isozyme loci currently being studied. *Maize Genet. Coop. News Letter* 59:90.

_____, _____, _____, and J. B. Beckett. 1988. New isozyme systems for maize (Zea mays L.) aconitate hydratase, adenylate kinase, NADH dehydrogenase, and shikimate dehydrogenase. *Biochem. Genet.* 26: 421–445.

_____, C. W. Stuber, M. D. Edwards, and M. M. Goodman. 1986. Duplicated chromosome segments in maize (Zea mays L.): further evidence from hexokinase isozymes. *Theor. Appl. Genet.* 72: 178–185.

CHAPTER 11

Bryophyte Isozymes: Systematic and Evolutionary Implications

Robert Wyatt
Ann Stoneburner
Ireneusz J. Odrzykoski*

Department of Botany, University of Georgia, Athens, GA 30602

Traditionally, bryophytes have been treated as a single division comprising three classes: Musci (mosses), Hepaticae (liverworts), and Anthocerotae (hornworts). The relative sizes of these groups are estimated as 700 genera and 10,000 species of mosses; 330 genera and 8,000 species of liverworts; and four genera and 360 species of hornworts (Schofield, 1985). Estimates of the total number of bryophyte species vary from 16,000 to 22,000 (Bold et al., 1986). Much of the unity of bryophytes as a group derives from their uniform possession of an alternation of generations unique among land plants. The sexual cycle involves a dominant, free-living, haploid gametophyte alternating with a reduced, dependent, diploid sporophyte (Fig. 11.1). Despite the elaborate specialization of its tissues in some groups, the sporophyte remains attached to, and nutritionally dependent on, the gametophyte throughout its lifetime. Bryophytes have successfully exploited a diversity of habitats in which their direct competition with seed plants is minimized (Anderson, 1980). According to Anderson (1980), this may explain why the number of bryophyte species is nearly twice that of pteridophytes. Morphologically and anatomically, bryophytes are considered structurally simple compared to vascular plants, but they have proven to be unexpectedly rich in chemical variation (Giannasi, 1978).

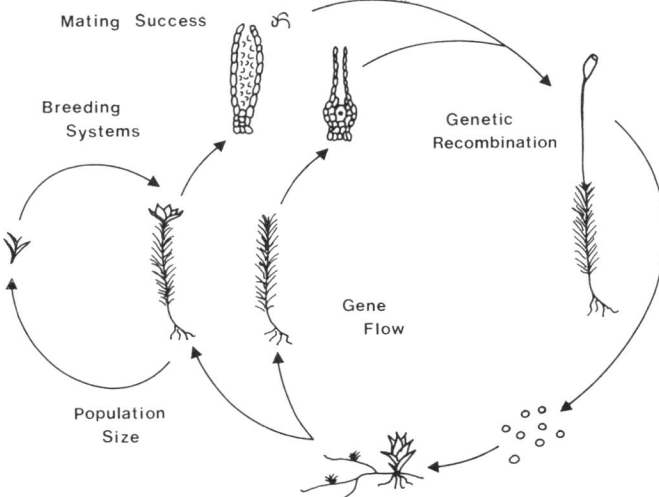

Fig. 11.1. Diagrammatic life cycle of a typical bryophyte, showing stages at which various processes may affect genetic variation. Both sexual and asexual reproduction are depicted. Effects of population size are expressed during the dominant, free-living haplophase. Breeding systems and mating success are important at the time of fertilization. Genetic recombination occurs during meiosis when spores are produced in the capsule. Gene flow has two components: sperm dispersal and spore dispersal. Not shown are the evolutionary processes of mutation and natural selection, which may operate at any stage of development of either generation.

*Permanent Address: Department of Genetics, Institute of Biology, Adam Mickiewicz University, Dabrowskiego St. 165, Poznan 60-594, Poland.

LEVELS OF ELECTROPHORETICALLY DETECTABLE GENETIC VARIATION

There has been much discussion in recent years of the "evolutionary potential" of bryophytes. The traditional view of these plants as evolutionary failures, lacking genetic variability, is based on the premise that dominance of a gametophytic generation precludes heterozygosity and permits selection to act directly on the haploid genotype (Anderson, 1963; Crum, 1972). The general inefficacy of sexual reproduction also is thought to lead to reduced levels of genetic variation (Anderson, 1963; Khanna, 1964; Schuster, 1966; Crum, 1972). Bryophytes rely on the availability of water for gamete dispersal, and sperm dispersal distances are short (Wyatt and Anderson, 1984). The 40% of mosses and 20% of liverworts that are bisexual are assumed to be predominantly self-fertilizing (Wyatt and Anderson, 1984). The second component of gene flow, effective dispersal distances of spores, is not known. Despite the production of hundreds of thousands of haploid spores by an individual sporophyte, most species migrate only over short distances and in a stepwise fashion. Bryophytes demonstrate restricted geographical distributions and species disjunctions that reflect tectonic and climatic events and are comparable to the patterns displayed by genera of vascular plants (Anderson, 1963; Crum, 1972; Zanten and Pocs, 1981; Schuster, 1983).

The prevalence of asexual reproduction among bryophytes has been cited as an additional factor reducing genetic variation within a species (Crum, 1972). Asexual reproduction can occur by growth and branching of individual plants, by regeneration of new gametophytes from fragments of haploid or diploid plants, or by the production of specialized propagules. The restriction of bryophytes to microhabitats with comparatively uniform conditions also may contribute to genetic uniformity (Anderson, 1963). According to Crum (1972), most species of mosses are distinct and unvarying; the size of genera in mosses is small and the number of varieties is few. Of the two major groups, liverworts are regarded as having even less genetic variability than mosses, due in part to their reduced capacity for sexual reproduction (Khanna, 1964; Schuster, 1966).

The prevailing view that bryophytes maintain little genetic variation has been seriously challenged in the past decade by evidence from electrophoretic studies. Three measures of intrapopulational variation that have been reported or could be calculated from reported data are given in Table 11.1: the percentage of loci polymorphic per population (P), the mean number of alleles per locus (A), and, from Nei (1973), the mean expected heterozygosity per locus (H_S). Although Nei (1973) preferred the term "gene diversity" rather than "heterozygosity" to describe genic variation in a nonrandomly mating population, for the comparative purposes of this review we will treat the terms interchangeably. Measures of genetic variability based on the analysis of only one enzyme system were not included in Table 11.1 (e.g., Krzakowa and Szweykowski, 1979) nor were those in which the genetic interpretations are questionable (e.g., Daniels, 1982, 1985a, 1985b). Because we calculated weighted averages by populations, values for some of the species are lower than reported in the literature.

For vascular plant species, the range of means reported for P, A, and H_S are 22.0–75.3%, 1.35–2.56, and 0.079–0.354, respectively (Hamrick, 1979). With the exception of *Plagiomnium insigne*, which exhibits low levels of genetic variability, values for moss populations fall within the ranges for vascular plants (Table 11.1). *Plagiomnium insigne*, confined to the coastal ranges of northwestern North America (Koponen, 1971), has the most restricted geographic distribution of the mosses listed in Table 11.1, but it is not known to differ significantly from *P. ciliare* or *P. ellipticum* in such life history variables as generation time, mode of reproduction, mating system, fecundity, chromosome number, or habitat. Wyatt et al. (1989a) sampled 14 enzyme loci from 13 populations of the dioecious moss *P. ciliare* from the southeastern United States. Gene diversity was much reduced in populations from disturbed, secondary forests in the Piedmont ($H_S = 0.04 \pm 0.02$) relative to those from undisturbed, primary forests in the Appalachian Mountains ($H_S = 0.11 \pm 0.01$). For the species as a whole, 71% of the loci surveyed were polymorphic. Spatial heterogeneity, perhaps multiple niche selection, was suggested to explain the observed variation (Wyatt et al., 1989a). *Plagiomnium ciliare* maintains about two-thirds as much variation as is found in *P. ellipticum*, based on comparisons of the percentages of polymorphic loci and proportions of expected heterozygosity (Table 11.1). The two species are similar in most life history characteristics, but *P. ellipticum* exhibits greater variation in

Table 11.1. Levels of genetic variability in species of bryophytes estimated by starch gel electrophoresis. Abbreviations: N_p, number of populations studied; N_e, number of enzymes studied; N_l, number of loci studied; P, percentage of loci polymorphic per population; A, mean number of alleles per locus per population; H_S, mean expected heterozygosity per locus per population

Species	N_p	N_e	N_l	P	A	H_S	Reference
Mosses							
Plagiomnium ciliare	13	9	14	31.1	1.40	0.079	Wyatt et al. (1989)
P. ellipticum	4	13	18	50.0	1.80	0.123	R. Wyatt et al. (unpubl. data)
P. insigne	4	13	18	16.7	1.17	0.065	R. Wyatt et al. (unpubl. data)
Plagiothecium curvifolium	8	—	20	48.0	1.90	0.190	Hofman (1988)
P. denticulatum	2	—	20	37.0	1.70	0.170	Hofman (1988)
P. ruthei	2	—	20	35.0	1.60	0.160	Hofman (1988)
P. latebricola	2	—	20	25.0	1.30	0.110	Hofman (1988)
P. nemorale	4	—	20	39.0	1.70	0.160	Hofman (1988)
P. undulatum	3	—	20	16.0	1.20	0.090	Hofman (1988)
Racopilum capense	1	—	7	29.0	1.29	0.069	Bramen (1986)
R. convolutaceum	2	—	8	25.0	1.43	0.102	Bramen (1986)
R. cuspidigerum	2	8	10	24.3	1.37	0.093	Vries et al. (1983)
R. cuspidigerum	2	—	8	50.0	1.71	0.242	Bramen (1986)
R. intermedium	1	—	7	43.0	1.43	0.093	Bramen (1986)
R. robustum	1	—	7	29.0	1.29	0.127	Bramen (1986)
R. spectabile	3	8	10	52.3	1.93	0.198	Vries et al. (1983)
R. spectabile	3	—	8	45.0	1.62	0.168	Bramen (1986)
R. strumiferum	3	—	8	56.0	1.57	0.180	Bramen (1986)
R. tomentosum	3	—	8	43.0	1.62	0.174	Bramen (1986)
Liverworts							
Conocephalum conicum (S)[a]	16	15	20	4.4	1.04	0.012	Odrzykoski (1986)
C. conicum (L)	24	15	20	8.5	1.09	0.025	Odrzykoski (1986)
C. conicum (A)	5	21	28	13.6	1.15	0.044	Odrzykoski (1986)
C. conicum (J)	2	6	11	82.7	2.81	0.167	Yamazaki (1984)
C. conicum (S or L)	—	8	—	—	—	0.026	Gliddon (1983)
Plagiochila asplenioides	20	13	18	2.5	1.02	0.008	Wachowiak (1986)
P. porelloides	8	13	18	0.0	1.00	0.000	Wachowiak (1986)
Pellia epiphylla (S)	6	13	13	9.1	1.14	0.026	Zielinski (1987)
P. borealis	12	13	13	5.9	1.07	0.024	Zielinski (1987)
P. neesiana	4	11	11	9.1	1.09	0.031	Zielinski (1987)
Riccia dictyospora (A)	19	7	8	11.0	1.30	0.024	Dewey (1989)
R. dictyospora (B)	11	7	8	20.0	1.24	0.076	Dewey (1989)
R. dictyospora (C)	2	7	8	19.0	1.20	0.072	Dewey (1989)

[a] Cryptic species (indicated in parentheses) are discussed in the text.

morphology and habitat and has a wider, circumboreal distribution (Koponen, 1971).

Species of *Racopilum* are the only tropical taxa that have been studied electrophoretically. Vries et al. (1983) surveyed two populations of *R. cuspidigerum* and three of *R. spectabile* from the Philippines and found high levels of polymorphism and gene diversity (Table 11.1). The more geographically widespread of the two species, *R. cuspidigerum*, has only about one-half the polymorphism and gene diversity found in *R. spectabile*. The maintenance of higher levels of genetic variation by geographically widespread species, as in *Plagiomnium ellipticum*, could be easily confounded by other life history variables. According to Vries et al. (1983), *R. cuspidigerum* also reproduces by specialized asexual propagules. Nevertheless, the levels of polymorphism and expected heterozygosity reported for both species of *Racopilum* fall within the ranges of values reported for vascular plants (Hamrick, 1979). Vries et al. (1983) suggested that variation at allozyme loci is maintained by their association with other genes selectively important in the evolution of these species. Bramen's (1986) recent survey of enzyme variation in eight taxa of Racopilum reported much higher values of polymorphism and gene diversity in R. cuspidigerum than those found by Vries et al. (1983; Table 11.1). Gene diversity statistics reported by Bramen (1986) for an additional six species of Racopilum (Table 11.1) generally fall in the lower range of values for vascular plants (Hamrick, 1979).

Other electrophoretic studies of mosses, not included in Table 11.1, also have reported levels of polymorphism similar to those of vascular plants. Cummins and Wyatt's (1981) study of the dioecious moss Atrichum angustatum compared 15 populations from three localities in eastern Texas. Enzyme polymorphisms in three systems were found among localities and within and among populations. Atrichum angustatum is a widespread dioecious species with a broad habitat range (Crum, 1972). According to Cummins and Wyatt (1981), these and other considerations, including high fecundity, suggest that the species should have a relatively high level of genetic variability. In a discussion of possible mechanisms maintaining genetic variation in haploid mosses, Cummins and Wyatt (1981) suggested close adaptation to microhabitats in which selective advantages vary over short distances.

Isozymes of EST, ACPH, and PER have been studied in the peat moss, Sphagnum (Daniels 1982, 1985a, b). Daniels (1982) found high levels of variation both within and among British populations of Sphagnum pulchrum, a moss that rarely reproduces sexually and which has a disjunctive circumboreal distribution. To explain the unexpectedly high level of variation, Daniels (1982) suggested that present-day populations represent relicts of a once widespread, sexually reproducing species. In a comparison of two British and two Finnish populations of Sphagnum recurvum var. mucronatum, a dioecious species that produces sporophytes regularly, Daniels (1985a) found only moderate levels of polymorphism and no relationship between environmentally induced morphotypes and isozyme banding patterns. Four populations of monoecious Sphagnum compactum from Finland and England showed moderately high levels of variation (Daniels, 1985b). He found that the population at the most microenvironmentally diverse site had the largest amount of polymorphism and related this to differential selection acting in response to specific local factors. There are problems with the methods used by Daniels (1982, 1985a, b) to study Sphagnum that complicate the interpretation of his results, including the large number of bands produced by polyacrylamide gel electrophoresis, the use of variable substrate enzymes, and the assumption that each position on the gel represents a locus that expresses either an active or a null allele.

Shaw et al. (1987) electrophoresed plants of Climacium americanum, a dioecious moss that rarely produces sporophytes, from three sites in the Piedmont of North Carolina. For six of the eight loci surveyed, they found genetic variation within and among populations. Plants from each site were separable into morphotypes corresponding to either var. kindbergii or var. americanum. Shaw et al. (1987) reported enzymatic differentiation between the morphotypes and suggested that linkage disequilibrium between genes coding for electrophoretically detectable enzymes and those controlling morphology developed from an initial founder effect and the absence of gene exchange between the varieties.

Szweykowski and Zielinski (1983) found four different PER phenotypes in Polish populations of the widespread moss Plagiothecium undulatum. At the most intensively sampled site, each population appeared genetically homogeneous; however, adjacent populations, separated by several hundred meters, were fixed for different PER phenotypes. Szweykowski and Zielinski (1983) concluded that the different populations had arisen independently from separate diaspores. A survey of variation at 20 enzyme loci in six species of Plagiothecium from the Netherlands (Hofman, 1988) reported values of P, A, and H_S (Table 11.1) comparable to those for species of vascular plants (Hamrick, 1979), except for P. undulatum. The populations of P. undulatum that were sampled were asexually reproducing, and both P and A were lower than reported for vascular plants, although H_S fell within the usual range (Hamrick, 1979).

Studies of enzyme polymorphisms in liverworts suggest that levels of genetic variability in natural populations of these plants are greater than traditionally expected. Those liverworts for which measures of genetic variability are available (Table 11.1), however, typically show much lower values than those of mosses. One of the best studied is the thallose liverwort *Conocephalum conicum* in Europe (Krzakowa, 1977; Szweykowski and Krzakowa, 1979; Odrzykoski et al., 1981; Szweykowski et al., 1981a; Gliddon, 1983; Odrzykoski, 1986). Generally regarded as morphologically uniform throughout its wide distribution in temperate areas of the Northern Hemisphere, *C. conicum* is genetically quite diverse and composed of five distinct races (Table 11.1). The levels of polymorphism and gene diversity in the European "small" ("S") and "large" ("L") forms and the "American" ("A") form are among the lowest reported for liverworts (Table 11.1; Odrzykoski, 1986). A conflicting view of variation in *C. conicum* was described by Yamazaki (1981, 1984) for plants from Japan. He reported much higher levels of polymorphic loci and expected heterozygosity than those found in European populations (Table 11.1). In fact, the high values he reported were comparable to those reported for trees and woody shrubs (Hamrick, 1979). Yamazaki (1981, 1984) argued that, because of their haploid-dominant life cycle, the amount of genetic polymorphism in bryophytes should be much less than in diploid organisms under most selection models. He suggested that the model of selective neutrality of enzyme polymorphisms best explains the high genetic polymorphism in haploid *C. conicum*.

It is likely that Yamazaki's (1981, 1984) data overestimate the levels of genetic variability in *C. conicum*, because he compared only two populations and three of the six enzymes he assayed are often highly variable in plants (SOD, PER, and EST). It is also possible that the values for *C. conicum* in Europe (Odrzykoski, 1986) are underestimates of the variation of the species, because the majority of sampled populations were from glaciated regions that may be genetically depauperate (see earlier discussion of *Plagiomnium ciliare*). The best indication of the amount of enzyme polymorphism and gene diversity in *C. conicum* may be those values reported by Odrzykoski (1986) for the "A-form" (Table 11.1). The American populations, from gorges in the Appalachian Mountains, occupy relatively ancient and undisturbed sites and show considerably greater gene diversity than samples from other areas.

Krzakowa and Szweykowski (1977b) found that the PER pattern of *Plagiochila asplenioides* differed from that of *P. porelloides*. Furthermore, Krzakowa and Szweykowski (1979) detected PER variation within and among five populations of *P. asplenioides* in Poland comparable to that reported for angiosperms. A contrasting view of variation in *Plagiochila* is presented by Wachowiak (1986; Table 11.1), who found very low levels of polymorphism and expected heterozygosity in these same two species. Wachowiak's data appear to be more representative, because more enzymes and populations were studied. These values, however, may underestimate overall species variability because the sampled populations again came only from glaciated regions of Poland.

The liverwort *Pellia* also has been studied extensively (Krzakowa, 1977, 1981; Krzakowa and Szweykowski, 1977a; Szweykowski et al., 1981b; Zielinski, 1984a, b, 1986; Zielinski et al., 1985). Zielinski (1987) reported values of P, A, and H_S in species of *Pellia* (Table 11.1) that are comparable to those found in *Conocephalum* by Odrzykoski (1986). These same measures of enzyme variability in moss populations are typically much greater than those for liverwort populations. It is difficult to detect a trend among life history characteristics of mosses and liverworts that can account for their apparent differences in genetic variability. Wyatt (1985a) noted a number of similarities between the liverwort *Conocephalum* and the moss *Plagiomnium*, including generation time, mating system, fecundity, and habitat, that preclude any simple explanation of the large differences in genetic variation.

Dewey (1989) surveyed eight enzyme loci in the monoecious liverwort *Riccia dictyospora* from the southeastern United States and reported a pattern of variation similar to that described for *Conocephalum conicum* (Szweykowski et al., 1981a; Odrzykoski, 1986) and *Pellia epiphylla* (Zielinski, 1984a). Three distinct sets of genotypes were found among 38 localities. Linkage tests provided no evidence of disequilibrium, and Dewey (1989) suggested that internal reproductive barriers prevented gene exchange among these sympatric sibling species. The mean proportion of loci polymorphic and mean expected heterozygosity for these incipient species of *Riccia* (Table 11.1) are somewhat higher than those reported for *Conocephalum* in Europe, *Plagiochila*, and *Pellia*, but considerably less than those found for *Conocephalum* in Japan and for most species of mosses. In *Riccia*, enzyme polymorphism and expected heterozygosity are comparable to levels reported for biennials (*P*

= 15.78%, H_S = 0.060: Hamrick et al., 1979); mosses typically have the higher values associated with annuals (P = 39.47%, H_S = 0.132: Hamrick et al., 1979) and perennials (P = 28.09–65.77%, H_S = 0.123–0.267: Hamrick et al., 1979). In a discussion of the maintenance and significance of allozyme variation, Dewey (1989) noted that "evolution in these organisms may center about physiological, rather than morphological, adaptations and divergence."

EVOLUTIONARY RELATIONSHIPS AND SYSTEMATICS

As noted by Anderson (1974), "mosses and liverworts are regarded by most phylogenists as evolutionary misfits or failures." Most bryologists agree further that bryophytes are an ancient group in which rates of evolution and speciation are extremely slow (Anderson, 1963, 1980; Schuster, 1966; Crum, 1972). The arguments in favor of this view are those enumerated earlier with respect to predictions of low genetic variation in bryophyte populations.

Given this degree of unanimity regarding evolutionary processes in bryophytes, it is surprising that bryologists do not agree on the issue of species distinctness. Noting that most species of mosses are believed to have originated long ago, Crum (1972) characterized these plants as "unchanging, unmoving sphinxes of the past" and suggested that moss species are genetically distinct and unvarying. Thus, electrophoretic variation should be minimal within moss species, but genetic distances between them should be large. Furthermore, Crum (1972) argued that the liverworts are a smaller, more stable group that has even less potential for change than the mosses. Hence, we should expect even less electrophoretically detectable genetic variation in hepatics.

Taking a middle-of-the-road viewpoint, Anderson (1964) noted that species problems in mosses parallel those in other groups of plants: there is the same lack of discontinuities, the inevitable morphological variations, the puzzles of subspecific categories, and the problems of vicariousness, polyploidy, hybridization, and disjunct distributions. Schuster (1966) went even further in suggesting that species problems in liverworts are more vexatious than those in vascular plants. The extreme adaptive plasticity of liverworts may exist as a compensation for genetic rigidity; most of the problems of species intergradation may be due to phenotypic plasticity. Thus, Schuster's (1966) model of hepatic evolution predicts less genetic variation not only within, but presumably among, species than is commonly observed in flowering plants. He also agreed with Khanna (1964) and Crum (1972) that speciation occurs more slowly in hepatics than in mosses, but cited as exceptions certain genera of epiphytic, leafy liverworts.

Bryologists are in general agreement also that the adoption of "geographical species" has led to a senseless multiplication of species (Schuster, 1966). The fact that mosses are more broadly distributed than higher plants of similar disjunctions and tend to occur as species disjuncts rather than as congeneric species pairs may be due to the extreme antiquity and slow rate of evolution in the group (Crum, 1972). Anderson (1963) noted that long-continued geographical isolation of species has produced few, if any, changes and applauded the modern tendency to reunite proposed species that are recognized chiefly on the basis of geography. In terms of electrophoretically detectable genetic variation, therefore, we should expect even continentally disjunct species to be little differentiated throughout their ranges. Furthermore, electrophoresis might reveal vicariant species to be genetically similar and, therefore, prime candidates for "lumping."

Genetic Similarity Among Taxa

The electrophoretic evidence from mosses and liverworts makes it abundantly clear that bryophyte species usually are *not* genetically uniform. As discussed above, it appears that morphological characters used to distinguish species of bryophytes are more conservative than are biochemical characters. There is a broad range of genetic variability for different mosses and liverworts, but most species appear comparable to higher plants (Wyatt, 1985a).

Electrophoresis has also provided an unequivocal answer to the issue of genetic distinctiveness of bryophyte species: they are as distinct, or more distinct, than species of higher plants investigated using the same techniques. To express genetic similarity between populations or species, Nei's (1972) genetic identity (I) is widely used. Gottlieb (1981) calculated the mean genetic identity for 21 pairs of congeneric plant species: $\bar{I} = 0.67 \pm 0.04$ (mean ± standard error). Significantly, this represents a large decrease from the mean genetic identity between populations of a single species. For 13 self-fertilizing species, Gottlieb reported $\bar{I} = 0.975$, whereas for 14 outcrossers, $\bar{I} = 0.956$. Similarly, Crawford (1983) reported high genetic similarities between conspecific populations with genetic identities nearly always > 0.900. His tabulation of mean identities for 16 congeneric species studied since Gottlieb's (1981) summary showed $\bar{I} = 0.789$, again supporting the view that populations of different species are considerably more differentiated genetically than are conspecific populations.

Among the mosses, R. Wyatt et al. (unpublished data) found that populations of two species of *Plagiomnium* section *Rosulata*, *P. ellipticum* and *P. insigne*, had a mean genetic identity of 0.612. This value is somewhat lower than the averages reported by Gottlieb (1981) or Crawford (1983) for flowering plants, but falls well within the wide range detected by previous studies. It is possible that the value for these two species is higher than it might have been had populations from outside the United States been sampled for the circumboreal *P. ellipticum*. Nevertheless, the observation that these two species differ on average at nearly half of any 100 loci will surprise many bryophyte taxonomists, who regard them as very closely related (Koponen, 1971). Within the species, $I = 0.960$ for *P. ellipticum* and $I = 0.927$ for *P. insigne*. These values are close to the mean for outcrossing flowering plants.

Vries et al. (1983) calculated a mean genetic identity of 0.511 between *Racopilum spectabile* and *R. cuspidigerum*, two tropical mosses from the Philippines. Within *R. spectabile*, $I = 0.870$, whereas among populations of *R. cuspidigerum*, $I = 0.877$. Again, these values suggest that populations of these mosses and the species themselves are more strongly differentiated than most bryologists would expect. The only other moss for which data on genetic identities are available is *Sphagnum*, populations of which appear to be extremely distinct from each other electrophoretically. Genetic identities among populations of one variety (var. *mucronatum*) of *S. recurvum* ranged from $I = 0.430$ to $I = 0.864$ (Daniels, 1985a), values closer to the average expected from comparisons between different species. Several low intraspecific genetic identities were also reported for *S. pulchrum* ($I = 0.657$–0.967; Daniels, 1982) and *S. compactum* ($I = 0.856$–0.962; Daniels, 1985b). These genetic identities, however, are probably underestimates because of Daniels's (1982, 1985a, b) questionable genetic interpretations. He treated each position on a gel as a locus that either expresses activity (an active allele) or fails to express activity (a null allele). Nevertheless, there do appear to be large differences between conspecific populations.

Dewey (1986) compared 16 species of the liverwort *Riccia* on the basis of MDH and PER electrophoretic patterns. He found species-specific banding patterns for all 16. Indeed, for PER alone, of the 16 species only *R. bifurca*, *R. hirta*, and *R. eldeeniae* lacked diagnostic profiles. In a more comprehensive survey of *R. dictyospora*, Dewey (1989) detected three morphologically similar, but electrophoretically distinct, "sibling species." Based on eight loci, genetic identities between these taxa ranged from $I = 0.211$ to 0.454. Genetic identities between conspecific populations ranged from $I = 0.856$ to 0.939. Therefore, although conspecific values fall within the range reported for flowering plants, some of the interspecific comparisons yield genetic identities lower than those reported previously for any congeneric pair of angiosperms. This is especially surprising in that Dewey's (1987) work is the first to suggest that any heterogeneity exists within *R. dictyospora sensu lato*.

Similar situations have been discovered in the two other genera of thallose liverworts that have been analyzed by electrophoresis, *Conocephalum* and *Pellia*. Odrzykoski (1987) reported that *C. conicum* actually consists of five sibling species. In Europe two of these, called the "large" or "L-form" and the "small" or "S-form" can occur sympatrically, yet they do not interbreed. The forms are sharply differentiated with genetic identities averaging 0.40. In the case of *Conocephalum*, previous studies of variation in flavonoid profiles and other characters had already established a high degree of differentiation within an apparently morphologically homogeneous set of populations. In *Pellia epiphylla*, Zielinski (1987) found that northern and southern races exist that differ at nearly all of the loci surveyed. Again, this sharp genetic differentiation was totally unexpected in a seemingly morphologically uniform taxon.

Finally, Shaw et al. (1987) used electrophoretic data to argue that *Climacium kindbergii*, viewed by most bryologists merely as a variant of *C. americanum* induced by aquatic environments, is genetically distinct and should probably be distinguished either at the varietal or specific level. They showed significant associations between allelic and morphological variation for several loci, including *Pgi* and *Pgm*, both of which were absolutely diagnostic for the two species.

In summary, it appears that bryophyte species generally are more distinct genetically than most bryologists would have predicted. Genetic identities based on electrophoretic data are lower, on average, between moss and liverwort species than between angiosperm species. In this respect bryophytes are similar to homosporous pteridophytes, congeneric species of which typically are strongly differentiated (Haufler, 1987; see the chapter in this volume by Soltis and Soltis). Furthermore, contrary to the view of most bryologists (e.g., Crum, 1972), moss species are not genetically uniform, but rather show at least as much differentiation among conspecific populations as do flowering plants.

Data for liverworts suggest that Schuster's (1966) view that most of the morphological variation observed is due to phenotypic plasticity may be in error. Electrophoretic analyses show large genetic differences among populations. Indeed, in all three species studied in depth, intensive sampling has detected genetically well-differentiated sibling species. It does appear, however, that liverworts are less genetically variable than mosses, when the variation is apportioned by sibling species. These data therefore fit Schuster's (1966), Khanna's (1964), and Crum's (1972) predictions that mosses should show more genetic variation than liverworts.

A final question relating to species concepts in bryophytes is the distinctness of "geographical" species. As noted earlier, Anderson (1963), Schuster (1966), and Crum (1972) have deplored the recognition of such taxa solely on the basis of geographical separation on different continents. The evidence from electrophoretic surveys suggests, however, that such disjunct sets of populations may be highly differentiated genetically. For example, Odrzykoski (1986) found that *Conocephalum conicum*, a liverwort ordinarily treated by bryologists as a uniform north temperate species, actually consists of five strongly differentiated races. Two of these occur in Europe, two in North America, and one in Japan. Genetic distances between these "forms" are as large in some cases as are commonly seen between genera of vascular plants (e.g., Rieseberg and Soltis, 1987). It would seem unwise, therefore, to assume that all continentally disjunct bryophyte species are uniform across their ranges. This issue has been discussed previously by Wyatt (1985a).

There are several problems associated with these conclusions about genetic distinctness of bryophytes that render them tentative at best. Relatively few species have been studied: we are extrapolating from data for fewer than 10 species of mosses and liverworts. Furthermore, in only a few cases have adequate numbers of populations and individuals been surveyed, and typically fewer than 10 loci have been scored. Finally, only a very restricted range of taxa has been sampled. This is especially true of liverworts, in which only temperate, thallose species have been examined. Schuster (1966) predicted higher levels of genetic variation for epiphytic, leafy hepatics from the tropics, where sexual reproduction is common. It will be interesting to compare data from such species to see if they more closely approximate mosses in terms of genetic differentiation.

Polyploidy and Hybridization

In vascular plants, electrophoretic data have proved very useful in addressing questions related to polyploidy. Among liverworts and hornworts, there is little variation in basic chromosome numbers, and relatively few cases of polyploidy have been reported (Smith, 1978; Newton, 1983). For nearly 64% of the species of liverworts, the basic chromosome number is $n = 9$, and only about 12% of the taxa are polyploid (Newton, 1983). For the hornworts, counts of $n = 5$ and 6 generally are reported (Smith, 1978). Some authors suggest that modern liverworts are ancient polyploids derived from taxa with a base number of $x = 4$ or 5 (Schuster, 1966; Newton, 1983). Others believe that the base number in hepatics is $x = 8–10$ (Smith, 1978; Crosby, 1980).

The diversity of chromosome numbers detected in mosses contrasts sharply with the uniformity observed in liverworts, and polyploidy apparently has played a considerable role in the evolution of the group. For *Sphagnum*, the majority of counts are $n = 19$ or 38 and for *Andreaea*, 10 or 11. Within the majority of mosses, however, the haploid chromosome number varies considerably within and among the three major peristomate groups. Mosses with sporophytes bearing a single articulated peristome (Haplolepideae) have a high frequency of the numbers $n = 12–16$; those with a double articulated peristome (Diplolepideae) frequently have $n = 6$, 10, and 11; and those with a solid, nonarticulated peristome (Nematodonteae) generally have $n = 7$, 8, 9, and 14 (Smith, 1978; Anderson, 1980). For the majority of orders and families of mosses, a base number of $x = 6$ or 7 is most probable with primary, possibly ancient, polyploidy varying from 17–98%, and secondary polyploidy, stemming from haploid numbers of $n = 10$–15, ranging from 0–68%.

Electrophoretic evidence is potentially useful in establishing the existence of polyploidy and in determining its nature (i.e., whether autopolyploidy or allopolyploidy is involved). For example, Haufler and Soltis (1986) used Gottlieb's (1981, 1982) model, which assumes that isozyme numbers are conserved, to argue that homosporous ferns with high chromosome numbers behave genetically as diploids. Similarly, Soltis (1986) reported diploid expression of isozymes in the fern ally *Equisetum*, species of which have uniformly high chromosome numbers of $2n = 216$. Haufler and Soltis (1986) and Haufler (1987) suggested that the evolution of taxa characterized by high chromosome numbers and isozymic diploidy, such as homosporous pteridophytes, could occur through several cycles of polyploidy followed by chromosomal diploidization, gene silencing, and extinction of progenitor diploids. Soltis and Soltis (chapter in this volume) outlined the equally likely alternative that pteridophytes are not of polyploid origin, but were initiated with high chromosome numbers, perhaps via chromosomal fission.

A comprehensive overview of isozyme expression in bryophytes is not possible at this time, because so few taxa have been investigated in sufficient detail. Nevertheless, it is noteworthy that all studies to date have reported simple haploid patterns of isozymic expression, even in taxa which, on the basis of chromosome number alone, would be classified as "polyploid." For example, Yamazaki (1984) found no evidence of duplicate gene expression in *Conocephalum conicum* ($n = 9$), which would be considered a polyploid on the basis of Schuster's (1966) base number of $n = 4$ or 5 for liverworts. Yamazaki (1984) suggested that the doubling event was so ancient that haploid isozymic expression has been restored. Of 2,000 individuals he screened, only three showed apparent heterozygous expression. Presumably, as in Dewey's (1989) observation of rare heterozygous banding in *Riccia dictyospora,* even these cases merely involved mixtures of tissue from genetically distinct thalli and not true expression of additional isozymes.

Our electrophoretic studies of *Plagiomnium* section *Rosulata* have uncovered some enzymes for which additional isozymes appear to be present (Wyatt et al., 1989a). These include triosephosphate isomerase, isocitrate dehydrogenase, phosphoglucomutase, glutamate dehydrogenase, and 6-phosphogluconate dehydrogenase, enzymes that typically exist in plants in only one or two isozymic forms (Gottlieb, 1982; see the chapter by Weeden and Wendel in this volume). The complex banding patterns observed in *P. ciliare, P. ellipticum,* and *P. insigne* (all $n = 6$) may result from some other phenomenon, such as post-translational modification of enzymes. Because a substantial proportion of the loci surveyed are involved, it appears that a simple explanation of single gene duplication is less likely than that the entire genome may have been doubled in the distant past. This problem requires further study. Unfortunately, no genetic data documenting the inheritance of isozymic patterns exist at present for *Plagiomnium*.

Bryologists have assumed that the 79% of mosses, 11% of liverworts, and 2% of hornworts presumed to have polyploid chromosome numbers are the result of autopolyploidy (Smith, 1978; Anderson, 1980). Furthermore, most bryologists have assumed that apospory is the predominant mechanism by which polyploids have arisen (Wyatt and Anderson, 1984), although Smith (1978) argued that diplospory, rather than apospory, is more likely. Diploid gametophytes of mosses are reportedly easily produced by regeneration of wounded sporophytic tissue in culture (Wettstein, 1932). On the other hand, it is apparently difficult to produce diploid gametophytes aposporously from sporophytic tissue of liverworts, partly because the sporophytes are evanescent and lack chlorophyll. Nevertheless, diploid gametophytes have been successfully produced by apospory in a few genera, including *Marchantia* (Burgeff, 1937).

It is somewhat ironic, therefore, that the best electrophoretically documented case of autopolyploidy in bryophytes involves the thallose liverwort *Pellia*. Hepaticologists were divided in their views regarding the status of *P. borealis* ($n = 18$), with some considering it an autopolyploid of *P. epiphylla* ($n = 9$) and others speculating that it might represent an allopolyploid of *P. epiphylla* and an unknown second species. Electrophoretic analysis revealed extensive homology between the genomes of P. borealis and P. epiphylla from southern Poland (Zielinski, 1984b; Zielinski et al., 1985), supporting an autopolyploid origin. These taxa were identical at five of six loci surveyed and, furthermore, differed at all six loci from northern populations of P. epiphylla.

In contrast to other land plants, there has been only one report of allopolyploidy in bryophytes and scant discussion even of the possibility of its occurrence. Based on evidence from cytology, morphology, and ecology, Khanna (1960) suggested that Weissia exserta ($n = 26$) might be an allodiploid derived from a cross between W. (Astomum) crispa ($n = 13$) and W. controversa ($n = 13$). None of the irregularities in spore formation usually seen in other Weissia-Astomum hybrids were detected, and Khanna (1960) believed the spores to be germinable. Unfortunately, the spores were not tested for germination, and the hybrid has not been synthesized. Anderson and Lemmon (1972) and Anderson (1980) questioned the allopolyploid nature of W. exserta, noting that the evidence is entirely circumstantial.

Numerous bryophyte genera contain "species pairs," in which a diploid, monoecious species is paired with its haploid, dioecious progenitor (Lowry, 1948; Yano, 1957a, b; Schuster, 1966). The moss family Mniaceae, for example, includes at least four such species pairs (Lowry, 1948; Koponen and Nilsson, 1977; Wyatt, 1985b). Species pairs have also been postulated in Atrichum (Lowry, 1954) and Fissidens (Smith and Newton, 1968). Evidence in most of these cases is entirely circumstantial, based on morphological similarities between haploids and their presumed diploid derivatives.

Within Plagiomnium, section Rosulata includes one polyploid species (P. medium; $n = 12$) and six haploid species (Koponen, 1971). On the basis of morphology, Lowry (1948) concluded that P. ellipticum ($n = 6$) was the progenitor of autopolyploid P. medium. Koponen (1971), however, stressed different characters and argued that P. insigne ($n = 6$) was the only possible progenitor. Because the morphological evidence was ambiguous, Wyatt et al. (1988) used electrophoretic evidence to resolve the origin of P. medium.

Variation was scored at 18 enzyme loci, of which nine contributed no useful information regarding interspecific relationships because they were monomorphic or nearly so (Table 11.2). The other nine included alleles present in only one of the two haploids, P. ellipticum or P. insigne, which therefore could be used as genetic markers. Surprisingly, P. medium was discovered to show fixed heterozygosity at several loci, suggesting immediately that the species might be an allopolyploid. When species are monomorphic for alternative alleles, their allopolyploid derivatives display fixed heterozygosity because of nonsegregation of nonhomologous chromosomes (e.g., Roose and Gottlieb, 1976; Werth et al., 1985a, b). Furthermore, when one or both of the progenitors is polymorphic, different fixed heterozygous genotypes of the allopolyploid may be produced. The existence of such variability constitutes a strong case for multiple origins of the derivative species.

As described earlier, P. ellipticum and P. insigne are very distinct electrophoretically. Our data show that P. medium displays several different combinations of alleles, one in each case originating from P. ellipticum and one from P. insigne (Table 11.2). The simplest case is that of a locus such as Pgm-2, in which the haploid progenitors are characterized by different alleles (Fig. 11.2a). The allopolyploid P. medium expresses both alleles, producing a fixed heterozygous pattern (Fig. 11.2a). Hk-1 shows a similar pattern of variation (Fig. 11.2b). At loci such as Per-1, a single species-specific allele is contributed by P. insigne to the allopolyploid P. medium. The contribution of the polymorphic P. ellipticum, however, is variable (Fig. 11.2c). The existence of different Per-1 genotypes of P. medium is strong evidence for multiple origins of this allopolyploid. Other loci, including Aco-2 (Fig. 11.2d) and Tpi-1 (Fig 11.2e), further illustrate this pattern.

The discovery of allopolyploidy has wide-ranging implications for bryophyte phylogeny and evolutionary biology. This important mode of speciation in other land plants has only rarely been mentioned (e.g., Khanna, 1960; Anderson and Bryan, 1956) and never documented in mosses or liverworts. The data regarding multiple origins of P. medium join a growing body of literature that suggests that allopolyploids may originate independently numerous times and in different places (Roose and Gottlieb, 1976; Werth et al., 1985a, b). This implies that gene flow from progenitor species into derivative allopolyploid gene pools is possible over long time spans. Electrophoretic studies seem likely to reveal additional examples of allopolyploidy in bryophytes.

Table 11.2. Allele frequencies for 18 enzyme loci of the allopolyploid moss Plagiomnium medium and its two haploid progenitors, P. ellipticum and P. insigne. Arrows designate the presumed source of particular alleles. Loci: Aco, aconitase; Ald, aldolase; Dia, diapharase; G3p, glyceraldehyde-3-phosphate dehydrogenase; Hk, hexokinase; Idh, isocitrate dehydrogenase; Mdh, malate dehydrogenase; Per, peroxidase; Pgi, phosphoglucose isomerase; Pgm, phosphoglucomutase; and Tpi, triosephosphate isomerase

Locus	Allele	P. ellipticum ($\bar{N}$ = 112)[a]		P. medium ($\bar{N}$ = 292)		P. insigne ($\bar{N}$ = 90)
Aco-1	a	0.991	→	0.483		—
	b	—		0.500	←	1.000
	c	0.009	→	0.010		—
	d	—	→	0.007		—
Aco-2	a	—		0.500	←	1.000
	b	0.111	→	0.146		—
	c	0.889	→	0.354		—
Ald-1	a	1.000		1.000		1.000
Dia-1	a	0.714		1.000		1.000
	b	0.286		—		—
Dia-2	a	1.000		1.000		1.000
G3p-1	a	1.000		1.000		1.000
G3p-2	a	1.000		0.983		1.000
	b	—		0.017		—
Hk-1	a	1.000	→	0.500		—
	b	—		0.500	←	1.000
Idh-1	a	0.974		1.000		1.000
	b	0.026		—		—
Mdh-1	a	1.000	→	0.500		—
	b	—		0.500	←	1.000
Mdh-2	a	0.966		0.910		0.372
	b	0.034		—		—
	c	—		0.090	←	0.628
Mdh-3	a	1.000		1.000		1.000
Per-1	a	0.530	→	0.102		—
	b	0.385	→	0.398		—
	c	—		0.500	←	1.000
	d	0.085		—		—
Pgi-1	a	0.991		1.000		1.000
	b	0.009		—		—
Pgm-1	a	0.983	→	0.500		—
	b	0.017		0.242	←	0.149
	c	—		0.236	←	0.851
	d	—		0.022	←	—
Pgm-2	a	0.846	→	0.261		0.340
	b	0.136		—		—
	c	—		0.500	←	0.660
	d	0.009	→	0.189		—
	e	0.009	→	0.050		—
Tpi-1	a	0.103	→	0.040		—
	b	0.897		0.960		1.000
Tpi-2	a	0.205		—		—
	b	0.761		1.000		1.000
	c	0.034		—		—

[a] Mean number of haploid genomes screened per locus.

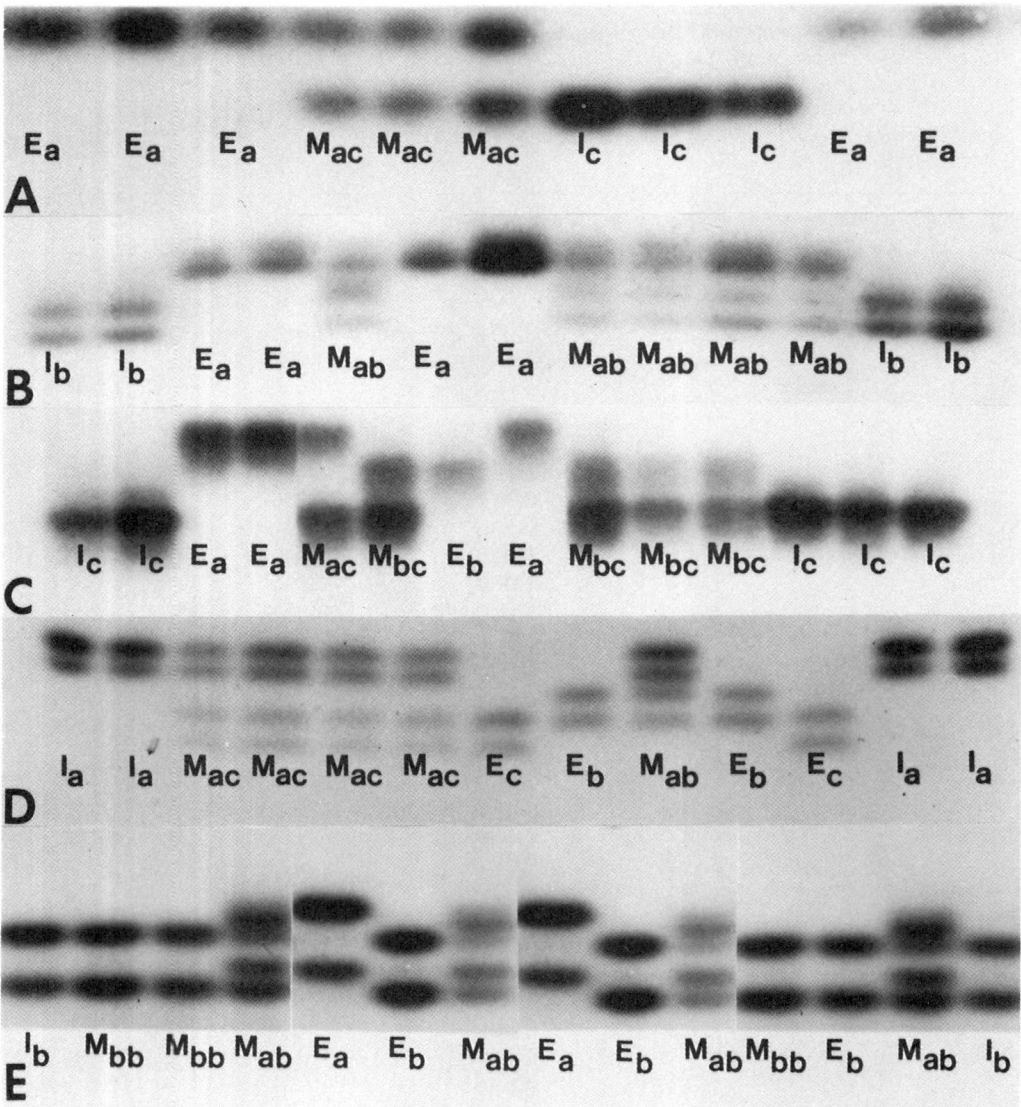

Fig. 11.2. Starch gels illustrating electrophoretic banding of individual gametophytic plants of the allopolyploid moss *Plagiomnium medium* (M) and its two haploid progenitors, *P. ellipticum* (E) and *P. insigne* (I), for five enzyme loci. Allelic designations match those in Table 11.2, but note that not all allelic variants appear on these gels. Subscripts to species designations indicate allelic combinations. Loci: A, *Pgm-2*; B, *Hk-1*; C, *Per-1*; D, *Aco-2*; and E, *Tpi-1*.

Hybridization appears to be rare in bryophytes. In contrast to vascular plants, in which hybridization is frequent and widespread, there are few reports of interspecific crosses in mosses (Anderson, 1980) and none in liverworts (Crundwell, 1970). This probably is due in part to spatial isolation of natural populations. Gamete dispersal distances are very short (Wyatt, 1982), so that even when species have similar ecological tolerances, they seldom occur sufficiently close together to interbreed. Anderson (1980) pointed out that most cases of hybridization involve mosses that grow in disturbed habitats where spores and/or gametophores can become intermixed.

Another reason why reports of hybrid mosses may be rare is that the products of interspecific crosses may not be recognized (Wyatt and Stoneburner, 1984). Nearly all of the well-documented hybrids are intergeneric and involve one parent with cleistocarpous capsules and another with stegocarpous capsules. Hybrids between such distinctly different plants are easily recognized. Gametophytes derived from hybrid sporophytes are possible and would be expected to show genetic segregation. Such hybrid gametophytes have been reported, but Anderson (1980) reviewed all of these and rejected each as based on insufficient evidence. Nevertheless, it seems likely that hybridization is more widespread in mosses than is presently believed but that it is often difficult to detect morphologically (Smith, 1979).

It is, therefore, reasonable to expect that electrophoretic analyses of natural populations will reveal additional cases of interspecific hybridization in bryophytes. The discovery that *Plagiomnium medium* is an allodiploid moss clearly indicates interspecific hybridization between *P. ellipticum* and *P. insigne*. Furthermore, this case suggests that hybridization may, at least sometimes, be of evolutionary importance. All previously reported instances of hybridization have involved the production of moss sporophytes whose spores are inviable (Wyatt and Stoneburner, 1984).

There have been only two other studies that used electrophoretic data to evaluate the extent of cross-fertilization between different taxa. I. Odrzykoski (1987) analyzed sporophytes collected from mixed colonies of the "L-" and "S-forms" of *Conocephalum conicum* in Poland. He found no evidence of any cross-fertilization, despite the fact that thalli of the two races were often completely intermixed. Similarly, Zielinski's (1987) observations of two races of *Pellia epiphylla* suggest that the two do not interbreed in Poland. Additional studies of sympatric taxa are sorely needed to determine if bryophytes are indeed very different from other land plants in having exceptionally low frequencies of hybridization. Among the pteridophytes, a group of land plants that have many reproductive and genetic features in common with bryophytes, hybridization is extremely common (Haufler, 1987).

BREEDING SYSTEMS AND POPULATION STRUCTURE

Among land plants, the breeding systems of bryophytes are most similar to those of pteridophytes. Both differ from seed plants in that the reproductive mechanisms that determine genetic diversity reside in a free-living, haploid gametophyte. In addition, because sperm are motile in both groups, gamete dispersal, fertilization, and hence sexual reproduction depend upon the availability of water. Bryophytes and pteridophytes are also similar in that self-fertilization results in a completely homozygous zygote. Finally, the two share a propensity for some species to reproduce asexually. Indeed, Anderson (1963) argued that reproduction in mosses is "almost entirely other than by spores." This regression of sexuality has been cited frequently as a major reason why bryophyte populations should be genetically uniform (Cummins and Wyatt, 1981).

Interestingly, electrophoretic data do not bear out these predictions (see above). It appears, therefore, that the importance of asexual reproduction has been overestimated or that genetic uniformity is not a necessary consequence of high levels of asexual reproduction. Lyman and Ellstrand (1984) found, for example, that populations of *Taraxacum officinale*, a species that habitually produces apomictic seeds, showed electrophoretically detectable genetic variation. In fact, if preexisting genotypes reproduce asexually at equal rates, one would expect no erosion of the original genetic variability that was present. Longton (1976) suggested that sexual reproduction in bryophytes is actually more common than many bryologists believe, although his recent observations of spore production, germination, and growth in field populations do not support this view (Miles and Longton, 1987). In any event, the electrophoretic evidence discussed earlier very clearly demonstrates that liverwort and especially moss populations are not genetically uniform.

Bryophytes differ strongly from pteridophytes in that gametophytes of about 57% of mosses and 68% of liverworts are unisexual (i.e., the species are dioecious) (Wyatt and Anderson, 1984; Wyatt, 1985c). Among the ferns and fern allies, nearly all species possess potentially bisexual gametophytes. In terms of electrophoretically detectable genetic variation, we expect populations of dioecious bryophytes, which are incapable of self-fertilization, to be more variable than populations of self-fertilizing monoecious or synoecious species. In fact, as noted earlier, self-fertilization in bryophytes results in a completely homozygous sporophyte. Spores produced on such plants, even though by normal meiosis, are genetically equivalent to asexual propagules of the gametophytic parent.

Unfortunately, electrophoretic data are lacking to test the prediction of greater variation in dioecious species. Studies of congeneric species that differ in sexuality are complicated by concomitant polyploidy. For example, synoecious *P. medium* is more variable than its dioecious congeners, *P. ellipticum* and *P. insigne,* but this is largely due to the unique phenomenon of fixed heterozygosity resulting from allopolyploidy. Ideally, comparisons should be made between congeneric dioecious and monoecious species that share a common ploidal level.

Generally, it has been assumed that monoecious bryophytes are largely self-fertilizing (Smith, 1978; Anderson, 1980). Stark's (1983) observations of sporophyte production and phenology of *Entodon cladorrhizans* suggested that sporophytes of this moss commonly result from self-fertilization. In addition, several monoecious species have been shown to be self-compatible (Wyatt and Anderson, 1984). On the other hand, Lazarenko and Lesnyak (1972) and Ashton and Cove (1976) reported self-incompatibility in monoecious *Desmatodon cernuus* and *Physcomitrella patens,* respectively. Even in self-compatible monoecious species, cross-fertilization may be promoted by mechanisms such as protogyny or protandry, as in *Splachnum ampullaceum* (Bauer, 1963) and *Funaria hygrometrica* (Crum, 1972). Nevertheless, what is sorely needed are direct estimates of outcrossing rates in natural populations, based most easily perhaps on electrophoretic analyses.

One such analysis was carried out in a Polish population of *Pellia borealis*, a monoecious liverwort polymorphic for two peroxidase alleles (Zielinski, 1986). Screening of 40 sets of progeny from field-collected sporophytes revealed only two instances of segregation for the peroxidase marker. Zielinski (1986) concluded that the frequency of cross-fertilization was low (5%) and that the predominant mode of reproduction in this monoecious species is by self-fertilization. These results contrasted sharply with those for *P. epiphylla*, the monoecious haploid progenitor of *P. borealis,* in which at least 25% of the sporophytes were the result of cross-fertilization.

Wyatt et al. (1988) detected 28 multilocus genotypes of the allodiploid *P. medium.* This abundance of variants implies that extensive genetic recombination has taken place among the original genotypes produced through allopolyploidy. An alternative hypothesis, that each of the multilocus genotypes represents a separate polyploid origin, seems less likely. Such an explanation would require a very large number of presumably rare events involving interspecific hybridization and polyploidization. It appears, therefore, that cross-fertilization is possible, if not common, in this synoecious moss.

Another hypothesized feature of the biology of mosses and liverworts that has led many bryologists to conclude that populations should be genetically uniform is severely limited gene flow (Wyatt and Anderson, 1984). In bryophytes, gene flow is effected through both sperm and spore dispersal. The data available, although largely inferential, indicate that sperm dispersal distances are short. Even in species with splash cups, which are generally interpreted as structures that have evolved to increase sperm dispersal distances, only rarely have sperm been demonstrated to travel more than 50 cm. In species without splash cups, fertilization distances are usually less than 10 cm.

Data regarding spore dispersal are even fewer (Anderson, 1980). Anecdotal evidence and one unpublished study discussed by Wyatt and Anderson (1984) suggest that spore dispersal distances are greater than sperm dispersal distances. Nevertheless, the pattern is strongly leptokurtic, with 97% of all spores of the moss *Atrichum angustatum* falling within 2.0 m of the source colony. It appears likely, therefore, that geographically separated populations of this moss and other bryophytes with similar gene flow distances can show a reasonable degree of differentiation. Electrophoretically, there is no reason to expect high genetic similarity among moss and liverwort populations, as proposed by some bryologists who emphasize the potential for long-range dispersal of these small-spored plants.

Wyatt et al. (1989) screened 14 enzyme loci from 13 natural populations of the dioecious moss *Plagiomnium ciliare*. Total gene diversity (H_T of Nei, 1973, 1975) based on mean frequencies of alleles over all populations was 0.144, a value comparable to those measured in pitch pine (*Pinus rigida*) by Guries and Ledig (1982). In pitch pine, however, there is almost no differentiation among populations (mean $G_{ST} = 0.023$). For *P. ciliare*, G_{ST} averaged 0.23 ± 0.06, reflecting strong differentiation among populations. This pattern is reinforced by close examination of geographical patterns in allele frequencies at particular loci, which differ sharply even between closely adjacent populations. Although gene diversities were much lower overall, a similar pattern was observed in each of the various geographical races of the liverwort *Conocephalum conicum* (Odrzykoski, 1986). Mean G_{ST} values were 0.276, 0.232, and 0.231 for the "S-," "L-," and "A-forms," respectively. Again this suggests that the genetic variation present in the species is apportioned such that populations are differentiated from each other genetically, presumably as a result of restricted gene flow.

On a smaller scale, Dewey (1986) analyzed variation in the liverwort *Riccia dictyospora* from 36 sites over an area of about two km². He discovered clear patterns in the frequencies of two polymorphic loci: allele *a* of *Mdh* occurred principally in the western half of the site, whereas high frequencies of the *d* allele of *Idh* were restricted to the center. The cause of these patterns was unclear. Occasional detection of sharply contrasting allele frequencies between adjacent sites was attributed to rare instances of spore dispersal from within the site. Spore dispersal of *R. dictyospora* is apparently very restricted. The large diameter of the spores and their enclosure in the gametophytic thallus appears to permit no agent, other than perhaps running water, to transport spores effectively. Combined with a lack of specialized mechanisms for asexual reproduction, this lack of dispersibility suggests that gene flow is particularly limited in this species. Dewey (1986) concluded that "while the allozymic data do not indicate a direct correspondence of separate drainages with discrete Mendelian populations, they do imply that even very limited distances (i.e., one kilometer or less) can effectively impede gene flow in *R. dictyospora*. They also indicate a considerable potential for genetic isolation and, in turn, genetic differentiation across the range of this species."

Using Wright's (1946) models for relating gene flow and population structure to genetic differentiation within populations. Wyatt (1977) calculated the effective neighborhood size (in essence, the number of individuals occupying that area of a colony within which mating is random) of the moss *Atrichum angustatum*. Neighborhood size was 225 individuals; neighborhood area, 0.005 m². When Wright's equation was expanded to include spore dispersal as well as sperm dispersal, neighborhood size increased to 371,780 individuals and neighborhood area, 8.08 m² (Wyatt, 1982). Because this moss usually occurs as widely separated colonies, the potential exists for microscale genetic differentiation. This prediction was, in fact, borne out by studies of Cummins and Wyatt (1981), who reported considerable variation among colonies and populations of this species in eastern Texas.

Wyatt et al. (1987) also found evidence of heterogeneity within colonies of the moss *Plagiomnium ciliare*. One particularly polymorphic population, consisting of 36 5 × 5 cm clumps, was sampled intensively. Five erect shoots were electrophoresed from each clump: one from the center and one from each corner of the square culture pots. Five of the 36 clumps were genetically heterogeneous (i.e., consisted of two or more plants that differed in multilocus electrophoretic phenotypes). Three of these clumps showed variability in more than two enzyme loci. Zielinski (1986) also found heterogeneity with respect to two alleles of a peroxidase locus in a Polish population of the liverwort *Pellia borealis*. Only three of 54 samples analyzed, however, were polymorphic, and Zielinski (1986) attributed the segregation of genotypes within the population to natural selection.

CONCLUSIONS

Isozyme studies of bryophytes are still in their infancy. The paucity of data urges caution in attempting to reach general conclusions. In many cases, authors have made sweeping generalizations on the basis of wholly inadequate samples. Usually fewer than 10 loci have been studied. Some studies have utilized only one or two loci, often those coding for nonspecific enzymes such as peroxidases, esterases, or phosphatases. In the absence of any firm genetic data from test crosses, the interpretation of electrophoretic banding patterns in mosses and liverworts remains open to question. In addition, few studies have been designed to include adequate numbers of populations and reasonable numbers of individuals per population.

Attempts to reach any general conclusions about the levels of genetic variation in bryophytes are hindered further by the fact that relatively few taxa have been studied. Of the 84 families (in 17 orders) of mosses (Watson, 1971), only taxa from six families (in five orders) have been examined electrophoretically. This represents only about 0.1% of the species. Of the 70 families (in eight orders) of liverworts, isozymes have been analyzed from only six families (in three orders), or about 0.5% of the species. Furthermore, the taxa that have been studied by no means represent a wide sampling of the biological diversity that exists. For example, most of the liverworts studied thus far are thallose species from temperate regions. There are no electrophoretic data for hornworts.

Nevertheless, it is already apparent from existing data that populations of mosses and liverworts, contrary to expectations based on their haploid-dominant life cycle, usually are not genetically uniform. Levels of genetic variability range from no electrophoretically detectable variation in the liverwort *Plagiochila porelloides* to levels as high as those seen in outcrossing diploid plants such as pines for species like the moss *Plagiomnium ellipticum*. Overall, mosses appear to be more genetically variable than liverworts. The reasons for such differences are not apparent at present.

Bryophyte species are as distinct, or more distinct, electrophoretically than species of higher plants investigated using the same techniques. Genetic identities are typically greater than 0.90 among populations within species but often considerably less than 0.67 (the mean for seed plants; Gottlieb, 1981) between congeneric species. Electrophoretic analysis has also uncovered numerous cases of "sibling species," many of which are separated by large genetic distances, a finding totally at odds with their morphological similarity. It appears either that mosses and liverworts evolve more rapidly biochemically than morphologically or that much morphological change in these organisms simply goes unnoticed (Wyatt, 1985a).

Using isozyme evidence, it has been shown that allopolyploidy does occur in mosses. The allodiploid *Plagiomnium medium* originated multiple times from hybridization between *P. ellipticum* and *P. insigne*. Isozyme analysis has the potential to reveal additional cases of interspecific hybridization and allopolyploidy. Electrophoretic data also have been used to demonstrate outcrossing in bisexual species, which generally were assumed to self-fertilize. Consistent with their limited capacity for sperm (and possibly, spore) dispersal, bryophyte populations show considerable genetic differentiation among populations. The moss *Plagiomnium ciliare*, for example, averages $G_{ST} = 0.23$, a value tenfold greater than that observed for pitch pine (*Pinus rigida*). Heterogeneity also occurs on small spatial scales within local populations of mosses and liverworts.

It is painfully apparent that we need more data regarding bryophyte isozymes. In particular, it would be most instructive to obtain hard evidence to support our genetic interpretations of electrophoretic banding patterns. Thus far, however, only Zielinski (1984a, 1986) and Odrzykoski (1986) have completed such analyses for one to three loci in *Pellia* and *Conocephalum*, respectively. In theory, obtaining segregation ratios in gametophytic progeny should be straightforward.

It would also be enlightening to determine if genetic variation in bryophytes can be linked to mating systems. In seed plants, predominantly outcrossing species show higher levels of genetic variation than do predominantly self-fertilizing species (Hamrick et al., 1979). In addition to this feature of overriding importance, Hamrick et al. (1979) also found greater variability on average for late successional species with wide geographical ranges, high fecundities, wind pollination, and long generation times. The body of data for relating life history traits to genetic variation in bryophytes is too meager at present to allow any generalizations to be made.

Another aspect of the biology of bryophytes that could be explored effectively using isozyme analysis is the rate of outcrossing in bisexual species. Models now exist for obtaining multilocus, as well as single locus, estimates, and these should be applied to natural populations of mosses and liverworts. A simple, direct estimate based on a peroxidase marker was obtained by Zielinski (1984a, 1986), who concluded that outcrossing was much higher in haploid *Pellia epiphylla* than in its diploid derivative, *P. borealis*. Isozymes should also prove of great benefit in resolving cases of autopolyploidy and allopolyploidy in bryophytes as shown by Wyatt et al. (1988). A final area in which we need additional electrophoretic investigations is the problem of genetic relatedness in continentally disjunct species.

Following electrophoretic analyses of additional species of bryophytes, it may be that mosses and liverworts are as genetically diverse as vascular plants. It is likely that the broad range of variation represented by the "*Conocephalum* model" and the "*Plagiomnium* model" denote endpoints of a continuum (Wyatt, 1985a). Isozyme techniques will allow us to detect cases in which previously unsuspected evolutionary processes, such as hybridization and allopolyploidy, have played a role in bryophyte evolution. At present, the prospects for such studies appear exceedingly bright.

ACKNOWLEDGMENTS

Our research on bryophytes has been supported by NSF Grants BSR-8408931 and BSR-8806386 (to RW and AS) and by CPBP-0404 II/1 (to IJO). Earlier drafts of portions of this manuscript were read by J. L. Hamrick, D. E. Soltis, P. S. Soltis, L. D. Gottlieb, L. E. Anderson, W. Schofield, J. Shaw, and T. R. Meagher. Unpublished manuscripts were kindly provided by R. Dewey, J. Shaw, and R. Zielinski.

LITERATURE CITED

Anderson, L. E. 1963. Modern species concepts: mosses. *Bryologist* 66: 107–119.

———. 1964. Biosystematic evaluations in the Musci. *Phytomorphology* 14: 27–51.

———. 1974. Bryology, 1947–1972. *Ann. Missouri Bot. Gard.* 61: 56–85.

———. 1980. Cytology and reproductive biology of mosses. In R. J. Taylor and A. E. Leviton [eds.], *The mosses of North America*, 37–76. Pac. Div., Amer. Assoc. Advancem. Sci., San Francisco.

———, and V. S. Bryan. 1956. A cytotaxonomic investigation of *Fissidens cristatus* Wils. and *F. adiantoides* Hedw. in North America. *Rev. Bryol. Lichenol.* 25: 254–267.

———, and B. E. Lemmon. 1972. Cytological studies of natural hybrids between species of the moss genera, *Astomum* and *Weissia*. *Ann. Missouri Bot. Gard.* 59: 382–416.

Ashton, N. W., and D. J. Cove. 1976. Auxotrophic and developmental mutants of *Physcomitrella patens*. *Bull. Brit. Bryol. Soc.* 27: 10.

Bauer, L. 1963. On the stabilization of the male sexual tendency in Musci. *Bot. J. Linn. Soc.* 58: 337–342.

Bold, H. C., C. J. Alexopoulos, and T. Delevoryas. 1986. Morphology of plants and fungi. Fifth Edition, Harper and Row, New York.

Bramen, J. P. J. 1986. *De systematiek van de Racopilaceae met behulp van electroforese.* M.S. thesis, University of Groningen, Haren, The Netherlands.

Burgeff, H. 1937. Über polyploidie bei *Marchantia*. *Z. Indukt Abstamm. Vererbungsl.* 73: 394–403.

Crawford, D. J. 1983. Phylogenetic and systematic inferences from electrophoretic studies. In S. O. Tanksley and T. J. Orton [eds.], *Isozymes in plant genetics and breeding*, Part A, 257–287. Elsevier, Amsterdam.

Crosby, M. R. 1980. Polyploidy in bryophytes with special emphasis on mosses. In W. H. Lewis [ed.], *Polyploidy: Biological relevance*, 193–198. Plenum Press, New York.

Crum, H. 1972. The geographic origins of the mosses of North America's eastern deciduous forest. *J. Hattori Bot. Lab.* 35: 269–298.

Crundwell, A. C. 1970. Infraspecific categories in Bryophyta. *Biol. J. Linn. Soc.* 2: 221–224.

Cummins, H., and R. Wyatt. 1981. Genetic variability in natural populations of the moss *Atrichum angustatum*. *Bryologist* 84: 30–38.

Daniels, R. E. 1982. Isozyme variation in British populations of *Sphagnum pulchrum* (Braithw.) Warnst. *J. Bryol.* 12: 65–76.

———. 1985a. Isozyme variation in populations of *Sphagnum recurvum* var. *mucronatum* from Britain and Finland. *J. Bryol.* 13: 563–570.

———. 1985b. Isozyme variation in Finnish and British populations of *Sphagnum compactum*. *Ann. Bot. Fenn.* 22: 275–279.

Dewey, R. M. 1986. *Taxonomic and populational studies of the thallose liverworts Riccia subgenus Riccia.* Ph.D. thesis, Texas A&M University, College Station.

———. 1989. Genetic variation in the liverwort *Riccia dictyospora* (Ricciaceae, Hepaticopsida). *Syst. Bot.* In press.

Giannasi, D. E. 1978. Systematic aspects of flavonoid biosynthesis and evolution. *Bot Rev.* (Lancaster) 44: 399–429.

Gliddon, C. J. 1983. Studies on population biology of four species of thallose liverwort. *Bull. Brit. Bryol. Soc.* 36: 14.

Gottlieb, L. D. 1981. Electrophoretic evidence and plant populations. *Prog. Phytochem.* 7: 1–46.

———. 1982. Conservation and duplication of isozymes in plants. *Science* 216: 373–380.

Guries, R. P., and F. T. Ledig. 1982. Genetic diversity and population structure in pitch pine (*Pinus rigida* Mill.). *Evolution* 36: 387–402.

Hamrick, J. L. 1979. Genetic variation and longevity. In O. T. Solbrig et al. [eds.], *Topics in plant population biology*, 84–113. Columbia University Press, New York.

———, Y. B. Linhart, and J. B. Mitton. 1979. Relationships between life history characteristics and electrophoretically detectable genetic variation in plants. *Ann. Rev. Ecol. Syst.* 10: 173–200.

Haufler, C. H. 1987. Electrophoresis is modifying our concepts of evolution in homosporous pteridophytes. *Amer. J. Bot.* 74: 953–966.

———, and D. E. Soltis. 1986. Genetic evidence suggests that homosporous ferns with high chromosome numbers are diploid. *Proc. Natl. Acad. Sci. USA* 83: 4389–4393.

Hofman, A. 1988. A preliminary survey of allozyme variation in the genus *Plagiothecium* (Plagiotheciaceae, Bryopsida). *J. Hattori Bot. Lab.* 64: 143–150.

Khanna, K. R. 1960. Studies in natural hybridization in the genus *Weisia*. *Bryologist* 63: 1–16.

———. 1964. Differential evolutionary activity in the bryophytes. *Evolution* 18: 642–670.

Koponen, T. 1971. A monograph of *Plagiomnium* section *Rosulata* (Mniaceae). *Ann. Bot. Fenn.* 8: 305–367.

———, and E. Nilsson. 1977. Flavonoid patterns and species pairs in *Plagiomnium* and *Rhizomnium* (Mniaceae). *Bryophyt. Biblioth.* 13: 411–425.

Krzakowa, M. 1977. Isozymes as markers of inter- and intraspecific differentiation in hepatics. *Bryophyt. Biblioth.* 13: 427–434.

———. 1981. Evolution and speciation in *Pellia*, with special reference to the *Pellia megaspora-endiviifolia* complex (Metzgeriales). IV. Isozyme investigations. *J. Bryol.* 11: 447–450.

———, and J. Szweykowski. 1977a. Peroxidases as taxonomic markers in two critical *Pellia* taxa (Hepaticae, Pelliaceae). *Bull. Acad. Polon. Sci., Ser. Sci. Biol.* 25: 203–204.

———, and ———. 1977b. Peroxidases as taxonomic characters. II. *Plagiochila asplenioides* (L.) Dum. *sensu* Grolle (= *P. maior* S. Arnell) and *Plagiochila porelloides* (= *P. asplenioides* aucti non Grolle; Hepaticae, Plagiochilaceae). *Bull. Soc. Sci. Letters Poznan Ser. D.* 17: 33–36.

———, and ———. 1979. Isozyme polymorphism in natural populations of a liverwort, *Plagiochila asplenioides*. *Genetics* 93: 711–719.

Lazarenko, A. S., and E. N. Lesnyak. 1972. A comparative study of two moss sibling species: *Desmatodon cernuus* (Hüb.) BSG—*D. ucrainicus* Laz. (Contribution to the problem of infrastructure of the moss species). *Zurn. Obscej. Biol.* 33: 657–667.

Longton, R. E. 1976. Reproductive biology and evolutionary potential in bryophytes. *J. Hattori Bot. Lab.* 41: 205–223.

Lowry, R. J. 1948. A cytotaxonomic study of the genus *Mnium*. *Mem. Torrey Bot. Club* 20: 1–42.

———. 1954. Chromosome numbers and relationships in the genus *Atrichum* in North America. *Amer. J. Bot.* 41: 410–414.

Lyman, J. C., and N. C. Ellstrand. 1984. Clonal diversity in *Taraxacum officinale* (Compositae), an apomict. *Heredity* 53: 1–10.

Miles, C. J., and R. E. Longton. 1987. Life history of the moss, *Atrichum undulatum* (Hedw.) P. Beauv. *Symp. Biol. Hung.* 35: 193–207.

Nei, M. 1972. Genetic distance between populations. *Amer. Naturalist* 106: 283–292.

———. 1973. Analysis of gene diversity in subdivided populations. *Proc. Natl. Acad. Sci. USA* 70: 3321–3323.

———. 1975. *Molecular population genetics and evolution*. North-Holland, Amsterdam.

Newton, M. E. 1983. Cytology of the Hepaticae and Anthocerotae. In R. M. Schuster [ed.], *New manual of bryology*, 117–148. Hattori Bot. Lab., Miyazaki, Japan.

Odrzykoski, I. J. 1986. *Genetic structure of natural populations of* Conocephalum conicum. Ph.D. thesis, Adam Mickiewicz University, Poznan, Poland.

———. 1987. Genetic evidence for reproductive isolation between two European "forms" of *Conocephalum conicum*. *Symp. Biol. Hung.* 35: 577–587.

———, M. A. Bobowicz, and M. Krzakowa. 1981. Variation in *Conocephalum conicum*—The existence of two genetically different forms in Europe. *In* J. Szweykowski [ed.], *New perspectives in bryotaxonomy and bryogeograpy,* 519–542. Adam Mickiewicz University, Poznan, Poland.

Rieseberg, L. H., and D. E. Soltis. 1987. Allozymic differentiation between *Tolmiea menziesii* and *Tellima grandiflora* (Saxifragaceae). *Syst. Bot.* 12: 154–161.

Roose, M. L., and L. D. Gottlieb. 1976. Genetic and biochemical consequences of polyploidy in *Tragopogon. Evolution* 30: 818–830.

Schofield, W. B. 1985. *Introduction to bryology.* Macmillan, New York.

Schuster, R. M. 1966. *The Hepaticae and Anthocerotae of North America.* Vol. I. Columbia University Press, New York.

———. 1983. Phytogeography of the Bryophyta. *In* R. M. Schuster [ed.], New manual of bryology, 463–626. *Hattori Bot. Lab.,* Miyazaki, Japan.

Shaw, J., T. R. Meagher, and P. Harley. 1987. Electrophoretic evidence of reproductive isolation between two varieties of the moss, *Climacium americanum. Heredity* 59: 337–343.

Smith, A. J. E. 1978. Cytogenetics, biosystematics and evolution in the Bryophyta. *Advances Bot. Res.* 6: 195–276.

———. 1979. Towards an experimental approach to bryophyte taxonomy. *In* G. C. S. Clarke and J. G. Duckett [eds.], *Bryophyte systematics,* 195–206. Academic Press, New York.

———, and M. E. Newton. 1968. Chromosome studies on some British and Irish mosses. III. *Trans. Brit. Bryol. Soc.* 5: 463–522.

Soltis, D. E. 1986. Isozyme number and enzyme compartmentalization in *Equisetum. Amer. J. Bot.* 73: 908–913.

Stark, L. R. 1983. Bisexuality as an adaptation in desert mosses. *Amer. Midl. Naturalist* 110: 445–448.

Szweykowski, J., and M. Krzakowa. 1979. Variation of four enzyme systems in Polish populations of *Conocephalum conicum* (L.) Dum. (Hepaticae, Marchantiales). *Bull. Acad. Polon. Sci., Ser. Sci. Biol.* 27: 37–41.

———, and R. Zielinski. 1983. Isoenzymatic variation in Polish populations of the moss *Plagiothecium undulatum* (Hedw.) BSG—A preliminary report. *J. Hattori Bot. Lab.* 54: 119–123.

———, I. J. Odrzykoski, and R. Zielinski. 1981a. Further data on the geographic distribution of two genetically different forms of the liverwort *Conocephalum conicum* (L.) Dum.: the sympatric and allopatric regions. *Bull. Acad Polon. Sci., Ser. Sci. Biol.* 28: 437–449.

———, R. Zielinski, and M. Mendelak. 1981b. Variation of peroxidase isoenzymes in central European taxa of the liverwort genus *Pellia. Bull. Acad. Polon. Sci., Ser. Sci. Biol.* 29: 9–19.

Vries, A. de, B. O. Van Zanten, and H. Van Dijk. 1983. Genetic variability within and between populations of two species of *Racopilum* (Racopilaceae, Bryopsida). *Lindbergia* 9: 73–80.

Wachowiak, M. 1986. *Enzyme polymorphism in populations of* Plagiochila asplenioides *and P. porelloides.* M.Sc. thesis, Adam Mickiewicz University, Poznan, Poland.

Watson, E. V. 1971. *The structure and life of bryophytes.* Hutchinson University Library, London.

Werth, C. R., S. I. Guttman, and W. H. Eshbaugh. 1985a. Recurring origins of allopolyploid species in *Asplenium. Science* 228: 731–733.

———, ———, and ———. 1985b. Electrophoretic evidence of reticulate evolution in the Appalachian *Asplenium* complex. *Syst. Bot.* 10: 184–192.

Wettstein, F. 1932. Genetik. *In* F. Verdoorn [ed.], *Manual of bryology,* 233–272. Nijhoff, The Hague.

Wright, S. 1946. Isolation by distance under diverse systems of mating. *Genetics* 31: 39–59.

Wyatt, R. 1977. Spatial pattern and gamete dispersal distances in *Atrichum angustatum,* a dioicous moss. *Bryologist* 80: 284–291.

———. 1982. Population ecology of bryophytes. *J. Hattori Bot. Lab.* 52: 179–198.

———. 1985a. Species concepts in bryophytes: input from population biology. *Bryologist* 88: 182–189.

———. 1985b. Chemosystematics of the Mniaceae. I. Identities of Lowry's species pairs. *Monogr. Syst. Bot. Missouri Bot. Gard.* 11: 187–194.

———. 1985c. Terminology for bryophyte sexuality: toward a unified system. *Taxon* 34: 420–425.

———, and L. E. Anderson. 1984. Breeding systems of bryophytes. *In* A. F. Dyer and J. G. Duckett [eds.], The experimental biology of bryophytes, 39–64. Academic Press, London.

———, and A. Stoneburner. 1984. Biosystematics of bryophytes: an overview. In W. F. Grant [ed.], Plant biosystematics, 519–542. Academic Press, Toronto.

———, I. J. Odrzykoski, and A. Stoneburner. 1987. Electrophoretically detectable genetic variation in Plagiomnium ciliare: a preliminary report. Symp. Biol. Hung. 35: 589–602.

———, ———, and ———. 1989. High levels of genetic variability in the haploid moss Plagiomnium ciliare. Evolution. In press.

———, ———, ———, H. W. Bass, and G. A. Galau. 1988. Multiple origins of Plagiomnium medium, an allopolyploid moss. Proc. Natl. Acad. Sci. USA. 85: 5601–5604.

Yamazaki, T. 1981. Genic variabilities in natural population of haploid plant, Conocephalum conicum. I. The amount of heterozygosity. Jpn. J. Genet. 56: 373–383.

———. 1984. The amount of polymorphism and genetic differentiation in natural populations of the haploid liverwort Conocephalum conicum. Jpn. J. Genet. 9: 133–139.

Yano, K. 1957a. Cytological studies on Japanese mosses. II. Hypnobryales. Mem. Takada Branch, Niigata Univ. 1: 85–127.

———. 1957b. Cytological studies on Japanese mosses. III. Isobryales, Polytrichinales. Mem. Takada Branch, Niigata Univ. 1: 129–159.

Zanten, B. O. Van, and T. Pocs. 1981. Distribution and dispersal of bryophytes. Adv. Bryol. 1: 479–562.

Zielinski, R. 1984a. Electrophoretic evidence of cross-fertilization in the monoecious Pellia epiphylla, $n = 9$. J. Hattori Bot. Lab. 56: 255–262.

———. 1984b. Electrophoretic and cytological study of the Pellia epiphylla and P. borealis complex. J. Hattori Bot. Lab. 56: 263–269.

———. 1986. Cross-fertilization in the monoecious Pellia borealis, $n = 18$, and spatial distribution of two peroxidase genotypes. Heredity 56: 299–304.

———. 1987. Genetic variation and evolution of the liverwort genus Pellia. Monograph. University of Szczecin Press, Poland.

———, J. Szweykowski, and E. Rutkowska. 1985. A further electrophoretic study of peroxidase isoenzyme variation in Pellia epiphylla (L.) Dum. from Poland, with special reference to the status of Pellia borealis Lorbeer. Monogr. Syst. Bot. Missouri Bot. Gard. 11: 199–209.

CHAPTER 12

Polyploidy, Breeding Systems, and Genetic Differentiation in Homosporous Pteridophytes

Douglas E. Soltis
Pamela S. Soltis

Department of Botany, Washington State University,
Pullman, WA 99164

The common name "pteridophyte" has been applied to plants composing four divisions (Bold et al., 1986): Psilotophyta, Microphyllophyta, Arthrophyta, and Pteridophyta. Most pteridophytes are homosporous, producing one type of spore that germinates to produce a potentially bisexual gametophyte. The homosporous pteridophytes are the whisk ferns, *Psilotum* and *Tmesipteris* (Psilotophyta); the lycopods, *Lycopodium* sensu lato and *Phylloglossum* (class Aglossopsida of the Microphyllophyta); the horsetails and scouring rushes, *Equisetum* (Arthrophyta); and most families of ferns (Pteridophyta). These are ancient lineages, some of which are regarded as living fossils (Bold et al., 1986). The lycopods, for example, can be traced back in the fossil record to the lower Devonian.

Homosporous pteridophytes are unique among vascular plants in possessing free-living gametophytes that are potentially bisexual. The unusual nature of the life cycle in these plants compared to all heterosporous plants, including seed plants, has important evolutionary implications, including: (1) the two free-living organisms, gametophyte and sporophyte, may often have distinct ecological requirements; (2) both the gametophyte and sporophyte experience separate selective forces; and (3) selfing of a bisexual gametophyte produces a completely homozygous sporophyte in a single generation.

These plants are also unusual in possessing the highest chromosome numbers known for any organisms. Whereas the average gametic chromosome number for angiosperms is $n = 15.99$, the mean for all homosporous pteridophytes is $n = 55.27$ (Klekowski and Baker, 1966). The average haploid number for homosporous ferns (Pteridophyta) is 57.05 (Klekowski and Baker, 1966). Using chromosome numbers reported by Löve et al. (1977) we have determined that the average chromosome number for lycopods (Aglossopsida, Microphyllophyta) is approximately $n = 86$. Chromosome numbers for Psilotophyta range from $n = 52$ to 208 in *Psilotum* and $n = 104$ to 208 in *Tmesipteris* (Löve et al., 1977). All species of *Equisetum* (Arthrophyta) for which chromosome counts are available have $n = 108$.

For the past two decades, the prevalent view of homosporous pteridophyte biology has been that these plants are fundamentally different from all other vascular plants in both levels of polyploidy and type of breeding system. Klekowski and Baker (1966) argued that the high chromosome numbers characteristic of homosporous pteridophytes reflected very high levels of polyploidy. For example, original base chromosome numbers of $x = 9$ and 12 have been proposed for *Equisetum* (Löve et al., 1977; Vida, 1976), suggesting that extant species are 24- or 18-ploid, respectively. Klekowski and Baker (1966) suggested that high levels of polyploidy were essential for maintaining genetic variation in all lineages of homosporous pteridophytes. The genetic variation stored through polyploidy could then be released as a result of pairing of chromosomes from these additional genomes (homoeologous pairing). Such a genetic mechanism was deemed critical because genetic variation would otherwise be eliminated by the presumed habitual self-fertilization of the bisexual gametophytes of these plants.

Until recently, the tenets of homosporous pteridophyte biology remained that: (1) these organisms maintain genetic variation via high levels of polyploidy; and (2) the gametophytes of these plants typically self-fertilize (Klekowski, 1972, 1973, 1976, 1979; Lloyd, 1974a; Hickok, 1978; Chapman et al., 1979), although some evidence to the contrary has been presented. Manton (1961) argued, for example, that the numerous interspecific hybrids reported between pteridophytes in nature suggested high levels of outcrossing. Furthermore, investigations of gametophytes in the laboratory indicated that some homosporous ferns possess mechanisms, such as production of the hormone antheridiogen, that should promote intergametophytic matings (Döpp, 1950; Voeller, 1964; Lloyd 1974a, b; Näf, 1979). Detailed field and laboratory studies of *Equisetum* showed that gametophytes of many species are initially either male or female, thus providing the opportunity for intergametophytic matings; however, in the absence of fertilization, gametophytes may become bisexual and self-fertilize (Hauke, 1967, 1969, 1977; Duckett, 1970, 1972; Duckett and Duckett, 1980).

During the past several years, genetic data compiled from enzyme electrophoresis have permitted investigators to test the traditional view that high levels of polyploidy and intragametophytic selfing (i.e., selfing of an individual gametophyte) are fundamentally intertwined aspects of homosporous pteridophyte biology. In fact, electrophoretic investigations have led directly to rejection of some of these traditional tenets (see recent reviews by Haufler and Soltis, 1986; Haufler, 1987; D. Soltis and P. Soltis, 1987a).

Genetic insights acquired via enzyme electrophoresis have accumulated rapidly for representatives of the remaining three lineages of homosporous pteridophytes (Psilotophyta, class Aglossopsida of Microphyllophyta, Arthrophyta), necessitating a reassessment of the basic biology of all of these organisms. Herein, we summarize the broad genetic data base now available for all groups of homosporous pteridophytes and extend the reassessments proposed recently for ferns (Haufler and Soltis, 1986; Haufler, 1987; D. Soltis and P. Soltis, 1987a) to all divisions containing homosporous vascular plants. We provide the following synthesis: (1) despite extremely high chromosome numbers, all groups of homosporous pteridophytes are functionally diploid organisms; there is no genetic evidence that species having the lowest chromosome numbers in any lineage of these plants are highly polyploid; neither is there convincing evidence that these organisms release genetic variation via homoeologous chromosome pairing; (2) in nature, the gametophytes of ferns, lycopods, and species of *Equisetum* frequently cross-fertilize; (3) inbreeding does not characterize homosporous pteridophytes; just as in seed plants, a variety of mating systems operates; (4) congeneric species of homosporous pteridophytes typically are more highly differentiated at genes coding for enzymes than are congeneric species of seed plants.

GENETIC DIPLOIDY

In diploid seed plants, the number of isozymes (different forms of an enzyme encoded by different loci) for a given enzyme is highly conserved (Gottlieb, 1981, 1982). For many of the enzymes commonly analyzed in electrophoretic investigations, diploid angiosperms and gymnosperms exhibit two nuclear-encoded isozymes, one catalyzing reactions in the chloroplast and the other in the cytosol. In allopolyploid angiosperms, however, it is well-known that isozyme number may increase due to the addition of divergent genomes (Roose and Gottlieb, 1976; Gottlieb, 1981, 1982; Crawford, 1983, 1985). It therefore seemed reasonable to postulate that if homosporous pteridophytes with high chromosome numbers are highly polyploid, as initially proposed by Klekowski and Baker (1966), they should exhibit a dramatic increase in isozyme number.

Electrophoretic studies have shown, however, that species having the lowest chromosome numbers within genera (hereafter referred to as diploid), representing all groups of homosporous pteridophytes, usually have the number of isozymes typical of diploid seed plants. These data are expressed (Table 12.1) as a percentage of loci surveyed that could possibly have arisen through polyploidy. In most lineages, rare examples of additional isozymes were detected. However, occasional

Table 12.1. *Genetic evidence for diploidy in homosporous pteridophytes*

Lineage	Total number of species studied	Total number of loci examined	Percentage of surveyed loci that possibly could have arisen via polyploidy
[a]Pteridophyta	39 (diploid species)	652	0.9%
	25 (polyploid species)	394	41.1%
[b]Arthrophyta	3	72	4.2%
[c]Psilotophyta	1	28	0.0%
[d]Aglossopsida (Microphyllophyta)	4	79	5.1%

[a] Haufler and Soltis (1986)
[b] D. Soltis (1986)
[c] P. Soltis and D. Soltis (1988c)
[d] D. Soltis and P. Soltis (1988)

gene duplications are seen in diploid angiosperms (Gottlieb, 1981, 1982; Tanksley, 1987). Thus, these rare duplications cannot by themselves be considered evidence of polyploidy. There certainly is no genetic evidence for the low base chromosome numbers proposed for these taxa. In ferns, species having multiples of the low, or diploid, chromosome number (true polyploids) for a genus do exhibit large increases in isozyme number (Table 12.1). The isozyme data for diploid pteridophytes and true polyploid pteridophytes are therefore similar to genetic data for diploid and polyploid angiosperms (Gottlieb, 1981, 1982).

These findings for all groups of homosporous pteridophytes are made more significant by isozyme data for woody angiosperms that similarly have been considered ancient polyploids because of their high chromosome numbers (Stebbins, 1971; White, 1978). Ongoing electrophoretic analyses of Salicaceae, Hippocastanaceae, Myristicaceae, Cercidiphyllaceae, and Magnoliaceae (with x = 19, 20, 19, 19, 19, respectively) indicate a significant increase in isozyme number relative to diploid angiosperms: in these lineages approximately 25–55% of the genes examined electrophoretically are duplicated (S. J. Brunsfeld, unpublished data; D. E. Soltis and P. S. Soltis, unpublished data; J. F. Wendel, personal communication). These data support the hypothesis that these angiosperms are ancient polyploids and contrast with data for homosporous pteridophytes (Table 12.1).

Three explanations for the ubiquitous isozymic diploidy in homosporous pteridophytes seem plausible (see also Haufler and Soltis, 1986; D. Soltis and P. Soltis 1987a, 1988): (1) repeated cycles of allopolyploidy followed by gene silencing; (2) ancient autopolyploidy, which would lead to duplicated, but not divergent, genomes; (3) initiation with relatively high chromosome numbers, possibly involving ancient events of chromosomal fission.

Evidence for and against these three hypotheses was discussed initially by Haufler and Soltis (1986) and at greater length by D. Soltis and P. Soltis (1988) (see also Wagner and Wagner, 1980, for an insightful earlier discussion before electrophoretic data were available). This evidence is also summarized in Table 12.2 and will therefore be discussed only briefly here. We stress that although hypotheses (1) and (3) seem most likely, it is impossible to differentiate unequivocally between them. However, the concept of ancient allopolyploidy is so pervasive among pteridologists that hypothesis (1) has been strongly advocated by some (e.g., Haufler, 1987) without definitive supporting evidence. Even before genetic data were available for pteridophytes, some authors cautioned against acceptance of the idea that high chromosome numbers in extant pteridophytes were necessarily derived from low numbers via polyploidy (Lovis, 1973; Walker, 1973; Duncan and Smith, 1978; Wagner and Wagner, 1980). The only existing evidence for hypothesis (1) is from implication based on the situation in extant plants (Table 12.2). This line of reasoning suggests that since allopolyploidy is prevalent among modern pteridophytes, it must have been common in the past, as well. Such reasoning can be applied mainly to the ferns; however, it is not applicable to *Equisetum* because all species have the same high chromosome number. With regard to the lycopods, Tryon and Tryon (1982) noted that aneuploidy is so prevalent in these plants that it is hard to distinguish episodes of polyploidy based on cytological data.

Table 12.2. Summary of supportive (above dotted line) and non-supportive (below dotted line) arguments and experimental evidence for the three possible explanations for genetic diploidy in homosporous pteridophytes

	Hypotheses		
	Ancient allopolyploidy followed by gene silencing	Ancient autopolyploidy followed by gene silencing	Initiation with high chromosome numbers
Time	Due to antiquity of taxa there has been sufficient time for complete silencing of duplicated genes.	Due to antiquity of taxa there has been sufficient time for complete silencing of duplicated genes.	
Inference	Allopolyploidy is prevalent in some modern lineages. Therefore, it must have been common in the past as well.		
Genetics			DNA reassociation kinetics indicate that taxa behave as diploids. They are isozymically diploid.
Cytology			There is no evidence for multiple chromosome sets. Aneuploidy is prevalent in some lineages (e.g., lycopods).
Time	Gene silencing has not occurred to completion in angiosperms of ancient polyploid origin, which at least approach homosporous pteridophytes in antiquity.	Gene silencing has not occurred to completion in angiosperms of ancient polyploid origin, which at least approach homosporous pteridophytes in antiquity.	
Inference		Autopolyploidy is apparently rare among extant pteridophytes. Therefore, it must have been rare in the past as well.	

	Hypotheses		
	Ancient allopolyploidy followed by gene silencing	Ancient autopolyploidy followed by gene silencing	Initiation with high chromosome numbers
Genetics	Gene silencing is a poorly understood and infrequently documented process in plants. It seems unlikely that gene silencing would have occurred to completion in all lineages; some lineages must be of more recent origin, yet none shows evidence of gene duplication.	Gene silencing is a poorly understood and infrequently documented process in plants. The genetic consequences of autopolyploidy (tetrasomic or higher level segregation) are not apparent.	
Cytology	Only univalents and bivalents are encountered, even in hybrids between closely related species.	Only univalents and bivalents are encountered, even in hybrids between closely related species.	

Most evidence (Table 12.2) supports hypothesis (3). DNA reassociation kinetic data indicate, for example, that *Equisetum* species have "one genome of high sequence complexity rather than many copies of a low complexity genome" (Bendich and Anderson, 1983). Perhaps the strongest evidence for hypothesis (3) is the critical analysis of karyotypes (Duncan and Smith, 1978), which provided no evidence of multiple chromosome sets in ferns. Furthermore, from a genetic standpoint it seems improbable that wholesale silencing of duplicated genes would occur in so many lineages within four divisions. The repeated cycles of allopolyploidy proposed for pteridophytes [hypothesis (1)] are considered to be ongoing, with different genera at different points in the polyploid-gene silencing cycle (Haufler, 1987). It is curious, therefore, that there is no genetic evidence for allopolyploidy in any species having the lowest chromosome number in any genus of homosporous pteridophyte.

Additional evidence is needed to differentiate among the above hypotheses; most promising would be molecular genetic approaches to ascertain gene copy number. Regardless of which hypothesis is correct, homosporous pteridophytes are genetically diploid and should therefore be considered diploid organisms. Furthermore, until the origin of high chromosome numbers in these organisms is determined, their isozyme condition should be referred to as diploid rather than diploidized. Levels of polyploidy in homosporous pteridophytes may therefore be grossly overestimated. Grant (1981) suggested, for example, that perhaps 95% of these plants may be of polyploid origin. Interestingly, however, when the low chromosome numbers within genera of Pteridophyta are considered diploid and only higher multiples of those numbers are considered polyploid, ferns may be only 43.5% polyploid (Vida, 1976), a value comparable to estimates of levels of polyploidy in angiosperms (Grant, 1981).

Because electrophoretic studies indicate that homosporous pteridophytes are diploid organisms without large numbers of duplicated genes, there also is no evidence that they maintain genetic variation via polyploidy and release this variation through homoeologous chromosome pairing. There is, in fact, convincing evidence that homoeologous chromosome pairing does not occur in these plants. Electrophoretic analyses of segregation ratios in several ferns, including *Pellaea andromedifolia* (Gastony and Gottlieb, 1982), *Bommeria hispida* (Haufler and Soltis, 1984), *Pteridium aquilinum* (Wolf et al., 1987), and three species of Equisetum (D. Soltis, 1986) showed simple patterns of disomic inheritance. Electrophoretic analyses of gametophytic progeny arrays of lycopods and Psilotum have not yet been conducted because of the difficulty in germinating spores and culturing gametophytes of these plants (Freeberg and Wetmore, 1957; Whittier, 1973, 1977).

Hickok (1978) reported homoeologous pairing in Ceratopteris based on morphological markers. This putative example of homoeologous pairing should be reexamined using electrophoretic markers.

The only electrophoretic investigation that purported to demonstrate homoeologous chromosome pairing in a homosporous pteridophyte was the study of Chapman et al. (1979) of Pteridium aquilinum. Wolf et al. (1987), however, subsequently demonstrated disomic segregation in this species. The report of homoeologous chromosome pairing in Pteridium is probably the result of several factors, including misinterpretation of enzyme banding patterns and use of an inappropriate grinding buffer.

MATING SYSTEMS IN HOMOSPOROUS PTERIDOPHYTES

The mating system plays a fundamental role in the evolutionary biology of a species. Not only does it result in reproduction, but the mating system determines the distribution of genetic variation within and among populations, thus providing an array of genotypes on which natural selection can act and influencing the genetic structure of populations and species. The mating systems of angiosperms and gymnosperms have been well characterized (see reviews by Jain, 1976; Brown, 1979; Hamrick et al., 1979; Clegg, 1980) and typically combine selfing and outcrossing components. The mating systems of homosporous plants, however, have received less attention. Because homosporous plants reproduce sexually via haploid, free-living, potentially bisexual gametophytes, their mating systems differ from those of seed plants. In fact, three types of matings are possible in homosporous plants (Lloyd, 1974a; Klekowski, 1979): (1) intergametophytic crossing, the cross-fertilization of gametophytes produced by different sporophytes (analogous to outcrossing in seed plants); (2) intergametophytic selfing, the cross-fertilization of gametophytes produced by the same sporophyte (analogous to selfing in seed plants); and (3) intragametophytic selfing, a mating system component, unique to homosporous plants, that involves the self-fertilization of a single gametophyte.

Because archegonia and antheridia may be borne on the same thallus, it has long been maintained that intragametophytic selfing is the predominant mode of reproduction in natural populations of homosporous pteridophytes (Klekowski and Baker, 1966; Klekowski, 1973, 1979), although at least occasional intergametophytic crossing was suspected in homosporous ferns because of their high rates of interspecific hybridization (Manton, 1950, 1961). Recent electrophoretic analyses challenge the traditional view that homosporous pteridophytes undergo high levels of intragametophytic selfing. In this section we provide analyses of genetic equilibrium and rates of intragametophytic selfing in natural populations of homosporous ferns, lycopods, and a single horsetail, Equisetum arvense.

Hardy-Weinberg Equilibrium in Homosporous Pteridophytes

Analysis of conformance of a population to Hardy-Weinberg genotypic proportions can be informative regarding evolutionary processes within that population. Several factors may contribute to observed departures from equilibrium expectations, including nonrandom mating, drift, selection, migration, and mutation. Of these, nonrandom mating plays a major role; therefore, the mating system of a population or species can be inferred by the distribution of allelic variation into genotypes and by comparison of observed genotypic proportions with those expected at Hardy-Weinberg equilibrium. Deviation from Hardy-Weinberg expectations can be represented by the fixation index, F, which ranges from -1, indicating an excess of heterozygotes, to 1, signifying a deficiency of heterozygotes, relative to Hardy-Weinberg expectations. If homosporous pteridophytes are highly inbred, populations should exhibit values of F approaching unity.

Using data from enzyme electrophoresis, we have calculated F for populations of several species of homosporous ferns and lycopods (Table 12.3). Values of F for species of *Botrychium* approach 1.0 in all populations examined, consistent with expectations based on a model of extreme inbreeding in homosporous ferns. Populations of *Asplenium platyneuron* and *A. montanum* are also inbreeding (Werth et al., 1985). However, populations of most other fern species investigated indicate little, if any, deviation from Hardy-Weinberg expectations. Most populations of homosporous ferns are therefore panmictic, exhibiting considerable levels of intergametophytic crossing, or outcrossing in the sense of seed plants.

Genotypic frequencies of several species of lycopods were also examined for conformance to Hardy-Weinberg expectations (Table 12.3). Populations of *Huperzia lucidula* and *Lycopodium annotinum* have fixation indices near zero, indicating that they are in Hardy-Weinberg equilibrium and are panmictic. Interpopulational variation in F was observed for *H. miyoshiana* and *L. clavatum*, ranging from significant heterozygote excesses to significant heterozygote deficiencies. Deviation from Hardy-Weinberg equilibrium was greater in populations of *L. clavatum* than in any other lycopod species, possibly reflecting the extensive vegetative reproduction of this species. Although $\bar{F}$ for *L. clavatum* was −0.064, only one population approached Hardy-Weinberg expectations; all other populations exhibited heterozygote excesses or deficiencies. Multiple sampling of a single genet could result in significant departures from expected Hardy-Weinberg genotypic proportions even in populations that experience random mating. Alternatively, some populations of *H. miyoshiana* and *L. clavatum* may exhibit mixed mating systems with both selfing and outcrossing components.

Table 12.3. *Fixation indices for species of homosporous pteridophytes*

Species	Number of Populations	$\bar{F}$	Range	Reference
Ferns				
Blechnum spicant	6	0.132	0.022–0.403	P. Soltis and D. Soltis, 1988a
Botrychium virginianum	4	0.962	0.948–0.977	D. Soltis and P. Soltis, 1986
Dryopteris expansa	8	0.335	−0.014–0.745	D. Soltis and P. Soltis, 1987b
Polystichum imbricans	8	0.033	−0.088–0.052	D. Soltis and P. Soltis, 1987a
P. munitum	4	0.052	−0.019–0.102	P. Soltis and D. Soltis, 1987
Cheilanthes gracillima	5	0.163	0.026–0.296	P. Soltis et al., 1989
Lypcopods				
Huperzia lucidula	2	0.017	0.008–0.017	P. Soltis and D. Soltis, unpubl. data
H. miyoshiana	4	−0.004	−0.197–0.296	P. Soltis and D. Soltis, unpubl. data
Lycopodium clavatum	6	−0.064	−0.590–0.672	P. Soltis and D. Soltis, unpubl. data
L. annotinum	1	0.016	—	P. Soltis and D. Soltis, unpubl. data
L. obscurum	1	−0.141	—	P. Soltis and D. Soltis, unpubl. data

Rates of Intragametophytic Selfing in Homosporous Pteridophytes

High levels of heterozygosity in populations of homosporous pteridophytes indicate extensive outcrossing. However, because of the differences between the mating systems of homosporous pteridophytes and those of seed plants, quantitative estimates of the relative contributions of selfing and outcrossing cannot be obtained using traditional methods of analysis (e.g., Clegg, 1980). Furthermore, the estimation procedure developed by Hedrick (1987) for homosporous plants is based on the analysis of sets of progeny derived from a single sporophyte and can only be used if a progeny array from a single maternal parent can be identified, a difficult, if not impossible, process in natural populations of homosporous pteridophytes. Estimates of the rate of intragametophytic self-fertilization can be obtained, however, from a recently developed procedure that uses multilocus genotypic frequencies of sporophytes (Holsinger, 1987).

Rates of intragametophytic selfing in homosporous ferns vary among species and among populations within a species (Table 12.4). In *Botrychium virginianum* (D. Soltis and P. Soltis, 1986) and *B. dissectum* (McCauley et al., 1985) intragametophytic matings are prevalent, possibly due to restricted sperm movement imposed by the subterranean gametophytes of these species. In a majority

Table 12.4. Estimates of intragametophytic selfing in homosporous pteridophytes, following the estimation procedure of Holsinger (1987). Because lycopods and Equisetum arvense undergo extensive vegetative reproduction, estimates for these species were calculated in two ways: 1) each ramet was treated as a distinct individual, and 2) each multilocus genotype was treated as a distinct individual. The results are presented following the format Method 1/Method 2

Species	Number of Populations	Number of Loci	Maximum-Likelihood Estimate (Range)	Bootstrap Estimate (Range)
Ferns				
Blechnum spicant[a]	5	12	0.092 (0–0.403)	0.095 (0–0.427)
Botrychium dissectum[b]	3	5	0.95	—
B. virginianum[c]	4	18	0.948 (0.930–0.960)	—
Dryopteris expansa[d]	8	12	0.336 (0–0.583)	0.338 (0.022–0.600)
Pellaea andromedifolia[e]	2	8	0	0
Polystichum imbricans[f]	8	12	0.021 (0–0.169)	0.022 (0–0.176)
P. munitum[g]	4	12	0.014 (0–0.029)	0.014 (0–0.29)
Lycopods				
Huperzia lucidula	2	18	0.010/0.003 (0–0.020/0–0.007)	0.002/0 (0–0.003/—)
H. miyoshiana[h]	4	8	0.063/0.033 (0–0.251/0–0.0131)	0.067/0.035 (0–0.267/0–0.138)
Lycopodium clavatum[h]	13	5	0.105/0.063 (0–0.709/0–0.537)	0.105/0.072 (0–0.690/0–0.575)
L. annotinum[h]	6	9	0.028/0.071 (0–0.167/0–0.427)	0.031/0.112 (0–0.184/0–0.674)
L. obscurum	1	19	0/0.229	0/0.269
			—	—
Diphasiastrum complanatum	2	17	0/0	0/0
D. digitatum	1	20	0/0	0/0
			—	—
Equisetum				
E. arvense[i]	17	14	0.044 (0–0.333)	0.063 (0–0.577)

[a]P. Soltis and D. Soltis, 1988a.
[b]McCauley et al., 1985. These estimates were not obtained following Holsinger (1987).
[c]D. Soltis and P. Soltis, 1986.
[d]D. Soltis and P. Soltis, 1987a.
[e]Gastony and Gottlieb, 1985; Holsinger, 1987.
[f]D. Soltis and P. Soltis, 1987a.
[g]P. Soltis and D. Soltis, 1987.
[h]P. Soltis and D. Soltis, 1988b.
[i]D. Soltis et al., 1988.

of other fern species examined, intragametophytic selfing is rare, ranging from 0 to 10%. Interpopulational variation in rates of intragametophytic selfing was observed most notably in *Dryopteris expansa*, where selfing estimates ranged from 0 to 60% among populations, and *Blechnum spicant*, where estimates varied from 0 to 40% among populations. Variation in rates of intragametophytic selfing may reflect varying ecological conditions among populations, such as population density and safe-site availability for spore germination. Therefore, in many cases, the mating system of a single population cannot be viewed as representative of the species as a whole.

Rates of intragametophytic self-fertilization in most lycopods approach zero (Table 12.4). Interpopulational variation in selfing levels was again observed, especially in *Lycopodium clavatum*, where 11 populations exhibited no intragametophytic selfing and two populations had substantial selfing rates, approaching 60–70%. Similarly, in five populations of *L. annotinum* no intragametophytic selfing was observed, whereas a single population exhibited high rates of selfing. One population of *Huperzia miyoshiana* experienced low levels of intragametophytic selfing, whereas in the other three populations examined intergametophytic matings predominated. One or two populations were examined from the following species: *L. obscurum*, *H. lucidula*, *Diphasiastrum complanatum*, and *D. digitatum*. In all cases, intragametophytic selfing was rare. Intergametophytic matings also predominate in populations of *Equisetum arvense* (D. Soltis et al., 1988). In all but two of the 15 populations for which estimates were obtained, intragametophytic selfing rates were near 0%; two populations experienced selfing rates of approximately 25% and 33%.

Electrophoretic evidence clearly indicates that members of all major lineages of homosporous pteridophytes (i.e., homosporous ferns, Lycopodiaceae *sensu lato,* and Arthrophyta) experience high levels of intergametophytic mating despite their potential for intragametophytic selfing. However, the mechanisms responsible for preventing intragametophytic selfing and promoting intergametophytic matings are not well understood. Several factors may be involved, including asynchronous initiation and/or maturation of gametangia, inbreeding depression, and the hormone antheridiogen, which effects the production of unisexual gametophytes. Laboratory and field studies have demonstrated that gametophytes of *Equisetum* species, including *E. arvense*, are typically unisexual (Hauke, 1967, 1969, 1977; Duckett, 1970, 1972; Duckett and Duckett, 1980), although the mechanisms promoting unisexuality are unknown. Furthermore, ecological factors may promote intergametophytic matings. For example, gametophytes of lycopods typically grow in groups, showing a preference for disturbed habitats (Bruce and Beitel, 1976; Lellinger, 1985; Wagner et al., 1985), and the occurrence of many gametophytes in a small area may facilitate intergametophytic matings.

Evolutionary Consequences of Outcrossing in Homosporous Pteridophytes

Electrophoretic data indicate that intergametophytic matings are common in most species of homosporous pteridophytes and that populations of most species are in Hardy-Weinberg equilibrium, exhibiting high levels of outcrossing. These data have significant implications regarding population subdivision and the genetic structure of species of homosporous pteridophytes. The traditional view that homosporous pteridophytes are highly inbred predicts that populations should exhibit significant genetic substructuring. Furthermore, inbreeding species typically partition relatively more of their genetic variation among rather than within populations than do outcrossing species (Brown, 1979; Loveless and Hamrick, 1984), resulting in greater differentiation among populations of the former than the latter. To test these predictions, we compiled data on the following aspects of the population biology of homosporous pteridophytes: (1) population genetic structure; (2) genetic structure of species as indicated by Wright's F-statistics; and (3) genetic structure of species as indicated by Nei's (1972) genetic identities among conspecific populations.

Population Genetic Structure: The genetic structure of populations can be analyzed using Wright's (1965) F-statistics, a hierarchical series of fixation indices. Genetic differentiation among population subdivisions can be evaluated with the F_{ST} statistic, which ranges from 0, indicating no genetic divergence among subpopulations, to 1, indicating fixation for different alleles among subpopulations. Populations of four species of homosporous pteridophytes were analyzed for genetic subdivision, three species of ferns and *Equisetum arvense* (Table 12.5). *Polystichum munitum*, a highly outcrossing species, exhibited substructure in neither of the two populations examined (P.

Soltis and D. Soltis, 1987). Similarly, the outcrossing populations of Blechnum spicant were genetically homogeneous. A single population of B. spicant that exhibited moderate levels of intragametophytic selfing (Mt. Rainier) showed evidence of genetic heterogeneity among subpopulations, possibly due to family structure (P. Soltis and D. Soltis, 1988a). In Cheilanthes gracillima, populations are highly structured despite the fact that sporophytes apparently arise via intergametophytic matings (P. Soltis et al., 1989). Subpopulations of C. gracillima typically are panmictic units, approaching Hardy-Weinberg equilibrium genotype frequencies, yet they differ substantially from neighboring subpopulations in allele frequencies. Therefore, factors other than the mating system, perhaps including ecological conditions and the availability of safe-sites for spore germination, must determine population structure in this species. A single population of Equisetum arvense exhibited significant genetic structure, yet population subdivision was attributed to the clonal nature of this species rather than to inbreeding (D. Soltis et al., 1988).

Genetic Structure of Species: The genetic structure of species can also be analyzed using F-statistics. Here, levels of interpopulational differentiation are measured by F_{ST}. F-statistics have been calculated for seven species of homosporous pteridophytes: five homosporous ferns and two lycopods. With the exception of Dryopteris expansa and Lycopodium clavatum, all species exhibit high levels of similarity among populations (Table 12.6), regardless of mating system. These low levels of interpopulational genetic divergence are also reflected in high genetic identity values among conspecific populations (Table 12.7). Previous studies have demonstrated high spore dispersibility (Tryon, 1970, 1986; Page 1979) and interpopulational gene flow (P. Soltis and D. Soltis, 1987; D. Soltis and P. Soltis, 1987a) in homosporous ferns, factors that may act as cohesive forces in the genetic structure of fern species, regardless of mating system.

Electrophoretic data therefore reveal substantial levels of intergametophytic matings in most species of homosporous pteridophytes examined, contrary to traditional views of mating systems in homosporous plants. These data have furthermore demonstrated that populations of outcrossing species are typically genetically homogeneous, unless ecological or developmental factors result in population substructure. Finally, in contrast to the genetic structure typical of inbreeders, species of homosporous pteridophytes maintain most of their genetic variation within rather than among populations. Interpopulational genetic divergence is relatively low, probably reflecting high levels of gene flow via spore dispersal.

Table 12.5. *Analysis of genetic structure within populations of homosporous pteridophytes. F_{ST} values were corrected for sample size following Nei and Imaizumi (1966)*

Species	Population	F_{IS}	F_{ST}	F_{IT}	Reference
Ferns					
Blechnum spicant	Russian Gulch	0.059	0.032	0.131	P. Soltis and
	Mill Creek	−0.654	0.000	−0.016	D. Soltis,
	S. Fork Smith	0.030	0.016	0.090	1988a
	Twanoh	−0.073	0.029	−0.005	
	Mt. Rainier	0.206	0.175	0.382	
Cheilanthes gracillima	Kamiak	0.006	0.225	0.267	P. Soltis
	Hughes Ridge	0.064	0.185	0.283	et al.,
	Tahoe	−0.175	0.286	0.206	1989
	Moscow Mtn.	−0.460	0.332	0.062	
	Steptoe	−0.357	0.292	0.079	
Polystichum munitum	Gorge-1	−0.100	0.056	0.017	P. Soltis and
	Gorge-2	−0.083	0.007	−0.018	D. Soltis, 1987
Equisetum					
E. arvense	Wenatch	−0.530	0.191	−0.163	D. Soltis et al., 1988

Table 12.6. Genetic structure of species of homosporous pteridophytes, as indicated by F-statistics

Species	F_{IS}	F_{ST}	F_{IT}	Number of Populations	Reference
Ferns					
Blechnum spicant	0.097	0.068	0.134	5	P. Soltis and D. Soltis, 1988a
Botrychium virginianum	0.957	0.080	0.960	4	D. Soltis and P. Soltis, 1986
Dryopteris expansa	0.396	0.235	0.522	8	D. Soltis and P. Soltis, 1987b
Polystichum munitum	0.033	0.054	0.101	4	P. Soltis and D. Soltis, 1987
P. imbricans	0.047	0.024	0.075	8	D. Soltis and P. Soltis, 1987a
Lycopods					
Huperzia miyoshiana	0.002	0.055	0.103	4	P. S. Soltis and D. E. Soltis, unpubl. data
Lycopodium clavatum	0.107	0.188	0.506	6	P. S. Soltis and D. E. Soltis, unpubl. data

Table 12.7 Genetic identities among conspecific populations of homosporous pteridophytes

Species	$\bar{I}$	Number of Populations	Range	Reference
Bommeria hispida	0.879	12	0.747–0.989	Haufler, 1985
Polystichum imbricans	0.982	8	0.930–0.995	D. Soltis and P. Soltis, 1987a
Polystichum munitum	0.995	4	0.990–0.999	P. Soltis amd D. Soltis, 1987
Blechnum spicant	0.996	6	0.988–1.000	P. Soltis and D. Soltis, 1988a
Pellaea andromedifolia	0.943	12	0.781–1.000	Gastony and Gottlieb, 1985
Dryopteris expansa	0.989	9	0.967–1.000	D. Soltis and P. Soltis, 1987a
Botrychium virginianum	0.992	4	0.986–1.000	D. Soltis and P. Soltis, 1986
Equisetum arvense	0.962	18	0.903–1.000	D. E. Soltis and P. S. Soltis, unpubl. data
Adiantum pedatum (eastern woodland diploid)	0.965	6	0.936–0.989	Paris and Windham, 1988
Adiantum pedatum (serpentine diploid)	0.986	2	0.986	Paris and Windham, 1988
Hemionitis palmata	0.823	13	0.520–1.000	Ranker, 1987
Hemionitis elegans	0.79	3	0.76–0.84	Ranker, 1987
Gymnopteris rufa	0.81	3	0.72–1.00	Ranker, 1987
Gymnopteris subcordata	0.90	2	0.90	Ranker, 1987
Hemionitis pinnatifida	0.78	3	0.67–0.76	Ranker, 1987
Pteridium aquilinum	0.78	4	0.723–0.968	Wolf et al., 1988

GENETIC DIFFERENTIATION

Several investigators have reviewed Nei's (1972) genetic identity values (I) for conspecific populations and congeneric species of seed plants (Gottlieb, 1977, 1981; Crawford, 1983). Because electrophoretic analyses of pteridophytes, as well as non-vascular plants (see chapter by Wyatt, Stoneburner, and Odrzykoski), have lagged behind investigations of seed plants, sufficient data have not previously been available to make definitive statements regarding genetic differentiation among taxa in these groups. However, a substantial electrophoretic data base is rapidly accumulating for pteridophytes, facilitating a meaningful synthesis of data.

The mean genetic identity values ($\bar{I}$) computed for conspecific populations of seed plants by Gottlieb (1977, 1981) and later by Crawford (1983), using a larger data base, are very high and remarkably consistent. Most values of $\bar{I}$ range from 0.95 to 1.00, with 0.87 the lowest value reported. These values represent an ecologically, as well as taxonomically, diverse group of species and include both selfers and outcrossers (Crawford, 1983).

In general, genetic identity values among subspecies or varieties of angiosperms are also high and comparable to those reported for populations of the same taxon (Crawford, 1983). However, genetic identity typically is lower for congeneric species of seed plants than for conspecific populations. Gottlieb (1981) reported a value of $\bar{I}$ of 0.67 for congeneric species, and Crawford (1983) reported the same mean for 23 additional species, with values ranging from 0.35 to 0.97.

Conspecific populations of pteridophytes in most instances similarly have very high genetic identity values (Table 12.7). This has been found for taxa occupying a wide range of habitats, such as desert ferns of the genus *Bommeria* (Haufler, 1985) and mesic ferns, including *Polystichum munitum* (P. Soltis and D. Soltis, 1987). The species examined (Table 12.7) also have a variety of mating systems from inbreeding to outcrossing and are taxonomically diverse. The high genetic identity values generally encountered are not surprising given the high levels of gene flow reported for some homosporous pteridophytes (P. Soltis and D. Soltis, 1987; D. Soltis and P. Soltis, 1987a). P. Soltis and D. Soltis (1987) determined, for example, that estimates of gene flow in *Polystichum munitum* (calculated following Slatkin, 1985) are comparable to those reported for gymnosperms.

The work of Ranker (1987), however, has revealed remarkably low I values among conspecific populations within *Gymnopteris rufa* and several species of *Hemionitis* (Table 12.7). Some of these values are much lower than values for conspecific populations in seed plants and, in fact, are more similar to values for congeneric species of seed plants. In *Hemionitis palmata*, for example, some pair-wise comparisons yielded values as low as 0.52, although most values are much higher (> 0.90). Ranker (1987, personal communication) attributes these low values to long periods of geographic isolation of one or a few populations. Furthermore, because of rarity and isolation, gene flow among populations in some taxa (e.g., *Gymnopteris rufa*) may be quite low, which could also contribute to low I values. Similar explanations can be proposed for the I values determined by Wolf et al. (1988) for populations of the cosmopolitan *Pteridium aquilinum*. Comparison of two populations from England yielded a high I value (> 0.95), but comparison of populations from Europe and North America yielded an I value of only 0.723.

In striking contrast to results for angiosperms, the several reports of I for pteridophytes originally treated as varieties are remarkably low. For example, Ranker et al. (1986) calculated an $\bar{I}$ value of only 0.030 for two varieties of *Notholaena candida*. Similarly, Paris and Windham (1988) found an $\bar{I}$ value of 0.495 for two varieties of *Adiantum pedatum*. Crawford (1983) suggested that the very high genetic identity values for subspecies or varieties of angiosperms were probably the result of either or both of the following: (1) recent divergence with insufficient time for divergence at isozyme loci; and (2) possible hybridization between the subspecific taxa, thus preventing divergence. The extremely low genetic identity values for varieties of pteridophytes indicate that hybridization and gene flow between these taxa probably are not occurring and further suggest that these taxa are very old. These data also indicate that high morphological similarity between pteridophyte taxa may, in some instances, belie considerable genetic differentiation. Such extreme genetic divergence raises the question as to whether specific status is more appropriate for these varieties. The data are also in striking contrast to morphologically diverse angiosperm congeners that exhibit extremely high genetic identities, such as species of Bidens and Tetramolopium from the Hawaiian Islands (Lowrey and Crawford, 1985; Helenurm and Ganders, 1985; see also the chapter by Crawford in this volume).

In most instances, *I* values for congeneric pteridophyte species are extremely low (Table 12.8; included in this table are genera representing Pteridophyta and Microphyllophyta; see also Haufler, 1987). Werth et al. (1985), for example, reported a mean Rogers's genetic similarity value of only 0.399 for three diploid species of *Asplenium*. C. H. Haufler (unpublished data) observed *I* values of 0.00 between several species of *Cystopteris*. The overall $\bar{I}$ value for congeneric species of pteridophytes based on the data in Table 12.8 is only 0.33, much lower than the mean value of 0.67 for congeneric species of seed plants.

Table 12.8. *Genetic identities between congeneric species of homosporous pteridophytes*

Species	I	Reference
Asplenium platyneuron—A. montanum	0.296	Werth et al., 1985*
A. platyneuron—A. rhizophyllum	0.403	
A. montanum—A. rhizophyllum	0.498	
Bommeria hispida—B. subpaleacea	0.205	Haufler, 1985
B. hispida—B. ehrenbergiana	0.099	
B. hispida—B. pedata	0.303	
B. subpaleacea—B. ehrenbergiana	0.339	
B. subpaleacea—B. pedata	0.232	
B. ehrenbergiana—B. pedata	0.247	
Cystopteris protrusa—C. bulbifera	0.00	C. H. Haufler, unpubl. data
Hemionitis elegans—Bommeria subpaleacea	0.06	Ranker, 1987
H. elegans—B. ehrenbergiana	0.06	
Hemionitis palmata—H. levyi	0.23	Ranker, 1987
H. palmata—Gymnopteris rufa	0.26	
H. levyi—Gymnopteris rufa	0.04	
Huperzia lucidula—H. "occidentalis"	0.39	P. S. Soltis and D. E. Soltis,
H. lucidula—H. miyoshiana	0.50	unpubl. data
H. "occidentalis"—H. miyoshiana	0.22	
Polystichum munitum—P. imbricans	0.85	P. S. Soltis and D. E. Soltis,
P. lemmonii—P. imbricans	0.62	unpubl. data
P. lemmonii—P. munitum	0.71	
P. lonchitis—P. munitum	0.57	
P. lonchitis—P. imbricans	0.52	
P. lonchitis—P. lemmonii	0.36	

*Rogers's genetic similarity

In some instances, these extremely low $\bar{I}$ values may be due, in part, to poorly understood generic concepts. Haufler (1985) cautioned, for example, that the genus *Bommeria*, for which the $\bar{I}$ value for the four species is only 0.238, may be polyphyletic. That is, due to considerable convergence in morphological characters, the species may not be as closely related as traditionally maintained. This, however, may only partially explain the low values obtained for *Bommeria*. Certainly some of the species analyzed by Haufler (1985) do represent part of a monophyletic assemblage (see Ranker, 1987), yet the highest pair-wise *I* value for any two species is only 0.336. More recently, for example, Ranker (1987) proposed that *Bommeria ehrenbergiana*, *B. subpaleacea*, and *Hemionitis elegans* form a monophyletic assemblage; the $\bar{I}$ value for these three species is only 0.08 (Ranker, 1987).

A likely explanation for the very low genetic identity values typically encountered in electrophoretic analyses of pteridophytes is the great antiquity of these plants. The data presented here support the view of Crawford (presented in this volume) that time of divergence plays a critical role in determining *I* values. Because most modern fern genera are considered to be at least 100 million years old (T. Delevoryas and J. Skog, personal communication), there has been considerable time for genetic divergence to occur among species. On the other hand, certainly not all pteridophyte species are of ancient origin; many undoubtedly represent more recent speciation events. The much higher genetic identity values reported for some congeneric species may therefore reflect a more recent origin. In *Polystichum*, for example, there is a complex of species restricted to the Pacific Northwest of North America, and some of these taxa may represent a relatively recent radiation. Genetic identity values among the diploids of this complex are generally high. P. S. Soltis and D. E. Soltis (unpublished data) found a very high genetic identity value between two of these species, *P. imbricans* and *P. munitum* (0.85). Comparison of *P. lemmonii*, also of this complex, to *P. imbricans* and *P. munitum* also yielded relatively high I values (0.62 and 0.71, respectively), relative to mean values for most pteridophyte congeners. However, when western North American Polystichums are compared to more widespread taxa, such as the circumboreal *P. lonchitis*, lower I values are observed (Table 12.8).

CONCLUSIONS

Genetic data obtained via enzyme electrophoresis have greatly changed our views of pteridophyte biology. The two traditional tenets of homosporous pteridophyte biology, (1) maintenance and release of genetic variability through polyploidy and homoeologous chromosome pairing, and (2) high levels of intragametophytic self-fertilization, have been shown to be incorrect. Electrophoretic analyses have demonstrated, instead, that all groups of homosporous pteridophytes are isozymically diploid. There is, therefore, no genetic evidence for high levels of polyploidy in these organisms. Since there is no evidence of duplicate gene expression due to polyploidy, there is also no evidence that homosporous pteridophytes release genetic variation through homoeologous pairing. Electrophoretic studies of gametophytes grown from spores of individual sporophytes have, in contrast, revealed simple patterns of disomic inheritance.

Electrophoretic evidence for genetic diploidy in pteridophytes also has profound evolutionary implications. Although the fixed heterozygous genotypes of allopolyploids may permit new allopolyploids to be successful, the polyploid genetic system may also impose restrictions on the release of new variation and thereby limit the evolutionary potential of the polyploid lineage (Grant, 1981; see also Stebbins, 1950, 1980). As the trend to higher ploidal levels progresses in a lineage, the limitations on evolutionary potential must become even greater. Because homosporous pteridophytes have traditionally been considered highly polyploid, they have also been viewed as potential evolutionary dead ends, locked into a genetic system that restricts mutational and recombinational variability. The fact that these plants are genetic diploids means, however, that this view no longer holds.

Although some homosporous pteridophytes exhibit high levels of intragametophytic selfing, in most taxa analyzed intergametophytic matings predominate. A range of mating systems operates in homosporous pteridophytes, from inbreeding through mixed to outcrossing; most pteridophytes are outcrossing. Thus, a new synthesis has replaced the traditional views of pteridophyte biology: genetic diploidy despite high chromosome numbers, and maintenance of genetic variation through various levels of outcrossing. In contrast to previous concepts of pteridophyte biology which emphasized that striking differences exist between these plants and all seed plants in terms of polyploidy and breeding system, the picture that has emerged is one of similarity. Electrophoretic analyses have revealed one obvious difference between seed plants and pteridophytes: congeneric species of pteridophytes are typically much more highly differentiated genetically than are congeneric species of seed plants. This difference is probably attributable to the great antiquity of pteridophyte species.

Enzyme electrophoresis will continue to play an important role in the study of pteridophyte systematics (e.g., in ascertaining the parentage of polyploids). Several important evolutionary questions that require enzyme electrophoresis also remain to be addressed in pteridophytes. Additional breeding system analyses are needed, for example, to understand the breeding systems of allopolyploids relative to their diploid progenitors. Also, electrophoretic studies should be coordinated with field and laboratory analyses of gametophytes to elucidate the factors that control breeding systems in these plants. Electrophoretic analyses also have posed a new set of major evolutionary questions that can only be addressed using molecular genetic approaches. It is essential to resolve the pteridophyte paradox of genetic diploidy despite high chromosome numbers. Using molecular genetic methodology it should be possible to ascertain whether homosporous pteridophytes are truly polyploids, having experienced massive gene silencing, or whether they are truly diploid, having achieved high chromosome numbers via another mechanism. Regardless of which hypothesis is correct, homosporous pteridophytes offer the possibility of a genetic system that contrasts with that of seed plants. As a result, they offer exciting possibilities for future molecular genetic analyses.

ACKNOWLEDGMENTS

We thank Steve Brunsfeld, Dan Crawford, George Diggs, Chris Haufler, Ruth Kirkpatrick, Bryan Ness, Tom Ranker, Paul Wolf, and Robert Wyatt for helpful comments on the manuscript. This work was supported in part by NSF grants BSR-8516721 and BSR-8620444 and a grant from the Whitehall Foundation.

LITERATURE CITED

Bendich, A. J., and R. S. Anderson. 1983. Repeated DNA sequences and species relatedness in the genus *Equisetum*. *Pl. Syst. Evol.* 143: 47–52.

Bold, H. C., C. J. Alexopoulos, and T. Delevoryas. 1986. *Morphology of plants and fungi.* Harper and Row, New York.

Brown, A. H. D. 1978. Isozymes, plant population genetic structure and genetic conservation. *Theor. Appl. Genet.* 52: 145–157.

———. 1979. Enzyme polymorphism in plant populations. *Theor. Popul. Biol.* 15: 1–42.

Bruce, J. G., and J. M. Beitel. 1979. A community of *Lycopodium* gametophytes in Michigan. *Amer. Fern J.* 69: 33–41.

Chapman, R. H., E. J. Klekowski, and R. K. Selander. 1979. Homoeologous heterozygosity and recombination in the fern *Pteridium aquilinum*. *Science* 204: 1207–1209.

Clegg, M. T. 1980. Measuring plant mating systems. *BioScience* 30: 814–818.

Crawford, D. J. 1983. Phylogenetic and systematic inference from electrophoretic studies. In S. D. Tanksley and T. J. Orton [eds.], *Isozymes in plant genetics and breeding,* Part A, 257–287. Elsevier, Amsterdam.

———. 1985. Electrophoretic data and plant speciation. *Syst. Bot.* 10: 405–416.

Döpp, W. 1950. Ein die Antheridienbildung bei Farnen fordernde Substanz in den Prothallien von *Pteridium aquilinum* (L). Kuhn. *Ber. Dtsch. Bot. Ges.* 63: 139–147.

Duckett, J. G. 1970. Sexual behavior of the genus *Equisetum*, subgenus *Equisetum*. *Bot. J. Linn. Soc.* 63: 327–352.

———. 1972. Sexual behavior of the genus *Equisetum*, subgenus *Hippochaete*. *Bot. J. Linn. Soc.* 65: 87–108.

———, and A. R. Duckett. 1980. Reproductive biology and population dynamics of wild gametophytes of *Equisetum*. *Bot. J. Linn. Soc.* 80: 1–40.

Duncan, T., and A. R. Smith. 1978. Primary basic chromosome numbers in ferns: facts or fantasies? *Syst. Bot.* 3: 105–114.

Freeberg, J. A., and R. H. Wetmore. 1957. Gametophytes of *Lycopodium* as grown *in vitro*. *Phytomorphology* 7: 204–217.

Gastony, G. J., and L. D. Gottlieb. 1982. Evidence for genetic heterozygosity in a homosporous fern. *Amer. J. Bot.* 69: 634–637.

———, and ———. 1985. Genetic variation in the homosporous fern *Pellaea andromedifolia*. *Amer. J. Bot.* 72: 257–267.

Gottlieb, L. D. 1977. Electrophoretic evidence and plant systematics. *Ann. Missouri Bot. Gard.* 64: 161–180.

———. 1981. Electrophoretic evidence and plant populations. *Prog. Phytochem.* 7: 1–46.
———. 1982. Conservation and duplication of isozymes in plants. *Science* 216: 373–380.
Grant, V. 1981. *Plant speciation.* Columbia Univ. Press, New York.
Hamrick, J. L., Y. B. Linhart, and J. B. Mitton. 1979. Relationships between life history characteristics and electrophoretically detectable genetic variation in plants. *Ann. Rev. Ecol. Syst.* 10: 173–200.
Haufler, C. H. 1985. Enzyme variability and modes of evolution in the fern genus *Bommeria*. *Syst. Bot.* 10: 92–104.
———. 1987. Electrophoresis is modifying our concepts of evolution in homosporous pteridophytes. *Amer. J. Bot.* 74: 953–966.
———, and D. E. Soltis. 1984. Obligate outcrossing in a homosporous fern: field confirmation of a laboratory prediction. *Amer. J. Bot.* 71: 876–881.
———, and ———. 1986. Genetic evidence suggests that homosporous ferns with high chromosome numbers are diploid. *Proc. Natl. Acad. Sci. USA* 83: 4389–4393.
Hauke, R. L. 1967. Sexuality in a wild population of *Equisetum arvense* gametophytes. *Amer. Fern J.* 57: 59–66.
———. 1969. Gametophyte development in Latin American horsetails. *Bull. Torrey Bot. Club* 96: 568–577.
———. 1977. Experimental studies on growth and sexual determination in *Equisetum* gametophytes. *Amer. Fern J.* 67: 18–31.
Hedrick, P. W. 1987. Population genetics of intragametophytic selfing. *Evolution* 41: 137–144.
Helenurm, K., and F. R. Ganders. 1985. Adaptive radiation and genetic differentiation in Hawaiian *Bidens*. *Evolution* 39: 753–765.
Hickok, L. G. 1978. Homoeologous chromosome pairing and restricted segregation in the fern *Ceratopteris*. *Amer. J. Bot.* 65: 526–521.
Holsinger, K. E. 1987. Gametophytic self-fertilization in homosporous plants: development, evaluation, and application of a statistical method for evaluating its importance. *Amer. J. Bot.* 74: 1173–1183.
Jain, S. K. 1976. The evolution of inbreeding in plants. *Ann. Rev. Ecol. Syst.* 7: 469–495.
Klekowski, E. J., Jr. 1972. Genetical features of ferns as contrasted to seed plants. *Ann. Missouri Bot. Gard.* 50: 138–151.
———. 1973. Sexual and subsexual systems in the homosporous ferns: a new hypothesis. *Amer. J. Bot.* 60: 535–544.
———. 1976. Homoeologous chromosome pairing in ferns. *In* K. Kones and P. E. Brandham [eds.], *Current chromosome research*, 175–184. North-Holland, New York.
———. 1979. The genetics and reproductive biology of ferns. *In* A. F. Dyer [ed.], *The experimental biology of ferns*, 133–170. Academic Press, London.
———, and H. G. Baker. 1966. Evolutionary significance of polyploidy in the Pteridophyta. *Science* 135: 305–307.
Lellinger, D. B. 1985. *A field manual of the ferns and fern allies of the United States and Canada.* Smithsonian Institution Press, Washington, D.C.
Lloyd, R. M. 1974a. Reproductive biology and evolution in the Pteridophyta. *Ann. Missouri Bot. Gard.* 61: 318–331.
———. 1974b. Mating systems and genetic load in pioneer and non-pioneer Hawaiian Pteridophyta. *Bot. J. Linn. Soc.* 69: 232–25.
Löve, A., D. Löve, and R. E. G. Pichi Sermolli. 1977. *Cytotaxonomical atlas of the Pteridophyta.* Cramer, Vaduz.
Loveless, M. D., and J. L. Hamrick. 1984. Ecological determinants of genetic structure in plant populations. *Ann. Rev. Ecol. Syst.* 15: 65–95.
Lovis, J. D. 1973. A biosystematic approach to phylogenetic problems and its application to the Aspleniaceae. *In* The phylogeny and classification of ferns, 221–228. *Bot. J. Linn. Soc.* 67: Supplement 1.
Lowrey, T. K., and D. J. Crawford. 1985. Allozyme divergence and evolution in *Tetramolopium*

(Compositae: Astereae) on the Hawaiian Islands. *Syst. Bot.* 10: 64–72.

Manton, I. 1950. *Problems of cytology and evolution in the Pteridophyta.* Cambridge Univ. Press, London.

———. 1961. Evolution in the Pteridophyta. *Bot. Soc. Br. Isles Conf. Rep.* 6: 105–120.

McCauley, D. E., D. P. Whittier, and L. M. Reilly. 1985. Inbreeding and the rate of self-fertilization in a grape fern, *Botrychium dissectum. Amer. J. Bot.* 72: 1978–1981.

Näf, U. 1979. Antheridiogens and antheridial development. *In* A. F. Dyer [ed.], *The experimental biology of ferns,* 436–479. Academic Press, London.

Nei, M. 1972. Genetic distance between populations. *Amer. Naturalist* 106: 283–293.

———, and Y. Imaizumi. 1966. Genetic structure of human populations. II. Differentiation of blood group frequencies among isolated populations. *Heredity* 21: 183–190.

Page, C. N. 1979. Experimental aspects of fern ecology. *In* A. F. Dyer [ed.], *The experimental biology of ferns,* 552–589. Academic Press, London.

Paris, C. A., and M. D. Windham. 1988. A biosystematic investigation of the *Adiantum pedatum* complex in eastern North America. *Syst. Bot.*: 13: 240–255.

Ranker, T. A. 1987. Experimental systematics and population biology of the fern genera *Hemionitis* and *Gymnopteris* with reference to *Bommeria.* Ph.D. thesis, University of Kansas, Lawrence.

———, C. H. Haufler, and M. D. Windham. 1986. Biosystematic studies of gymnogrammoid/cheilanthoid ferns. *Amer. J. Bot.* 73: 731–732.

Roose, M. L., and L. D. Gottlieb. 1976. Genetic consequences of allopolyploidy in *Tragopogon. Evolution* 30: 818–830.

Slatkin, M. 1985. Rare alleles as indicators of gene flow. *Evolution* 39: 53–56.

Soltis, D. E. 1986. Genetic evidence for diploidy in *Equisetum. Amer. J. Bot.* 73: 908–913.

———, and P. S. Soltis. 1986. Electrophoretic evidence for inbreeding in the fern *Botrychium virginianum* (Ophioglossaceae). *Amer. J. Bot.* 73: 588–592.

———, and ———. 1987a. Polyploidy and breeding systems in homosporous Pteridophyta: a reevaluation. *Amer. Naturalist* 130: 219–232.

———, and ———. 1987b. Breeding system of the fern *Dryopteris expansa*: evidence for mixed-mating. *Amer. J. Bot.* 74: 504–509.

———, and ———. 1988. Are lycopods with high chromosome numbers ancient polyploids? *Amer. J. Bot.* 75: 238–247.

———, ———, and R. D. Noyes. 1988. An electrophoretic investigation of intragametophytic selfing in *Equisetum arvense. Amer. J. Bot.* 75: 231–237.

Soltis, P. S., and D. S. Soltis. 1987. Population structure and estimates of gene flow in the homosporous fern *Polystichum munitum. Evolution* 41: 620–629.

———, and ———. 1988a. Genetic variation and population structure in *Blechnum spicant* (Blechnaceae) in western North America. *Amer. J. Bot.* 75: 37–44.

———, and ———. 1988b. Estimated rates of intragametophytic selfing in lycopods. *Amer. J. Bot.* 75: 248–256.

———, and ———. 1988c. Electrophoretic evidence for genetic diploidy in *Psilotum nudum. Amer. J. Bot.*: 75: 1667–1671.

———, ———, and B. D. Ness. 1989. Population genetic structure in *Cheilanthes gracillima. Amer. J. Bot.*: In press.

Stebbins, G. L. 1971. *Chromosome evolution in higher plants.* Edward Arnold, London.

Tanksley, S. D. 1987. Organization of the nuclear genome in tomato and related diploid species. *Amer. Naturalist* 130: 546–561.

Tryon, R. M. 1970. Development and evolution of fern floras of oceanic islands. *Biotropica* 2: 76–84.

———. 1986. The biogeography of species, with special reference to ferns. *Bot. Rev.* 52: 117–156.

———, and A. F. Tryon. 1982. *Ferns and allied plants.* Springer-Verlag, New York.

Vida, G. 1976. The role of polyploidy in evolution. *In* V. J. A. Novák and B. Pacltová [eds.], *Evolutionary biology,* 267–204. Czechoslovak Academy of Sciences, Prague.

Voeller, B. R. 1964. Antheridogens in ferns. *Colloq. Int. Cent. Natl. Rech. Sci.* 123: 665–684.

Wagner, W. H., and F. S. Wagner. 1980. Polyploidy in pteridophytes. *In* W. H. Lewis [ed.],

Polyploidy: biological relevance, 199–214. Plenum, New York.

———, ———, and J. M. Beitel. 1985. Evidence for interspecific hybridization in pteridophytes with subterranean mycoparasitic gametophytes. *Proc. Roy. Soc. Edin.* 86B: 273–281.

Werth, C. R., S. I. Guttman, and W. H. Eshbaugh. 1985. Electrophoretic evidence of reticulate evolution in the Appalachian *Asplenium* complex. *Syst. Bot.* 10: 184–192.

White, M. J. D. 1978. *Modes of speciation.* Freeman, San Francisco.

Whittier, D. P. 1973. Germination of *Psilotum* spores in axenic culture. *Canad. J. Bot.* 51: 2000–2001.

———. 1977. Gametophytes of *Lycopodium obscurum* as grown in axenic culture. *Canad. J. Bot.* 55: 563–567.

Wolf, P. G., C. H. Haufler, and E. Sheffield. 1987. Electrophoretic evidence for genetic diploidy in the bracken fern (*Pteridium aquilinum*). *Science* 236: 947–949.

———, ———, and ———. 1988. Genetic variation and mating system of the clonal weed *Pteridium aquilinum* L. Kuhn (Bracken). *Evolution:* 42: 1350–1354.

Wright, S. 1965. The interpretation of population structure by F-statistics with special regard to systems of mating. *Evolution* 19: 395–420.

Index

A, mean number of alleles per locus, 89, 107, 112, 120, 222
AAT, see aspartate aminotransferase
Abies, 59
Abies balsamea, 132
Abutilon theophrasti, 113–114
acid phosphatase (ACP), 18, 38, 195, 197, 200
aconitase, see aconitate hydratase
aconitate hydratase (ACO), 18
ACP, see acid phosphatase
adaptive distance, 135, 137
adaptive importance, 121, 138
adenylate kinase (ADK), 19
ADH, see alcohol dehydrogenase
Adiantum pedatum, 251
agarose gels, 6
Ageratum, 118
Aglossopsida, 241–243
Agropyron, 53
ajara, 174
alanine aminotransferase (ALT), 19
alcohol dehydrogenase (ADH), 20, 116, 134, 140, 199–200, 213–214
aldolase (ALD), 20, 23
alfalfa, 59, 179
alkaline phosphatase, 200
allelic frequencies, 217
allogamy, 78
allopolyploidy, 56, 107, 109, 118–121, 152–154, 156–157, 159, 229–230, 236, 242–244
allozyme divergence, 150–151, 157
allozymes, 5, 146
alternation of generations, 221
Amaranthaceae, 118
amaranths, see also *Amaranthus*, 179
Amaranthus, 118
aminopeptidase (AMP), 20, 213
AMP, see aminopeptidase
Amsinckia, 118
amylase (AMY), 20, 198
ancient polyploidy, 245
Andreaea, 228
Annona cherimola, 201
Annona cherimola × *Annona squamosa*, 201
annual ryegrass, 136
Antennaria, 56, 153

antheridiogen, 242, 249
Anthocerotae, 221
Apera spica-venti, 112
apomixis, 74–76, 78, 84, 112
apospory, 229
apple, see also *Malus domestica*, 192–194
Arthrophyta, 240–243, 249
asexual reproduction, 222, 233
aspartate aminotransferase (AAT), see also glutamate oxaloacetate transaminase, 10, 21, 38
Asplenium, 152–153, 253
Asplenium abscissum, 153
Asplenium cristatum, 153
Asplenium curtissii, 153
Asplenium montanum, 247, 253
Asplenium platyneuron, 247, 253
Asplenium plenum, 153
Asplenium rhizophyllum, 253
Asplenium verecundum, 153
Astereae, 156, 158
atemoya, 201
Atrichum, 230
Atrichum angustatum, 224, 234–235
Atriplex, 157
autogamy, 78
autopolyploidy, 57, 107, 109, 118–121, 152–154, 156, 159, 229–230, 243–245
Avena, 1, 119
Avena barbata, 91, 98–99, 113, 131, 207
Avena fatua, 91
avocado, 179, 192–193, 199

β-galactosidase (GAL), 24
β-glucosidase (GLU), 25
banana, see also *Musa*, 167, 192, 201
barley, see also *Hordeum*, 1, 131, 167, 207
Bidens, 91–92, 119, 150, 253
Bidens menziesii, 75
biparental inbreeding, 74, 79
Blechnum spicant, 247–251
blue mussel, 130–131
Bommeria, 252
Bommeria ehrenbergiana, 253
Bommeria hispida, 245, 251, 253
Bommeria pedata, 253
Bommeria subpaleacea, 253

Botrychium dissectum, 248
Botrychium virginianum, 247–248, 251
bottlenecks during domestication, 183
bracken fern, see also *Pteridium aquilinum*, 112
breeding structure, 100
breeding system, 90, 218, 233, 241
Bromus mollis, 119, 140
Bromus tectorum, 113
buckwheat, 179, 184
bryophytes, 221
buffers
    electrode, 11–13
    extraction, 9
    gel, 11–13

Calocedrus, 60
Camellia, 56, 59
Camellia japonica, 54–55
Capsella bursa-pastoris, 116
Capsicum, 54, 56, 59, 61, 149, 170
Capsicum annuum var. *annuum*, 168, 170, 173
Capsicum annuum var. *aviculare*, 168, 173
Capsicum baccatum var. *baccatum*, 168, 170, 173
Capsicum baccatum var. *pendulum*, 168, 170, 173
Capsicum chinense, 168, 170, 173
Capsicum eximium, 168, 170, 173
Capsicum frutescens, 168, 170, 173
Capsicum pubescens, 168, 170, 173
Carduus nutans, 93, 96
Carica cauliflora, 201
Carica papaya, 201
cashew, 192
catalase (CAT), 10, 21
Catalpa speciosa, 83
catechol oxidase, 33
Ceanothus, 157
celery, 59
cellulose acetate membranes, 6
Celosia, 118
Ceratopteris, 246
Cercidiphyllaceae, 243
chalco teosinte, 181
Chamaelirium luteum, 76, 98, 100
Cheilanthes gracillima, 247, 250
Chenopodium, 2, 113, 119, 148, 158, 174
Chenopodium berlandieri subsp. *nuttallieae*, 174
Chenopodium hircinum, 174
Chenopodium incognitum, 148

Chenopodium quinoa subsp. *milleanum*, 168, 173–174
Chenopodium quinoa subsp. *quinoa*, 168, 173–174
Chenopodium quinoa var. *melanospermum*, 174
cherimoya, 192, 201
chili peppers, 170
chloroplast isozymes, 10, 48, 50
Citrullus, 54, 59
Citrullus colocynthis, 180
Citrullus lanatus, 180
Citrullus pepo, 180, 182
Citrullus texana, 180, 182
Citrus, 197
citrus, 179, 192–193, 197–199, 201
Clarkia, 3, 54–56, 62, 154–155, 158
Clarkia arcuata, 155
Clarkia cylindrica, 155
Clarkia epilobioides, 155
Clarkia exilis, 82
Clarkia franciscana, 55
Clarkia lewisii, 155
Clarkia rostrata, 154–155
Clarkia rubicunda, 55
Clarkia tembloriensis, 82
Clarkia williamsonii, 91
Claytonia perfoliata, 118
Climacium americanum, 224, 227
Climacium kindbergii, 227
codominance, 34, 36, 38, 47, 73, 87, 92, 107, 146, 151, 159, 193, 198, 207
cole, 179
Colias, 131
Collinsia sparsiflora var. *arvensis*, 82
colonizing species, 106
Compositae, 150, 156
conifers, 11, 149
Conocephalum, 227
Conocephalum conicum, 223, 225, 228–229, 233, 235
conserved number of isozymes, 38, 47–48, 154–159, 242–243
Coreopsis, 56, 149
Coreopsis grandiflora var. *longipes*, 120, 153
corn, see also maize and *Zea*, 134, 215
cotton, 179
crops
    centers of diversity, 184
    genetic structure, 183
    origin, 166
    progenitors, 166
cryptic variability, 6

cucumber, 179
Cucumis melo, 96, 173
Cucumis melo var. *agrestis*, 173
Cucurbita, 54, 59, 61
Cucurbita fraterna, 176
Cucurbita mixta, 168, 173, 176
Cucurbita pepo, 54, 173, 176
Cucurbita pepo var. *ovifera*, 168
Cucurbita sororia, 168, 173, 176
Cucurbita texana, 168, 173, 176
cultivars, 193
cultivated variety, 192
    identification, 195, 197, 201
cushaw, 176
Cynosurus, 55
Cystopteris, 253
Cystopteris bulbifera, 253
Cystopteris protrusa, 253
cytosolic isozymes, 48, 50

Dactylis glomerata, 153, 155
dark respiration, 138
date palm, 192–193, 200
Datura stramonium, 113–114
deer mouse, 131
Delphinium, 151
Delphinium nelsoni, 129
Dendroseris, 150
Deschampsia cespitosa, 118
Desmatodon cernuus, 234
developmental stability, 130
DIA, see diaphorase
diaphorase (DIA), see also NAD(P)H
    dehydrogenase, 21, 29, 195, 213, 224, 233
Diphasiastrum complanatum, 248–249
Diphasiastrum digitatum, 248–249
diphenol oxidase, 194
diploidization, 57, 229
diplospory, 229
disomic inheritance, 57, 254
domestication, 149, 182
    bottlenecks during, 183
dominance, 127
Drosophila, 1, 6, 206
Drosophila melanogaster, 97, 130
Dryopteris expansa, 247–251
D_{ST}, 89
Dubautia, 150

Echinochloa, 113, 119
Echinochloa crus-galli, 115–117
Echinochloa microstachya, 109–110
Echinochloa oryzoides, 115–117
Echium plantagineum, 82–83, 99, 111
Eichhornia, 118
Eichhornia paniculata, 110–111
electrode buffers, 11–13
electrophoresis, 15
    apparatus, 14
    buffers, 9, 11–13
    methodology, 5
    troubleshooting, 16
    wicks, 15
Elymus, 118
embryony, 198
Emex spinosa, 113
endopeptidase (ENP), 21
endosymbiont hypothesis, 48
Entodon cladorrhizans, 234
enzyme extraction, 7
Equisetum, 56, 156, 241–243, 245–246, 248–250
Equisetum arvense, 246, 248–251
Erharta, 18
Erodium, 113
EST, see esterase
esterase (EST), 22, 38, 194–195, 197–198, 200, 207
Eucalyptus, 82
Eucalyptus delegatensis, 83, 94
Eucalyptus obliqua, 93
Eucharidium, 154
Eupatorium microstemon, 118
extraction buffers, 9

F-statistics, 107, 249–250
female fertilities, 74–75, 80–81
ferns, 11, 56, 151, 156, 158, 241–243, 245–251
fertility
    female, 74–75, 80–81
    male, 74–75, 80–81
Festuca microstachys, 95
Ficus carica, 197
fig, 192–193, 197
F_{IS}, 251
Fissidens, 230
F_{IT}, 251
fitness, 99, 128, 135
    and heterozygosity, 128
fixation index, 107, 109, 246
fixed heterozygosity, 109, 119, 121, 153, 156, 230
fixing gels, 33
flax, 60
formate dehydrogenase (FDH), 22

fructose-1,6-diphosphatase (FDP), 23
fructose-bisphosphatase (FBP), 23
fructose-bisphosphatase aldolase (FBA), 23
fruit crops, 192–193
F_{ST}, 89, 250–251
fumarase, see fumarate hydratase
fumarate hydratase (FUM), 23
Funaria hygrometrica, 234
Fundulus heteroclitus, 130

Galax urceolata, 120, 153
gametic fertility, 74–75
gametophyte, 221
garden pea, 59
geitonogamy, 82–83
gel buffers, 11–13
gel mold, 14
gel slicing, 15–16
gel staining, 16
gene diversity, 89, 107, 109, 148, 222, 224–225, 234
gene duplication, 37–38, 55, 154, 156–159, 243
gene flow, 95–98, 149, 234–235, 250, 252
gene number conservation, see also isozyme, conserved number, 38, 47–48, 154–159, 242–243
gene silencing, 119, 154–158, 229, 243–255
genetic analysis, 38, 52
genetic differentiation, 235–236, 241, 252
genetic diploidy, 254–255
genetic distance, 110
genetic divergence, 254
genetic drift, 208
genetic identity, 148–151, 157–158, 166, 226–228, 236, 250–254
genetic interpretation, 5
genetic load, 128
genetic similarity, 156, 159, 226
genetic structure, 87, 100, 249–250
  of crops, 183
  of populations, 83–84, 107, 109, 112, 115, 120, 233, 235, 246, 249–250
  of species, 249–250
genetic studies, 52
genetic substructure, 93
genetic variability, 222, 236
genetic variation, 88–89, 235
genotroph, 60
genotype, 137
geographical model of speciation, 151
geographical species, 226, 228
Gilia achilleifolia, 82, 95

glucose-6-phosphate dehydrogenase (G6PDH), 24, 213
glucose-6-phosphate isomerase (GPI), 10, 25, 56, 155, 195–196, 198–199
glucosephosphate isomerase, see glucose-6-phosphate isomerase
glutamate dehydrogenase (GDH), 26
glutamate oxaloacetate transaminase (GOT), see also aspartate aminotransferase, 21, 26, 195, 197–200
glutamate-ammonia ligase (GS), 25–26
glutamate-pyruvate transaminase (GPT), 19
glutamine synthetase (GS), see glutamate-ammonia ligase
glyceraldehyde-3-phosphate dehydrogenase (G3PDH), 26
glycerate dehydrogenase, 33
Glycine argyrea, 74, 83
Glycine max, 168, 173
Glycine soja, 168, 173
Godetia, 155
Gossypium, 59, 61
Gossypium arboreum, 54
Gossypium herbaceum, 54
GOT, see glutamate oxaloacetate transaminase
gourds, 176
GPI, see glucose-6-phosphate isomerase
grape, 179, 192
grapefruit, 198
growth rate, 130, 132
G_{ST}, 89–90, 110, 112
Gymnopteris rufa, 251–253
Gymnopteris subcordata, 251
gymnosperms, 78, 80

Haplopappus spinulosus, 119
Hardy-Weinberg equilibrium, 249–250
Helianthus annuus, 54, 83, 92
Hemionitis, 252
Hemionitis elegans, 251, 253
Hemionitis levyi, 253
Hemionitis palmata, 251–253
Hemionitis pinnatifida, 251
hemoglobin, 130–131
Hepaticae, 221
Heterogaura heterandra, 154
heterosis, 127, 129, 208
heterozygosity, 99, 107, 131–134, 222, 247
  and development, 130
  and growth, 130
  and physiology, 130
  and viability, 130, 137–138

fixed, 109, 119–121, 153, 156, 230
mean (H), 2, 89
observed (H_O), 109–110, 112, 120
heterozygote advantage, 99
heterozygote excess, 132
Heuchera, 2, 150
Heuchera americana, 150
Heuchera grossulariifolia, 120, 153
Heuchera micrantha, 153
Heuchera parviflora, 150
Heuchera pubescens, 150
Heuchera villosa, 150
hexokinase (HEX), 27
Hippocastanaceae, 243
hitchhiking, 99
H_O, see heterozygosity, observed (H_O)
homoeologous chromosome pairing, 241–242, 245–246, 254
homosporous ferns, 249–250
homosporous pteridophytes, 250, 252
Hordeum, 53, 167
Hordeum bulbosum, 167
Hordeum jubatum, 91, 119
Hordeum murinum, 91, 113, 167
Hordeum spontaneum, 55, 91, 98
Hordeum vulgare, 91, 95, 98, 131, 207
Hordeum vulgare subsp. *spontaneum*, 167–168, 173
Hordeum vulgare subsp. *vulgare*, 167–168
hornworts, 221
horsetails, 241, 246
H_S, 110, 112, 120, 222
H_T, 89, 110, 112
huauzontle, 174
Huperzia lucidula, 247–249, 253
Huperzia miyoshiana, 247–249, 251, 253
Huperzia "occidentalis", 253
hybrid origin, 157, 198–199
hybrid speciation, 151, 157, 159
hybridization, 148, 152, 156, 158, 228, 232–233, 242, 252
Hydrocharis morsus-ranae, 113–114

I, see genetic identity
IDH, see isocitrate dehydrogenase
inbreeding, 92, 127, 242, 254
depression, 121, 127, 129, 249
indoleacetic acid oxidase, 195, 197
infraspecific taxa, 149
intergametophytic mating, 249–250, 254
intragametophytic self-fertilization, 248–249, 254
intrapopulational variation, 88

introgression, 179
invading species, 106
Ipomoea purpurea, 93, 107
isocitrate dehydrogenase (IDH), 27, 195, 213–214
isoelectric focusing, 196
isozyme, 5, 47, 146
analysis, 52
conserved number, 38, 47–48, 154–159, 242–243
cytosolic, 48, 50
mitochondrial-specific, 10
multiplicity, 51
plastid-specific, 10, 48, 50
subcellular location, 10, 34, 36, 47–50

killifish, 130–131
Knautia, 118

laccase, 33
lactate dehydrogenase (LDH), 28, 131, 140
Lactuca, 54
Lactuca sativa, 169, 173
Lactuca serriola, 169, 173
LAP, see leucine aminopeptidase
Lasthenia, 151
Lasthenia burkei, 3, 152
Lasthenia conjugens, 152
Lasthenia fremontii, 152
Layia, 56, 91, 151
Lens, 59, 61, 171
Lens culinaris, 54
Lens culinaris subsp. *culinaris*, 169, 171–172
Lens culinaris subsp. *orientalis*, 169, 171–172
Lens ervoides, 172
Lens nigricans, 171–172
Lens odomensis, 172
lentil, see also *Lens culinaris* subsp. *culinaris*, 54, 171
lettuce, see also *Lactuca sativa*, 172
leucine aminopeptidase (LAP), 20, 131, 140, 158, 200
Liatris cylindracea, 99, 131
lime, 198
Limnanthes, 54, 149
linkage, 195–196, 198–199, 201, 207, 218
conservation, 59
disequilibrium, 111, 135, 138, 207–209
equilibrium, 209
map, 52–53, 201
relationships, 38
Liriodendron tulipifera, 131
Lisianthius skinneri, 91

liverworts, 221, 223, 225
lodgepole pine, 132
Lolium multiflorum, 136
Lolium perenne, 138
Lupinus albus, 80
Lupinus nanus, 82
Lupinus succulentus, 82
Lupinus texensis, 96
Lycopersicon, 59, 61, 177, 208
Lycopersicon esculentum, see also tomato, 53, 177, 207–208
Lycopersicon esculentum var. *cerasiforme*, 169, 177, 180
Lycopersicon esculentum var. *esculentum*, 169, 177, 180
Lycopersicon pennellii, 208
Lycopersicon pimpinellifolium, 82, 177, 180, 184, 208
Lycopersicon cheesmanii, 177
Lycopodiaceae, 249
Lycopodium, see also lycopods, 241
Lycopodium annotinum, 247–249
Lycopodium clavatum, 247–251
Lycopodium obscurum, 247–249
lycopods, 156, 241, 243–244, 246–248, 250–251

Mabrya, 151, 158
Madiinae, 150
Magnoliaceae, 243
maize, see also corn and *Zea*, 11, 53, 56, 59, 61–62, 172, 181, 184, 207–209, 211, 215, 218
malate dehydrogenase (MDH), 10, 28
malate dehydrogenase (oxaloacetate-decarboxylating) (NADP$^+$) (ME), 28
male and female fertility, 76
male fertility, 74–75, 80–81
malic enzyme (ME), see malate dehydrogenase (oxaloacetate-decarboxylating) (NADP$^+$)
Malus, 59
Malus domestica, 194
Malus pumila, 194
mandarins, 199
Mangifera indica, 200
mango, 192, 200
mannose phosphate isomerase (MPI), 29
marker loci, 210–211, 213–215, 217–218
mating parameters of populations, 73
mating systems, 73, 92, 95, 236, 246, 254
  of individuals, 73–74
  of populations, 78

MDH, see also malate dehydrogenase, 194, 196, 198, 213–214
ME, see malate dehydrogenase (oxaloacetate-decarboxylating) (NADP$^+$)
Medicago, 57, 119
melons, 179
menadione reductase (MNR), 29
microbodies, 10
Microphyllophyta, 241–243, 253
mitochondrial-specific isozymes, 10
mixed mating, 78–79, 83, 92, 96, 247
Mniaceae, 230
modifiers, 60
Modiolus demissus, 130
morphometric traits, 90–91
Morus, 196
Morus alba, 196
Morus bomysis, 196
Morus latifolia, 196
mosses, 221
mulberry, 192, 196
Musa, 167, 201
Musa acuminata, 167, 201
Musa balbisiana, 167, 201
Musa paradisiaca, 201
Musci, 221
Myristicaceae, 243
Mytilis edulis, 130
Myxocarpa, 155

N-acetyl-β-glucosaminidase (NAG), 18
NAD(P)H dehydrogenase (NAD(P)DH), 29
naranjilla, 179
nectarine, 193
neighborhood size, 235
Nicotiana, 176
Nicotiana otophora, 177
Nicotiana sylvestris, 176
Nicotiana tabacum, 176
Nicotiana tomentosiformis, 177
nitrate reductase, 33
non-genetic phenomena, 36
Notholaena candida, 252
null alleles, 36–37, 61

oats, 179
Oenothera, 54
Oenothera biennis, 82
Olea europaea, 197
olive, 192–193, 197
Onagraceae, 56
organelle-specific isozymes, 50, 193

Oryza, 119, 175
Oryza breviligulata, 169, 175, 180–181
Oryza glaberrima, 169, 173, 175, 180
Oryza rufipogon, 169, 173–175
Oryza sativa, see also rice, 53, 175, 184
Oryza sativa subsp. *indica*, 169, 173, 175
Oryza sativa subsp. *japonica*, 169, 173, 175
outbreeding depression, 129
outcrossing, 92–96, 112, 236, 242, 247, 249
    spatial and temporal variation, 95
overdominance, 89, 107, 127, 214–215

P, proportion of polymorphic loci per population, 1, 110, 112, 120, 222
Panicum miliaceum, 113–114
papaya, 192, 201
paternity, 75–76, 80, 83
paternity analysis, 94, 97, 99–100
pea, 53
peach, 192–193, 196
pear, 192–193, 195
Pellaea andromedifolia, 245, 248, 251
Pellia, 225, 227, 229
Pellia borealis, 223, 229–230, 234–235
Pellia epiphylla, 223, 227, 229–230, 233–234
Pellia medium, 234
Pellia neesiana, 223
PEP carboxylase, 33
peptidase (PEP), 30
perennial ryegrass, 138
Peripetasma, 154–155
Peromyscus maniculatus, 131
peroxidase (PRX), 10, 30, 38
Persea, 56
Persea americana, 199
PGD, see 6-phosphogluconate dehydrogenase
PGI, see glucose-6-phosphate isomerase
PGM, see phosphoglucomutase
Phalaris, 118
Phaseolus, 56
Phaseolus vulgaris, 54
phenolase, 33
Phlox drummondii, 91
Phoenix dactylifera, 200
phosphoenolpyruvate carboxykinase, 33
phosphoglucoisomerase (PGI), 25, 30, 131, 140, 154
phosphoglucomutase (PGM), 10, 30, 38, 155, 196, 198, 213
phosphogluconate dehydrogenase (PGD), 31
phosphoglucose isomerase, see phosphoglucoisomerase

phosphoglycerate kinase, 33
photographic equipment, 33
Phylloglossum, 241
phylogeny, 154, 157–159
Physcomitrella patens, 234
physiology, 130, 138, 140
Picea mariana, 132
pine, 207
pineapple, 192
Pinus, 59–60, 99, 207
Pinus attenuata, 132
Pinus banksiana, 83, 94, 132
Pinus contorta, 91, 93–94
Pinus ponderosa, 61, 99, 132
Pinus radiata, 132
Pinus resinosa, 61
Pinus rigida, 99, 132
Pinus sylvestris, 60, 96
Pinus taeda, 97–98
pistachio, 179
Pisum, 59
Pisum sativum, 53
pitch pine, 135
Plagiochila asplenioides, 223, 225
Plagiochila porelloides, 223, 225
Plagiomnium, 229–230
Plagiomnium ciliare, 222–223, 229, 234–235
Plagiomnium ellipticum, 222–223, 227, 229–231, 233–234
Plagiomnium insigne, 222–223, 227, 229–231, 233–234
Plagiomnium medium, 230–231, 233–234
Plagiomnium section *Rosulata*, 227
Plagiothecium curvifolium, 223
Plagiothecium denticulatum, 223
Plagiothecium latebricola, 223
Plagiothecium nemorale, 223
Plagiothecium ruthei, 223
Plagiothecium undulatum, 223–224
plant breeding, 218
Plantago lanceolata, 54
plantain, 192
plastid-specific isozymes, 10, 48, 50
plum, 196
pollen carryover, 95–96
pollen loads, 92
pollination, 82–90
polyacrylamide gel electrophoresis (PAGE), 6, 195, 198
polyembryony, 197, 200–201
polygenic, 206
Polygonum lapathifolium, 113
polymorphism, 6, 224–225

polyphenol oxidase, 33
polyploid, 55, 107, 117–118, 121, 151–152, 154, 156–159, 195, 201, 228–229, 241–245, 254
polysomic inheritance, 57, 107, 109, 119
Polystichum, 254
Polystichum imbricans, 247–248, 251, 253–254
Polystichum lemmonii, 253
Polystichum lonchitis, 253–254
Polystichum munitum, 247–254
ponderosa pine, 132
population differentiation, 89
population genetic structure, 83–84, 107, 109, 112, 115, 120, 233, 235, 246, 249–250
Populus tremuloides, 132
post-translational modification, 36, 38
potato, 171
POX, 197
private alleles, 96
progenitor-derivative relationships, 149, 151, 157, 159
protandry, 234
protogyny, 234
Prunus persica, 196
Prunus salicina, 196
Pseudotsuga menziesii, 91, 132
Psilotophyta, 241–243
Psilotum, 156, 241, 246
Pteridium aquilinum, 112, 245–246, 251–252
Pteridophyta, 241, 243, 253
pteridophytes, 233, 241–247, 252
Puccinellia × *phryganodes*, 113
pummelo, 198
pumpkin, 176
PX, see peroxidase, 194–198
Pyrus, 195
Pyrus calleryana, 195
Pyrus communis, 195
pyruvate kinase, 33

quantitative trait loci (QTLs), 207–211, 213, 215, 218
quantitative traits, 206, 214
quantum speciation, 151
quaternary structure of enzymes, 34, 49–50
quinoa, 174

Racopilum, 223–224
Racopilum capense, 223
Racopilum convolutaceum, 223
Racopilum cuspidigerum, 223–224, 227
Racopilum intermedium, 223

Racopilum robustum, 223
Racopilum spectabile, 223, 227
Racopilum strumiferum, 223
Racopilum tomentosum, 223
Raphanus sativus, 98, 169
recombination frequency, 210, 213
respiration, 138
Rhodanthos, 155
ribbed mussel, 130
ribulose-bisphosphate carboxylase (RBC), 31
Riccia, 227
Riccia bifurca, 227
Riccia dictyospora, 223, 225, 227, 229, 235
Riccia eldeeniae, 227
Riccia hirta, 227
rice, see also *Oryza sativa*, 53, 175, 184
R_{ST}, 110
rubisco, see ribulose-bisphosphate carboxylase
rye, 59

Salicaceae, 243
Salicornia, 2, 113–115, 148
Salicornia europaea, 114
scouring rushes, 241
Scrophulariaceae, 151, 158
Secale, 53
seed dispersal, 90
segregation, 201, 245
segregation distortion, 61, 195, 198–199, 201
selection, 73, 94, 98–99, 127, 131–132, 135, 215–218
  truncation, 128
self-fertilization, 94–96, 112, 233–234, 242
self-incompatibility, 234
Senecio viscosus, 113
Setaria faberi, 113–114, 119
Setaria italica subsp. *italica*, 169–170, 174
Setaria italica subsp. *viridis*, 169–170, 174
Shannon-Weaver information index, 107
sibling species, 227–228, 236
Silene diclinis, 91
silver-seeded squash, 176
silverswords, 150, 157
6-phosphogluconate dehydrogenase (6PGD), 10, 154
6PGD, see 6-phosphogluconate dehydrogenase, 155, 194–195
slender wild oats, 207
Solanum, 57, 149, 171
Solanum goniocalyx, 171
Solanum group *Andigena*, 171
Solanum group *Tuberosum*, 171

Solanum sparsipilum, 171
Solanum stenotomum, 171
Solanum tuberosum, 119, 171
sorbitol dehydrogenase, 33
Sorghum, 60, 114
Sorghum bicolor, 54–55, 92
Sorghum halepense, 113, 119
sorghum, 11, 54, 59
soybean, 54, 179
spatial autocorrelation, 107
speciation, 146, 149, 151, 156–157, 159
   geographical, 151
   quantum, 151
sperm dispersal, 234–235
Sphagnum, 224, 227–228
Sphagnum compactum, 224, 227
Sphagnum pulchrum, 224, 227
Sphagnum recurvum, 227
Sphagnum recurvum var. *mucronatum*, 224
Splachnum ampullaceum, 234
spore dispersal, 234–235
sporophyte, 221
squash, 176
   silver-seeded, 176
starch, 6
starch gels, 13
Stephanomeria, 54, 151
Stephanomeria diegensis, 2, 152
Stephanomeria exigua, 55, 152
Stipa, 118
strawberry, 179
subcellular location of isozymes, 10, 34, 36, 47–50
subunit structure, 36, 49, 193
superoxide dismutase (SOD), 10, 32
sweet orange, 198
synteny, 59
systematics, 146, 156–157, 159, 226

Tachigali versicolor, 93, 97
tandem duplication, 155
Taraxacum obliquum, 113
taxonomic markers, 146
tea, 54
teosintes, 172
Tetramolopium, 150, 253
tetrasomic inheritance, 57, 120, 153, 156, 159, 245
Tmesipteris, 241
tobacco, 176
Tolmiea menziesii, 57, 120, 153
tomato, 53, 59–60, 177, 184, 207–208, 218
TPI, see triose-phosphate isomerase

tracker dyes, 15
Tragopogon, 56, 152–153
Tragopogon mirus, 119
Tragopogon miscellus, 119
transketolase, 33
tree fruit crops, 192–193, 201
Trifolium hirtum, 91
Trillium, 83
triose-phosphate isomerase (TPI), 10, 26, 32
Triticum, 56
Triticum aestivum, 177, 206–207
Triticum bicorne, 178
Triticum dicoccoides, 98
Triticum longissimum, 178
Triticum monococcum var. *boeoticum*, 177–178
Triticum speltoides, 178
Triticum tauschii, 177–178
Triticum tauschii subsp. *strangulata*, 178
Triticum timopheevii, 178
Triticum timopheevii var. *araraticum*, 177
Triticum timopheevii var. *timopheevii*, 177–178
Triticum turgidum, 177–178
Triticum turgidum var. *dicoccoides*, 177
Triticum turgidum var. *dicoccum*, 177
trout, 130
truncation selection, 128
Turnera ulmifolia, 54, 75, 119
Turnera ulmifolia var. *angustifolia*, 120
Turnera ulmifolia var. *elegans*, 120
Turnera ulmifolia var. *intermedia*, 120
Typha, 113–114
tyrosinase, 33

vegetative reproduction, 78
Veronica peregrina, 119
viability, 131, 136–137

watermelon, 54, 179
weeds, 106
Weissia controversa, 230
Weissia crispa, 230
Weissia exserta, 230
wheat, see also *Triticum*, 52–53, 177, 206–207
whisk ferns, 241
wild oats, 131
wild rice, 178

xanthine dehydrogenase, 33
Xanthium, 113

Zea, see also corn and maize, 56, 149, 172

Zea diploperennis, 172, 180
Zea luxurians, 180
Zea mays, 132, 207
Zea mays subsp. *mays*, 169, 172, 174, 180–181
Zea mays subsp. *mexicana*, 172, 180–181
Zea mays subsp. *parviglumis*, 172
Zea mays var. *huehuetenangensis*, 172
Zea mays var. *parviglumis*, 169, 172, 174, 180
Zea perennis, 153, 172
Zizania, 149, 178
Zizania aquatica, 179
Zizania palustris, 169, 174, 179
Zizania palustris var. *interior*, 169, 174, 179
Zizania palustris var. *palustris*, 179